Organic Chemical Methods

Daniel Krois

Organic Chemical Methods

in Biochemistry, Molecular Biology, Medicine, and Pharmacy

Springer

Daniel Krois
Faculty of Chemistry Universität Wien
Vienna, Austria

ISBN 978-3-662-73572-5 ISBN 978-3-662-73573-2 (eBook)
https://doi.org/10.1007/978-3-662-73573-2

Translation from the German language edition: "Organisch-chemische Methoden" by Daniel Krois, © Der/die Herausgeber bzw. der/die Autor(en), exklusiv lizenziert an Springer-Verlag GmbH, DE, ein Teil von Springer Nature 2026. Published by Springer Berlin Heidelberg. All Rights Reserved.

This book is a translation of the original German edition "Organisch-chemische Methoden," 2nd edition, by Daniel Krois, published by Springer-Verlag GmbH, DE in 2026. The translation was done with the help of an artificial intelligence machine translation tool. A subsequent human revision was done primarily in terms of content, so that the book will read stylistically differently from a conventional translation. Springer Nature works continuously to further the development of tools for the production of books and on the related technologies to support the authors.

Planung/Lektorat: Sinem Toksabay
This Springer imprint is published by the registered company Springer-Verlag GmbH, DE, part of Springer Nature.
The registered company address is: Heidelberger Platz 3, 14197 Berlin, Germany

Preface

On the Origin of this Textbook

This textbook has evolved from a lecture entitled "Organic Chemical Methods in Molecular Biology and Biochemistry," which I taught at the University of Vienna for students of Chemistry, Molecular Biology, Microbiology and Genetics, as well as Biological Chemistry, from 1999 to 2021. For this reason, you will occasionally find that I address the readers directly or use the formulation "we...".

As an organic chemist with a strong interest in biochemistry, as well as in physical-chemical and spectroscopic methods, I have long been fascinated by the field of bioorganic chemistry, as it might be called—certainly also strongly influenced by an extended research stay at the Max Planck Institute of Biochemistry in Martinsried during the years 1986–1987. Therefore, I decided to design a lecture that highlights the significance of organic chemistry for biological and medical topics and that could be suitable for students from both "camps" (chemistry and biology).

I would like to highlight another aspect that has significantly shaped the development of this lecture and, consequently, this book. Ever since my studies in chemistry (1974–1981) at the University of Vienna, I have repeatedly become aware of a shortcoming in our university system: as a student, one is plagued for many years with acquiring long-accumulated knowledge, and only much later does one reach the "research frontier." At the beginning of the 20th century, things were probably quite different. At that time, groundbreaking experiments were still conducted in introductory lectures—or at least that is the impression one gets. But I have always been aware that it is necessary to possess a certain amount of basic knowledge before one can begin to think and discuss scientifically. As this foundational knowledge continues to grow, this dilemma also becomes increasingly pronounced. I have not found an ideal solution to this problem either. However, in this lecture—which can already assume a certain basic knowledge of chemistry, especially organic chemistry and biochemistry, since it was originally intended for the second phase of a diploma program and is now designed for master's degree programs—this opportunity should be available. For this reason, it is divided into two parts: a methodological first part, which already integrates the latest developments, and a second part in which selected topics—application examples of organic-chemical methodology—are examined and discussed using original literature.

About the 2nd Edition

For this new edition, I have updated the methods discussed in the first part to reflect the latest developments. Section 4.2 has been somewhat expanded—originally, it was kept very brief in my lecture, as there was a separate course on this topic at our institution. Entirely new is Chap. 8, a subject that was missing from my lecture because it was thoroughly covered by a colleague. In addition, I have incorporated several application examples into the chapters. As for the second part, if you compare it with the first edition, you will notice that the fundamental reference examples have remained the same—you might say I took the easy way out... But that is not quite the case, as I still consider the methodology and findings of these works to be very important and instructive, and, as you will notice, I have followed the further development of the specific topic in the literature and included the relevant citations, so that one can also see what additional consequences have arisen. I find this somewhat "historical" perspective particularly stimulating and interesting for students.

At the end of each chapter, a short set of questions is appended, for which the corresponding solutions with explanations can be found in Chap. 15. New as well is Chap. 14, which deals with fluorescence and fluorophores. Fluorescence methods have become increasingly important, and I have endeavored, after a brief physical introduction, to present primarily the types of organic-chemical fluorophores and, by way of example, to list their photophysical properties—not an easy task, as these data are difficult to extract from the literature. And although this is not directly related to organic-chemical laboratory methodology, two examples of naturally occurring fluorescent proteins are discussed, and the formation of their fluorophores from the amino acid sequence is examined mechanistically—after all, these are also organic-chemical reaction methods, albeit employed by nature.

Overall, the number of literature citations has increased significantly, extending to the most recent publications from 2024 and, in some cases, even 2025. What has always set this textbook apart—and is even more pronounced in the new edition—is that virtually all methods and facts are drawn from the scientific primary and review literature, not from other textbooks. In this respect, this book is also suitable for PhD students as a monograph, and indeed for fully trained scientists for whom a topic covered in this book may be a peripheral area of their research. For most literature citations, the DOI code is now also provided to facilitate easier and faster access to the source.

On the Subject Matter of the Textbook

The chemistry underlying life processes has lost none of its fascination, even though we understand much more about it today than we did 50 years ago. One could even say, on the contrary—it has become even more exciting in many ways, and we are certainly far from the end of this thrilling journey of discovery. Newly acquired insights must subsequently be translated into new medical therapies or new approaches in agriculture, and the elucidation of brain functions will also have an impact on computer technology. All of this together will continually reshape our understanding of nature as a whole and of humanity. It is no wonder, then, that studies in biochemistry, bioorganic chemistry, biological chemistry, molecular biology, microbiology and genetics, pharmaceutical chemistry, medicinal chemistry—and whatever other names they may go by—are attracting ever more interest. In 2003, for example, Herbert Waldmann articulated the fascination of the field of "biological chemistry" in a review [Waldmann, H.: At the crossroads of chemistry and biology. *Bioorganic & Medicinal Chemistry* **11**, 3045–3051 (2003).]

The foundation of all these branches of knowledge is, on the one hand, chemistry—particularly organic chemistry (the chemistry of hydrocarbon derivatives)—and, on the other hand, biology. This textbook aims to present and examine in greater detail the essential concepts and methods of organic chemistry that are relevant to these research areas (Part 1), and then to discuss exemplary applications of these methodologies from the literature (Part 2). The goal is, on the one hand, to demonstrate and convey to chemists the possibilities of method development for these biological questions, and on the other hand, to provide users on the more biologically oriented side with insight into the background and capabilities of these methods. Since scientific research today, especially that which concerns biological topics, can only be conducted interdisciplinarily as teamwork, it is particularly important that every specialist also possesses a certain understanding of their partner's field. This book is intended to make that possible.

Finally, a provocative thought: Someone might say, why do we even need textbooks today—when we have a question, when we are looking for a method, when we don't know something?—We can just ask AI. In this sense, I am "old-fashioned"—I also practically did not use AI for this book. But it is clear that AI can and should be used for research purposes. AI can provide inspiration, can make lexical knowledge accessible more quickly, can analyze large amounts of data much faster than we humans can. But I am convinced that one needs a fundamental education, must develop one's own personal knowledge base, in order to be able to develop new ideas and to use the results of AI advantageously. And this book is intended to provide such a foundation, as already mentioned above, so that one can also have constructive discussions with colleagues—AI cannot help with that.

Vienna, Austria — Daniel Krois
18th of August 2025

Contents

Brief Review of Fundamental Concepts of General and Organic Chemistry

1

Abstract

This chapter reviews and emphasizes some fundamental concepts of general and organic chemistry that I consider particularly important. However, a basic knowledge of organic chemistry is assumed—only certain key points will be highlighted. The concepts of acid-base properties, nucleophilic and electrophilic reactivity, resonance, tautomerism, aromaticity, and the terms "configuration" and "conformation" are defined and summarized.

1.1 Acidity—Basicity ↔ Electrophilicity—Nucleophilicity

According to Brønsted, acids donate H^+ ions, and bases can accept H^+ ions. According to the more general Lewis definition, acids possess an electron deficiency—that is, they are electron pair acceptors, while bases are electron pair donors. The terms "**acidity**" and "**basicity**" describe the ***thermodynamic*** equilibrium of an acidic or basic compound with water as the solvent (in principle, this can and must in some cases be extended to other solvents), as described by the general reaction equations. From these, the acidity constant K_a and the basicity constant K_b can be derived (Fig. 1.1). From these relationships, the so-called buffer equation also follows quite simply, showing that an equimolar mixture of an acid (HA) with its conjugate base (A^-) provides the best buffering effect—the pH value is then numerically equal to the pK_a.

In contrast, "**electrophilicity**" and "**nucleophilicity**" are ***kinetic*** parameters. In a standard reaction of an electrophilic reagent with various nucleophiles, an increased reaction rate corresponds to greater nucleophilicity. Often, nucleophilicity increases in parallel with basicity—in both cases, a free electron pair is involved. However, this only applies to the same atom or to atoms within the same period, for example:

D. Krois, *Organic Chemical Methods*, https://doi.org/10.1007/978-3-662-73573-2_1

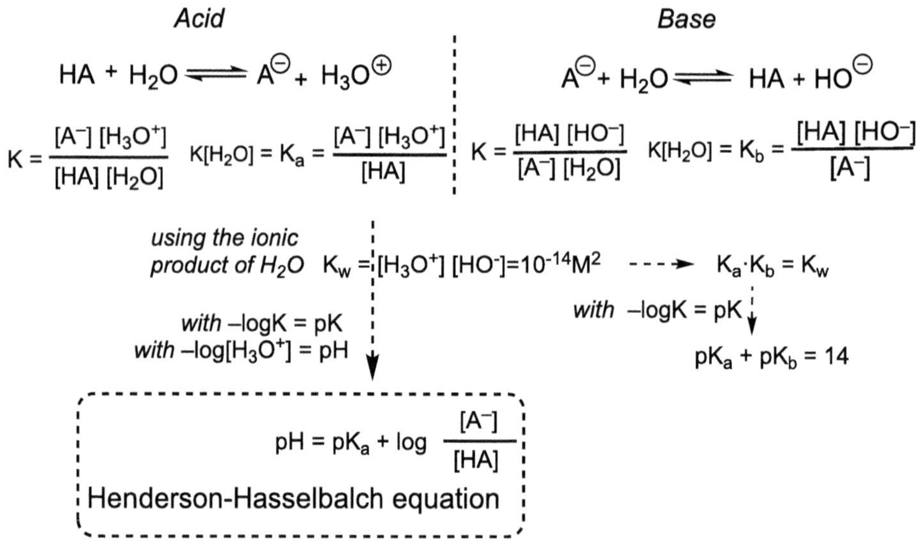

Fig. 1.1 Acidity and basicity as thermodynamic parameters, definition and relationship of the acidity constant K_a and the basicity constant K_b, as well as derivation of the buffer equation

- **ROH** is less nucleophilic (less basic) than **RO**$^-$ and
- **ROH** is less nucleophilic (less basic) than **RNH**$_2$.

But this is usually no longer the case when atoms from different periods are involved, for example:

- The ion **HS**$^-$ is significantly more nucleophilic (but less basic) than **HO**$^-$.

In kinetics, that is, in the transition state, other factors also play a role, such as the softness or flexibility of an atom or ion during collision.

1.2 Resonance and Tautomerism

"**Resonance**" (or mesomerism) refers to the description of the actual bonding situation in a molecule by means of several hypothetical limiting structures (=resonance structures), where only bonds and charges may be shifted, but the spatial arrangement of the atoms must of course not change.

We speak of "resonance stabilization" when several physically meaningful resonance structures can be drawn for a compound; these are connected by the characteristic double arrow ↔. Examples are given in Fig. 1.2. While the dipolar limiting structure

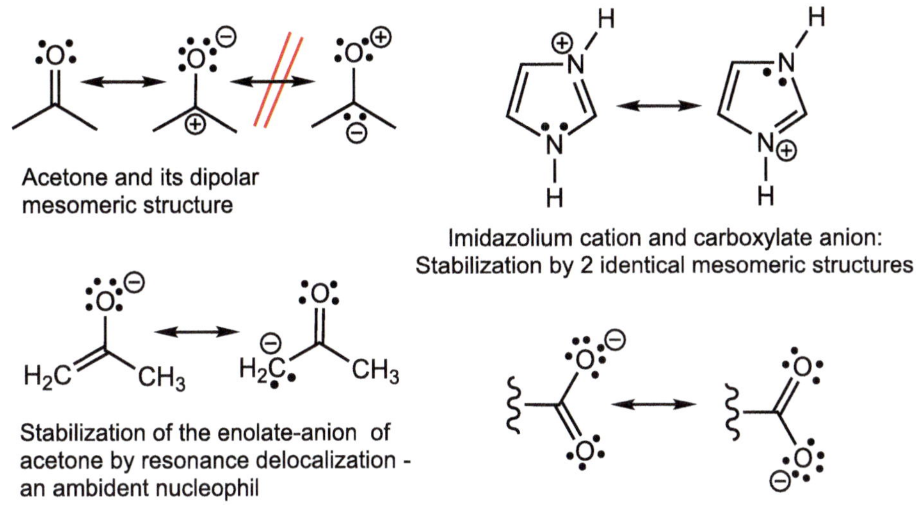

Fig. 1.2 Examples of the concept of resonance (mesomerism)

of carbonyl compounds with a positively charged carbonyl C atom (here using acetone as an example) makes a significant contribution to the overall bonding situation, the oppositely polarized form is, of course, meaningless, as it is physically nonsensical (O is the more electronegative atom). The acidity of the α-H atom of carbonyl compounds is well explained by the resonance stabilization of the enolate anion. The more delocalized a charge is, the lower its energy generally is, and thus the more stable the species. The enolate anion is therefore also an ambident nucleophile, having two (*amphi*) possible centers that can act as electron pair donors. In the imidazolium cation, the positive charge is distributed completely evenly over both N atoms, which is the reason for the increased basicity of imidazole, just as the distribution of the negative charge in the carboxylate anion is the reason for the increased acidity of carboxylic acids.

Tautomerism is a form of constitutional ***isomerism***, in which an atom (or a group of atoms) changes position with simultaneous shift of double bonds—with one important restriction: this process must also be able to occur rapidly under normal conditions, at room temperature. Almost always, heteroatoms are involved, and the migrating atom is usually a hydrogen. The equilibrium is therefore usually acid- and base-catalyzed. Examples are given in Fig. 1.3. The tautomerism of imidazole leads to an identical structure; one could say it is a "degenerate tautomerism," yet it is a kinetically relevant process. The enzymatic conversion of dihydroxyacetone to glyceraldehyde occurs via two keto-enol tautomerizations. The nucleobase thymine could also exist in a tautomeric form with an OH group, but under normal conditions this form is insignificant. Such forms have been detected in G–T mismatches (instead of the Watson-Crick pairing A–T) [1].

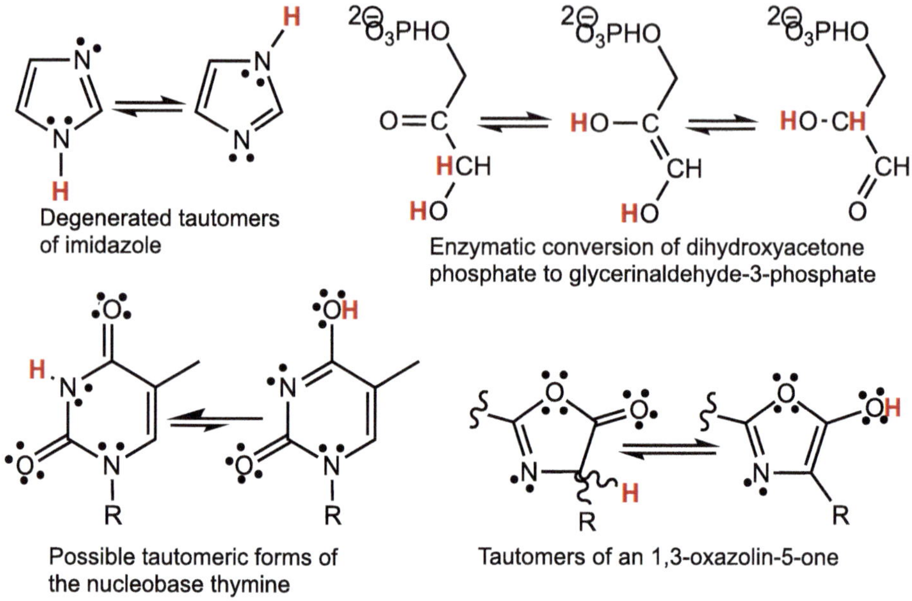

Fig. 1.3 Examples of tautomeric equilibria with biological relevance

Racemization observed during activation of the carboxyl group of peptides is, for example, based on the tautomerism of the intermediate 1,3-oxazolin-5-one (see Sect. 3.1.3).

1.3 Aromaticity and Heterocycles

The concept of aromaticity is very important for understanding organic chemistry, and the special properties of aromatic compounds also play a significant role in the biosphere. In particular, this refers to the special chemical reactivity and the delocalization of electrons in the aromatic molecular orbitals (MOs). These special electronic properties give rise to π–π interactions, which, for example, are partly responsible for the stability of the DNA double helix. For an unsaturated compound to be aromatic, that is, to experience a special additional resonance stabilization, certain *necessary conditions* must be met—this can already be checked in advance using the constitutional formula.

▶ Conditions for Aromaticity (necessary prerequisites for aromaticity that must be fulfilled in at least ***one*** realistic resonance structure)

- Cyclic, fully conjugated system, no “cross” conjugation
- $(4n + 2)$ π electrons: Hückel rule
- Planarity (only slight twisting allowed)

Whether a compound is actually aromatic can be recognized by certain properties, which we will call the *criteria for aromaticity*.

▶ Criteria for Aromaticity (how can it be recognized experimentally?)
- Reactivity: no electrophilic addition (as with –C=C–), but substitution; this preserves the energetic advantage due to aromaticity (addition would destroy the aromatic system)
- Above-average resonance energy compared to "normal" conjugated double bonds
- Planarity
- Ring current effect in nuclear magnetic resonance (NMR) spectra: This leads to unusual chemical shifts of protons located near an aromatic ring system. Explanation: *A fully conjugated aromatic system reacts in a strong magnetic field—as in the NMR experiment—like a conducting loop, i.e., a magnetic field is induced that opposes the external field, which influences the effective magnetic field experienced by the proton* (there is also a more precise quantum mechanical explanation).

Figure 1.4 shows examples of carbocyclic aromatic systems. Benzene, the aromatics naphthalene and pyrene, which are composed of two and four fused benzene rings, respectively, as well as the cyclopentadienyl anion, are examples of aromatic systems with 6 π electrons (Hückel rule $n = 1$). Cyclohepta-2,4,6-trienone is also an aromatic compound with 6 π electrons due to the dipolar resonance structure of the C=O group, whereas cyclopenta-2,4-dienone is not aromatic (for 6 π electrons, the oppositely polarized resonance structure would have to contribute—but it must be excluded, cf. Fig. 1.2).

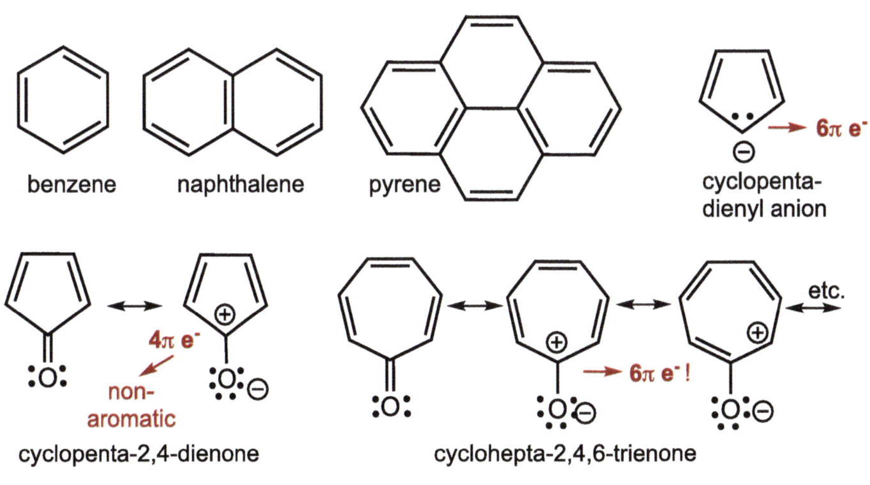

Fig. 1.4 Examples of carbocyclic aromatics

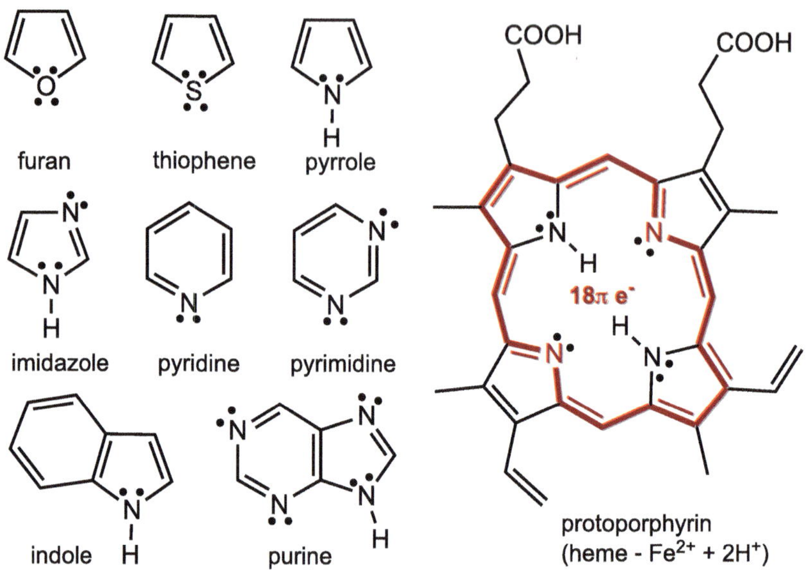

Fig. 1.5 Examples of heterocycles

Figure 1.5 shows examples of heterocycles, aromatic ring systems with at least one non-carbon atom in the ring. Here too, the ring system must be fully conjugated, with the heteroatom either contributing its non-bonding electron pair or, if possible, forming a double bond (π bond). In the heterocycles furan, thiophene, and pyrrole, a p orbital of the heteroatom (O, S, N) occupied by two electrons is included in the aromatic molecular orbital, so that the Hückel rule (with 6 π electrons) is fulfilled. Accordingly, there is no "free" electron pair on the N atom of pyrrole; the N atom is **not** basic. In fact, the N–H bond is even somewhat weakened, and the H atom is slightly acidic. Many heterocycles occur in natural products: indole in tryptophan and the indole alkaloids, imidazole in the amino acid histidine, pyridine in nicotine, NAD, etc., pyrimidine and purine in the nucleobases, and the porphyrin system in heme and also in other important structures with metal ion binding, to name just a few examples. In the porphyrin system, the bold "path" corresponds to the aromatic molecular orbital. If one were to go around the two five-membered rings with a C=N double bond on the outside, cross-conjugations would occur (double bonds extending out of the aromatic ring system), which should be avoided, since the π electrons must be assigned unambiguously in order to check the Hückel rule.

1.4 Configuration—Conformation

Configuration is the spatial arrangement of atoms and bonds, without considering rotation about single bonds, which occurs very rapidly at room temperature. **Conformation**, on the other hand, describes the spatial arrangement (of atoms and bonds), but also takes

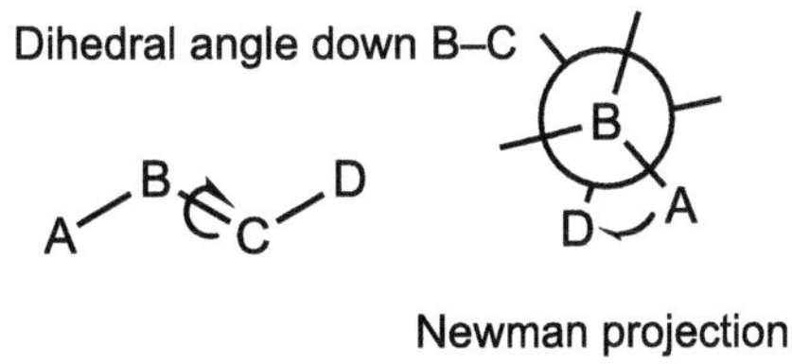

Fig. 1.6 Torsion angles

into account the torsion angles about single bonds. In other words: a defined compound can exist in a rapid dynamic equilibrium of several different conformations. While different configurations are described on the one hand by the *R*/*S* nomenclature and on the other hand by *Z*, *E* or *cis*, *trans*, different conformations, i.e., different torsion angles (Fig. 1.6) about single bonds, are designated as *syn(periplanar)*, *anti(periplanar)* or *synclinal* or also *gauche*. Of course, there are also borderline cases here, when rotation about single bonds is significantly slowed. In the peptide bond of proteins, for example, a *syn-* ⇆*anti* "isomerization" about the **C**(=O)–**N** bond, which is already half a double bond, can be so hindered by the partial need for overall reorientation of the remaining protein that one can already speak of a *cis*⇆*trans* isomerization.A configuration can be described by bond lengths (A–B, B–C, C–D) (Fig. 1.6), bond angles (A–B–C, B–C–D), and torsion angles (A–B–C–D) about multiple bonds. For chiral molecules, the relative and absolute spatial arrangement at chiral centers (or axes or planes) must also be distinguished (→ diastereomers, enantiomers). For conformation, all torsion angles about single bonds must also be considered.

The conformation of a molecule is influenced by various types of non-covalent interactions (int.):

- Van der Waals interactions (dispersive nonpolar interactions = "London forces"; interactions between dipole ↔ dipole, and dipole ↔ induced dipole; all proportional to r^{-6})
- ionic interactions (e.g., salt bridges = *salt bridges* in protein structure)
- hydrogen bonds (donor and acceptor sites) (a special case of dipole-dipole interaction)
- hydrophobic interactions: occur mainly in polar solvents, especially in H_2O, driven on the one hand by disruption of the water structure, on the other hand by solvation effects; salt effects ("salting out" = *salting-out* ≠ *salting-in*); amphiphilic interactions (keyword: amphiphilic helices in proteins)
- π–π electron interactions: especially interactions between aromatic systems (e.g., stacking of nucleic acid bases, intercalation)
- repulsive interactions: steric, electronic

1.5 Appendix

Basicities of some N-containing organic compounds for better understanding of the properties of biologically relevant compounds. (According to Refs. [2] and [3])

Base	Protonation at	Basicity constant pK_b
Aminocyclohexane (cyclohexylamine)	Primary amine N atom	3.34
Aminobenzene (aniline)	Aromatic primary amine N atom	9.37
Piperidine	Secondary amine N atom	3.88
Pyridine	Heterocyclic N atom	8.75
4-Methylpyridine	Heterocyclic N atom	7.9
Pyrimidine	Heterocyclic N atom	12.7
2,4,6-Trimethylpyrimidine	Heterocyclic N atom	10.05
2-Aminopyrimidine	Heterocyclic N atom	10.29
Imidazole	Heterocyclic N atom	7.05
Acetamide (CH_3–CO–NH_2)	Carbonyl O atom	14
Glycine	α-Amino N atom	4.22
Arginine	Guanidino =NH group	1.8
Lysine	α-Amino N atom ε-Amino N atom	4.8 3.2

1.6 Questions

Let's test our understanding of the content of this chapter!

1.1 Is there a relationship between the acidity constant K_a and the basicity constant K_b?

1.2 An alkyl halide R–Cl reacts faster with an alkoxide anion R′–O⁻ than with the corresponding alcohol R′–OH. Can this be explained by the higher basicity of the alkoxide?

1.3 Why is cyclopentadiene not aromatic?

True or False?

1.4 In a chemical reaction, only one resonance form reacts. ☐

1.5 Tautomers are constitutional isomers that can easily interconvert. ☐

1.6 A specific conformer of a compound cannot be separated from another at room temperature. ☐

References

1. Kimsey, I.J., Petzold, K., Sathyamoorthy, B., Stein, Z.W., Al-Hashimi, H.M.: Visualizing transient Watson-Crick-like mispairs in DNA and RNA duplexes. Nature. **519**(7543), 315–320 (2015). https://doi.org/10.1038/nature14227
2. Streitwieser, A., Heathcock, C.H.: Organische Chemie. Verlag Chemie, Weinheim/Deerfield Beach/Basel (1980)
3. Weast, R.C. (Hrsg.): CRC Handbook of Chemistry and Physics, 60. Aufl. CRC Press, Boca Raton (1980)

Part I

Organic Chemical Methods Commonly Used in Biological Research

Chemical Properties of "Biomolecules"—A Summary

2

Abstract

This chapter is intended solely as a review and, if necessary, a supplement to basic knowledge of organic chemistry. The most important biologically relevant classes of compounds are reviewed and their essential properties recalled. The first subsection covers the proteinogenic amino acids, discussing their structure and general properties, as well as the peptide bond and its significance for the structure and properties of proteins. This is followed by a section on carbohydrates, including the "family tree" of aldoses, the hemiacetal form of monosaccharides, and the glycosidic bond. The next subsection discusses nucleic acids, i.e., DNA and RNA, with Watson-Crick base pairing. This is followed by a brief selection from the field of natural lipids and, finally, a section on the significance of heterocycles in biomolecules.

2.1 Amino Acids, Peptides, Proteins

There are 20 (sometimes 21 are mentioned: including Sec, selenocysteine) amino acids encoded by DNA and incorporated into proteins via protein biosynthesis (so-called ***encoded proteinogenic*** amino acids) (Fig. 2.1). However, there are a number of post-translational modifications (PTMs) (which occur subsequently outside the ribosome), such as phosphorylation, sulfation, and glycosylation of hydroxy amino acids (at Ser, Thr, Tyr), hydroxylation (e.g., Pro → Hyp hydroxyproline), degradation (e.g., Arg → Orn, ornithine), *N*-glycosylation (at Asn). All proteinogenic amino acids are exclusively α-amino acids. With the exception of Gly, they are all chiral and display an L-configuration at the α-C atom in the Fischer projection (corresponding to the *S*-configuration, exception Cys: *R*). There has been and continues to be much speculation as to why all proteins in terrestrial life forms contain only L-configured amino acids. Here are just a

D. Krois, *Organic Chemical Methods*, https://doi.org/10.1007/978-3-662-73573-2_2

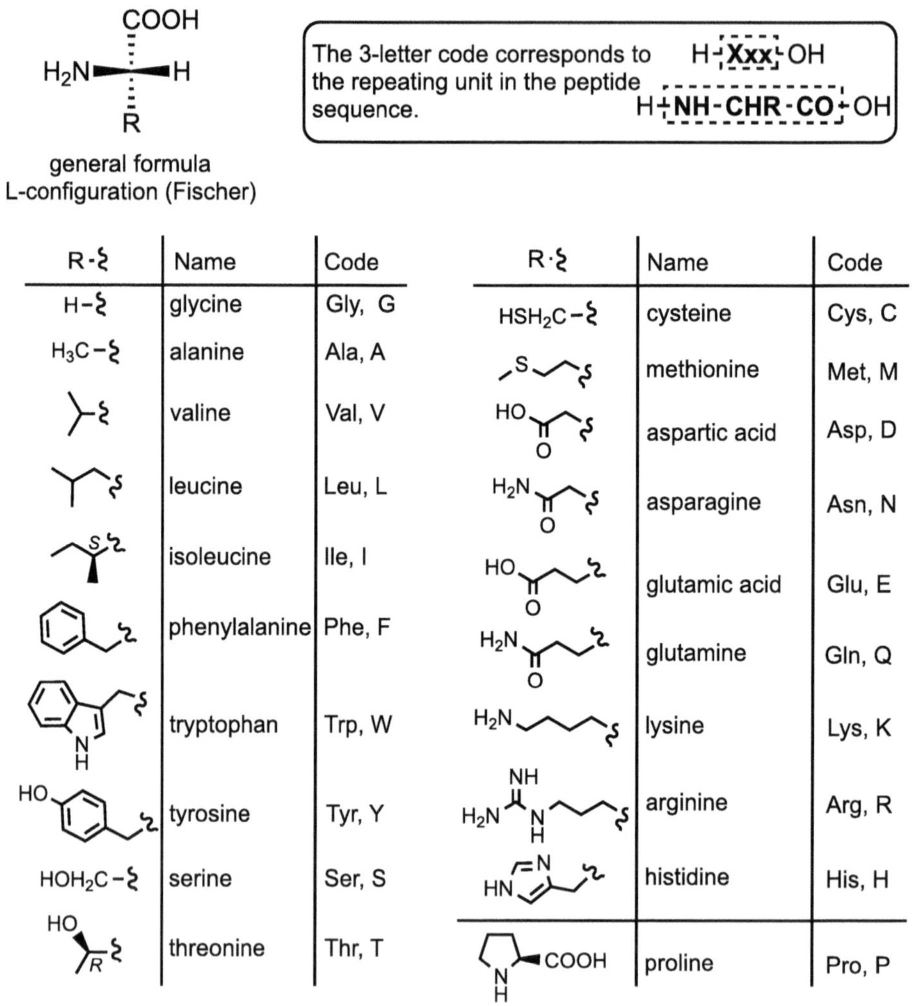

Fig. 2.1 Encoded proteinogenic α-amino acids

few references on this topic [1]. Two encoded amino acids possess an additional center of asymmetry at the β-C atom (C3) [L-isoleucine: 2*S*,3*S*; L-threonine: 2*S*,3*R*].

The amino acid proline occupies a special position, as it is the only one to possess a secondary amino group; formally, its side chain forms a ring with the amino nitrogen atom. This unique feature also has significant consequences for the conformational influence of proline in proteins.

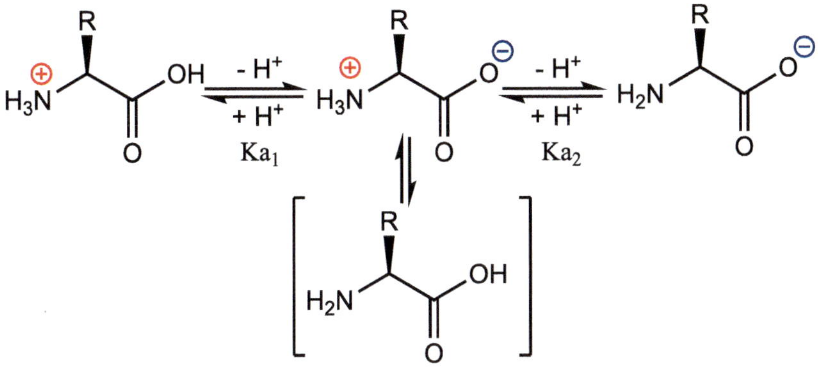

Fig. 2.2 Acid-base equilibria of an α-amino acid

Peptides produced by certain microorganisms may contain several non-encoded, non-proteinogenic amino acids; for example, β- or δ-amino acids may occur. In addition, amino acids of the D-configuration are often incorporated. However, this peptide synthesis usually proceeds via multi-enzyme complexes outside the ribosome [2].

All amino acids are amphoteric substances (Fig. 2.2); depending on the pH of the aqueous solution, they exist as protonated ammonium cations, neutral zwitterions, or deprotonated carboxylate anions (if the side chain contains additional basic or acidic functional groups, the situation becomes even more complex). Each amino acid has a specific pH at which it is electrically neutral overall (zwitterion)—the so-called isoelectric point (experimentally determined by titration with a pH electrode or by the electrophoretic method of "isoelectric focusing"). In principle, this form could also be written without charges, but in aqueous solution, the amino acid will always exist as a zwitterion.

Peptides and **proteins** are the oligo- and polymers of amino acids, formed by linking an amide bond between carboxyl and amino groups. In this context, the bond is referred to as a **peptide bond**. By convention, these oligomers are written with the N-terminus on the left and the C-terminus on the right. For example, the tripeptide H–Ala–Gly–Phe–OH; alanine is thus the N-terminal amino acid (with a free amino group), phenylalanine the C-terminal amino acid (with a free carboxyl group).

In the peptide bond, the "third" resonance structure of acyl compounds is particularly significant (Fig. 2.3). In fact, there is a partial double bond between the C and N atoms. This can be clearly demonstrated spectroscopically (IR and NMR spectra). This, in turn, has important consequences for the properties of peptides (and proteins). From this concept, important conclusions can be drawn for amides in general and for peptides and proteins in particular, which are clearly reflected in the actual properties of proteins.

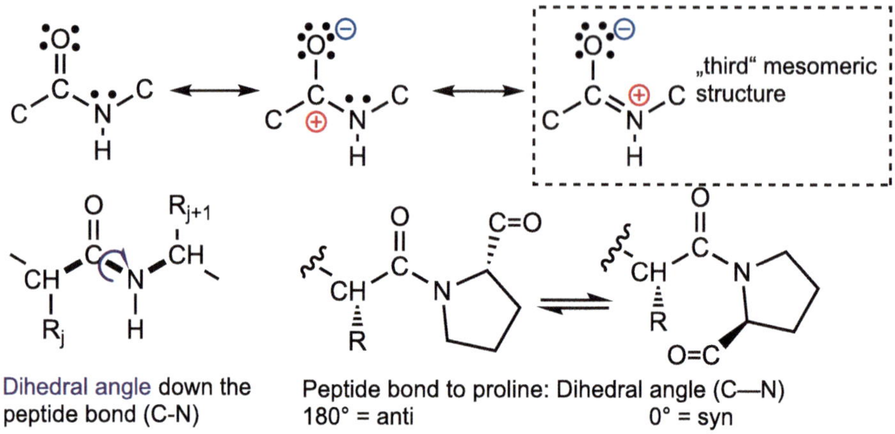

Fig. 2.3 The peptide bond—an amide bond: resonance structures; definition of the torsion angle around the peptide bond, *anti* and *syn* conformations in peptide bonds to proline

▶ Consequences of the significance of the "third" resonance structure of the amide bond = peptide bond for peptides and proteins

1. The N atom no longer has a free electron pair, it is **not** basic!
2. If the N atom carries an H atom (in all amino acids except Pro, Hyp), the H atom is a good hydrogen bond donor (in contrast to amino-H).
3. The C–N bond is **planar** (partially π-bonded), the torsion angle is about 0° (*syn* or *cis*) or 180° (*anti* or *trans*) ➔ significant restriction of the conformational possibilities of peptides and proteins.
4. Rotation around this bond is therefore restricted (the activation barrier is significantly increased), i.e., *syn-* ⇆ *anti* interconversions occur much more slowly in proteins. This has consequences for the kinetic process of protein folding, especially for peptide bonds to proline, which can exist in both *anti* and *syn* forms—in other amino acids, practically only the *anti* conformation occurs.
5. Among acyl compounds, the peptide bond is the **most stable** bond against nucleophilic attack ➔ peptides and proteins are therefore hydrolyzed only under very drastic conditions.

Primary, Secondary, Tertiary, and Quaternary Structure of Proteins

- **Primary structure:** amino acid sequence, by convention starting from the N-terminus on the left to the C-terminus on the right (3-letter code: e.g., H–Ala–Gly–Thr–[…]–Asp–Leu–OH); encoded by DNA
- **Secondary structure**: conformation of the peptide chain over shorter contiguous regions; main types: α-helix, β-sheet (β-*structure*), various loops and proline helix

- **Tertiary structure**: folding of the entire peptide chain in space
- **Quaternary structure**: spatial structure of the entire functional protein (possibly consisting of several peptide chains)

The native secondary and tertiary structure of the peptide chain is strictly determined by the primary structure; prediction is already possible in many cases today, see Sect. 3.2.5. The decisive factors are the interactions of the amino acid side chains with each other and with the peptide bonds [types of interactions: hydrogen bonds, dipolar interactions, ionic interactions, hydrophobic interactions, and—as the only covalent interaction—disulfide bridges formed by two cysteines → cystine], but also the folding process itself (kinetics!).

2.2 Carbohydrates

Carbohydrates are polyhydroxycarbonyl compounds (monosaccharides) and their oligo- and polymers. Depending on whether the monomers are aldehydes or ketones, they are referred to as "aldoses" or "ketoses." The most biologically important monosaccharides have five (pentoses) or six C atoms (hexoses), but those with seven, eight, or nine C atoms are also significant.

At the beginning of the 20th century, Emil Fischer elucidated the family tree of the aldoses (Fischer projection formulas Fig. 2.4). What does "elucidated" mean? – Well, starting from three C atoms, there is an increasing number of asymmetric C atoms. The aldopentoses have **3** centers of asymmetry and thus $2^3 = 8$ stereoisomers and therefore **4** diastereomers (each as image and mirror image, enantiomeric pair), and for the aldohexoses with **4** asymmetric C atoms, there are already $2^4 = 16$ stereoisomers and thus **8** diastereomers. The ketohexoses (carbonyl function at C2) have only three asymmetric C atoms, so there are only 4 diastereomeric ketohexoses (one of which is fructose). The elucidation of the relative configurations was achieved primarily through building-up and degradation reactions and based on the stereochemical relationships of the dicarboxylic acids (aldaric acids) obtained by oxidation. Emil Fischer (Nobel laureate 1902) arbitrarily (50:50 chance) assigned the absolute configuration for D-glyceraldehyde. Only around 1950 could this be indirectly proven by anomalous X-ray diffraction. He had guessed correctly. The designation "D-" refers to the position of the HO group at the last asymmetric C atom. In the D-series, this HO group is on the right (Latin: *dexter*). The corresponding L-aldoses are the mirror images with this HO group on the left (Latin: *laevus*). Since the naming, as with amino acids, is done using trivial names, the designations can be recalled using mnemonic words or phrases. For this, one only needs to know the order of the formulas in the Fischer projection, which follows logical criteria: For the D-series, all HO groups are on the right, then the first on the left, then the second on the left (all others right), now also the first on the left, then the third on the left (all others right), and so on.

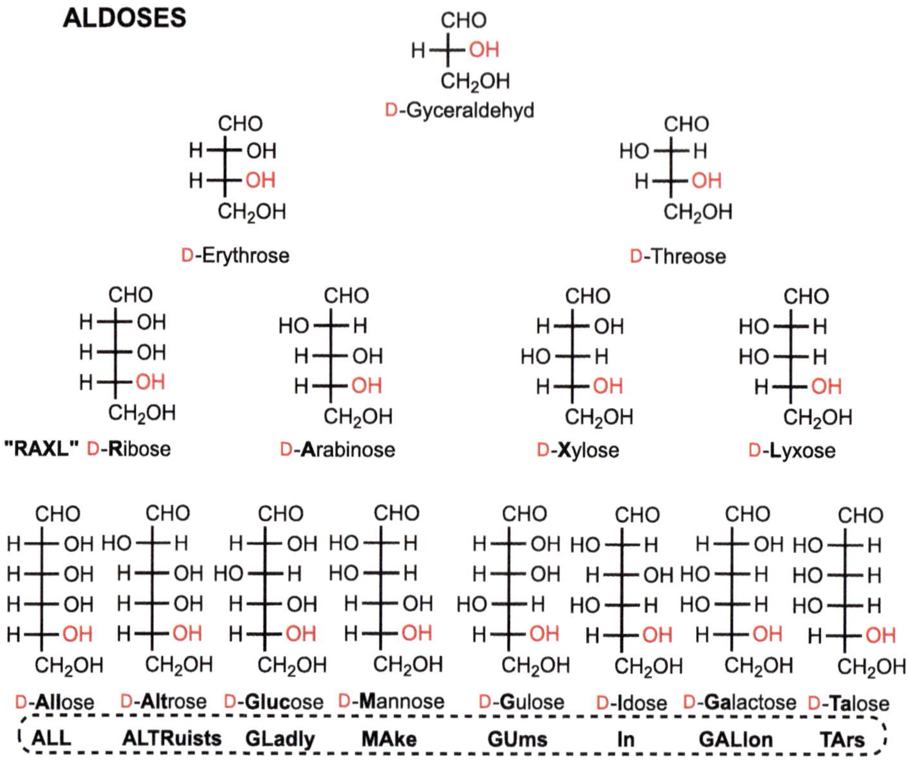

Fig. 2.4 "Family tree" of the aldoses

But are monosaccharides really free aldehydes or ketones? For monosaccharides with five or more C atoms, there is the possibility of forming cyclic hemiacetals (derivatives of ketones may be called hemiketals, but "hemiacetals" is also correct in this case). This involves an intramolecular addition of appropriate HO groups to the electrophilic carbonyl C atom. These products are both kinetically and thermodynamically favored in the case of 5- and 6-membered rings (see for D-glucose as an example Fig. 2.5). In fact, these monosaccharides exist mainly (>99%) as cyclic hemiacetals. Now, the original carbonyl C atom is sp^3-hybridized, with four different substituents, and thus becomes another asymmetric C atom. Thus, two different diastereomeric forms occur, also called "anomers," namely the α- and β-anomers. In aqueous solution, an equilibrium of all three species is established, since hemiacetal formation is both acid- and base-catalyzed. For D-glucose, as an example, 66% is present in the β-form, 33% in the α-form, and less than 1% as the aldehyde.

Mutarotation
D-glucose normally crystallizes as the pure α-anomer. If crystalline D-glucose is dissolved, the establishment of equilibrium can be monitored in a polarimeter, since the optical rotation values for the two anomers differ greatly. This phenomenon is called "mutarotation."

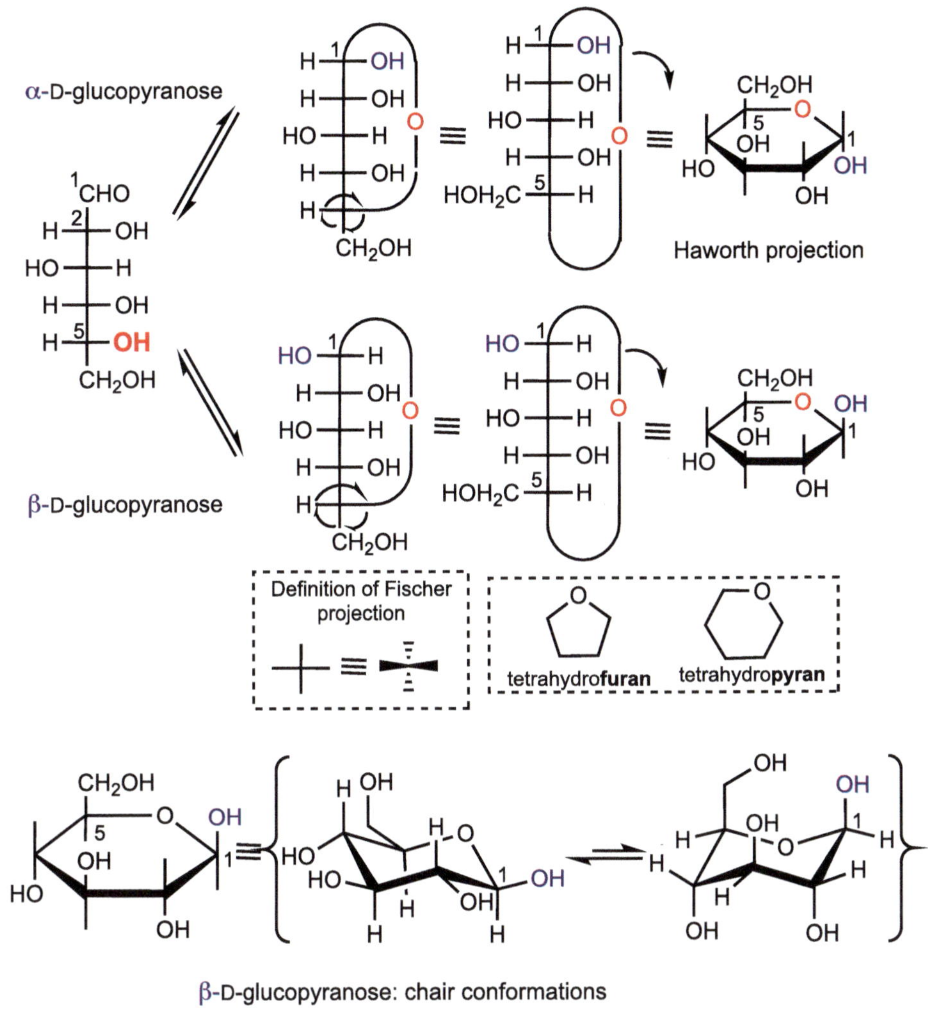

Fig. 2.5 Formation of the cyclic hemiacetal anomers using D-glucose as an example

How can one derive spatially accurate structural formulas of the cyclic hemiacetal forms from the Fischer projection formula? This is demonstrated in Fig. 2.5 using D-glucopyranose as an example. By convention, the HO group at the former carbonyl C atom in the Fischer projection of the α-anomer is on the right for the D-series (on the left for the L-series). However, the C–O–C bond is unrealistic in this form, considering the definition of the Fischer projection. By "rotation" around the C4–C5 bond, a more realistic representation is obtained—the backbone of the C chain curves behind the drawing plane and is linked by the O atom (at the very back). If this formula is then swung to the right, one arrives at the Haworth projection formula. Pyranose forms exist in chair conformations, of which there are two different types.

If the HO group at C4 were to react, a 5-membered ring, a furanose form, would result; however, this plays no role in unprotected aldohexoses and can only be enforced by introducing specific protecting groups. In ketohexoses, which have the carbonyl C atom at position 2, the furanose form is normally formed by reaction with the HO group at C5. Aldopentoses, such as ribose, form the cyclic hemiacetal with the HO group at C4 and generally exist in the furanose form.

If the hemiacetal HO group at C1 (at C2 in ketoses) reacts with another hydroxy group, a full acetal is formed, referred to as an ***O*-glycosidic** bond; reaction with N-derivatives yields ***N*-glycosidic** bonds. These glycosidic bonds can only be formed or cleaved non-enzymatically under acid catalysis and are therefore stable, in contrast to the hemiacetal forms. In this way, oligomeric and polymeric carbohydrates, oligosaccharides, and polysaccharides (Chap. 4) are formed.

Monosaccharide building blocks are also of great importance in other classes of natural products. In nucleosides, there is an *N*-glycosidic linkage of β-D-ribofuranose or β-D-2-deoxyribofuranose with the heterocyclic nitrogen atom of the nucleobases (Sect. 2.3). Glycoconjugates with lipids and/or proteins are of great importance for cell-cell recognition, immunological properties, etc. (Chap. 5).

2.3 Nucleic Acids

Nucleic acids, along with proteins and carbohydrates, represent the third major class of biopolymers. The monomeric unit is the nucleotide. This refers to the phosphoric acid ester at the 3′ or 5′ position of a nucleoside (Fig. 2.6). A nucleoside is a ribose or 2′-deoxyribose with an *N*-glycosidically linked nucleobase, which is a pyrimidine or purine derivative (Table 2.1). The positions in the sugar unit are indicated by numbers with a prime. As an example, 2′-deoxyadenosine-5′-phosphate (5′-dAMP) is shown. Phosphoric acid readily forms anhydrides, which in biochemistry are called di- or triphosphates. Here, 5′-ADP is shown as an example. In the polymer, the individual nucleotides are linked via a 3′–5′ phosphodiester bond. By convention, a nucleic acid sequence is given from the 5′ to the 3′ end. This is also the direction in which biological synthesis occurs.

The essential difference between RNA and DNA is the 2′-OH group of ribose, which is absent in 2′-deoxyribose. One reason for this is certainly the lower stability of phosphoric acid esters at the 3′ position of ribose. Formation of a 2′,3′-cyclophosphate, with elimination of the alcohol residue (in this case, the next nucleoside), occurs relatively easily (see Fig. 2.7). This property would, of course, be unsuitable for DNA, which is responsible for the reliable preservation of genetic information. In addition, RNA lacks the methyl group of the nucleobase thymine; this base is then called uracil.

The base pairings of DNA according to Watson and Crick are based on hydrogen bonds, as shown in Fig. 2.8 for the adenine–thymine (A–T) and guanine–cytosine (G–C) pairs. The numbering of the heterocycles is also indicated, which, confusingly, is done in

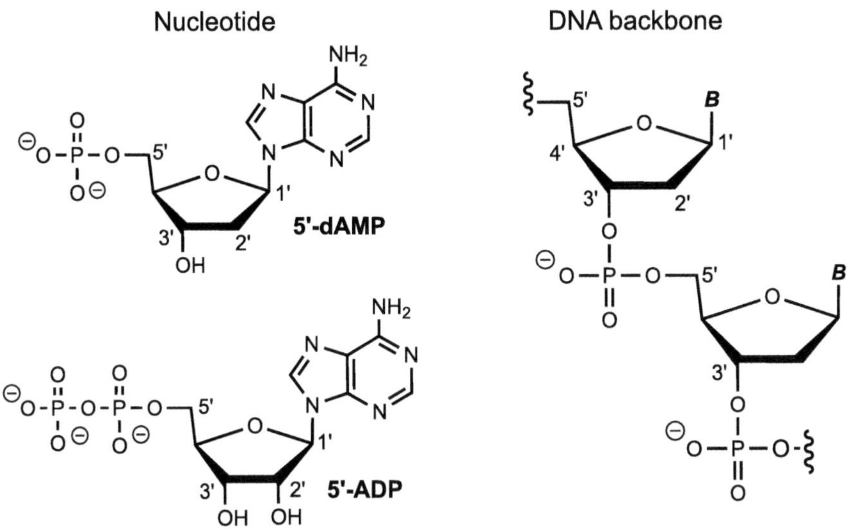

Fig. 2.6 Nucleotides and structure of DNA

Table 2.1 On the nomenclature of nucleobases, nucleosides, and nucleotides

Abbr.	Nucleobase	Nucleoside	Nucleotide (examples)
A	Adenine	Adenosine	5′-**D**eoxy**a**denosine**tri****p**hosphate (5′-dATP)
G	Guanine	Guanosine	5′-**G**uanosine**di****p**hosphate (5′-GDP)
C	Cytosine	Cytidine	5′-**C**ytidine**mono****p**hosphate (5′-CMP)
T	Thymine	Deoxythymidine (often simply called thymidine)	5′-**D**eoxy**t**hymidine**t**riphosphate (5′-dTTP)
U	Uracil	Uridine	5′-**U**ridine**tri****p**hosphate (UTP)

The abbreviations (Abbr.) thus refer to the nucleosides themselves. When discussing the properties of the bases in base pairing, often only the nucleobase itself is mentioned, even though the ribose or deoxyribose residue is, of course, always present. This also avoids specifying whether RNA or DNA is meant.

two different ways for purine derivatives in the literature (see Ref. [3]). A brief comment on these hydrogen bonds is necessary. From general chemistry, it is known that amino hydrogen atoms are very poor hydrogen bond donors. Evidence for this is the boiling point of NH_3, which is about 130 °C lower than that of H_2O, even though N is the better hydrogen bond acceptor. Why, then, should the hydrogen bonds from NH_2 groups to the carbonyl O atom contribute so much to stability? However, these are aromatic amines, in which the lone pair on the N atom is partially delocalized into the heterocycle (cf. the weak basicity of aniline, see table in the appendix to Chap. 1). As a result, the N atom carries a partial positive charge, making the hydrogen atom a good donor.

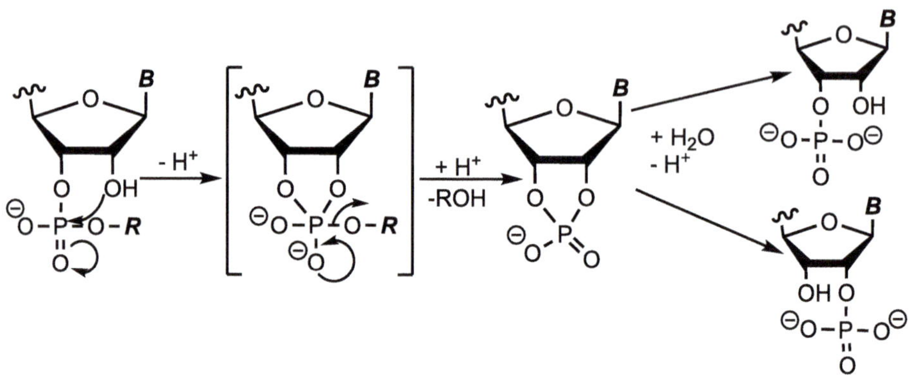

Fig. 2.7 Mechanism of cleavage of phosphodiesters at the 3′ position of ribose

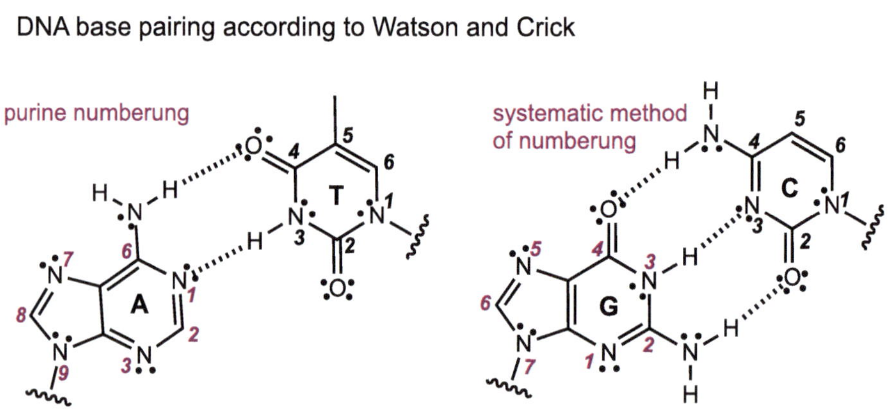

Fig. 2.8 Watson-Crick DNA base pairings and numbering of purines and pyrimidines

Apart from the polar sites of the nucleobases involved in A–T and G–C base pairing, there are additional free polar sites that could enable further dipolar interactions. These are called Hoogsteen hydrogen bonds and can contribute to the formation of triple helices, as well as to interactions with proteins.

In Chap. 6 we will take a closer look at this class of compounds and the biologically relevant methods that organic chemistry can contribute in this area.

2.4 Lipids

The class of "lipids" generally includes organic-chemical substances that are predominantly lipophilic, i.e., not hydrophilic and nonpolar. In these compounds, the hydrocarbon portion predominates, and there are few or no heteroatoms such as O and N. As a result, these compounds are usually not water-soluble. Here are just a few examples (Fig. 2.9):

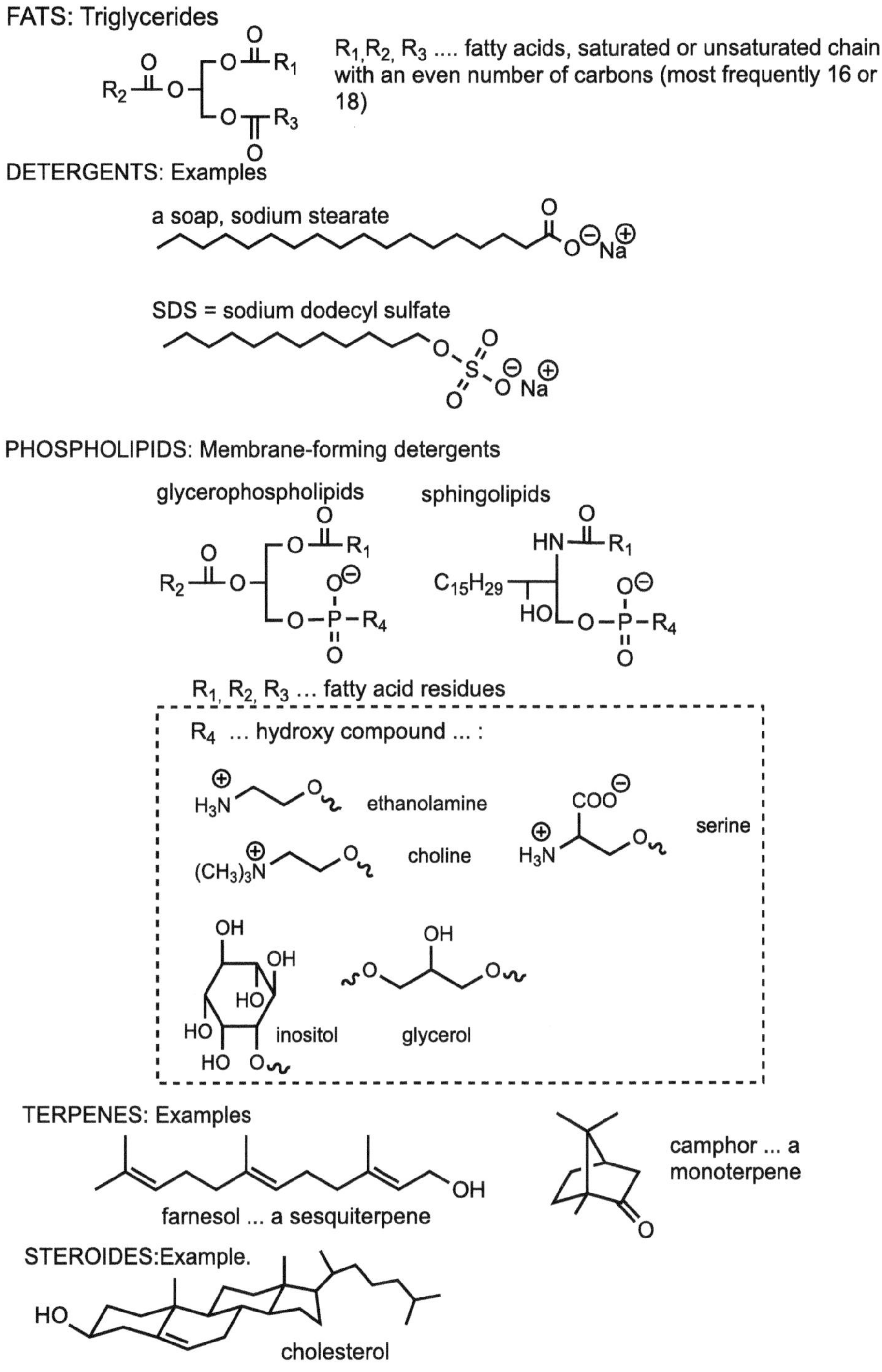

Fig. 2.9 Examples of lipids

- Fats: Triesters of glycerol with long-chain carboxylic acids (so-called fatty acids). The most biologically important representatives of these fatty acids are the saturated ones with 16 carbon atoms (palmitic acid = hexadecanoic acid) and 18 carbon atoms (stearic acid = octadecanoic acid), as well as the unsaturated fatty acids with 18 carbon atoms: oleic acid, linoleic acid, and linolenic acid, which contain one, two, or three isolated *cis*-[–HC=CH–] elements, respectively.
- Detergents: These have a very polar (usually ionic), but narrowly confined region and **at least one** long lipophilic hydrocarbon chain. They are used as cleaning agents, the oldest example being soaps such as sodium stearate [CH_3–$(CH_2)_{16}$–COONa], or as solubilizers for lipophilic substances in aqueous solutions such as SDS (*sodium dodecyl sulfate)* (e.g., by micelle formation). Detergents with two long hydrocarbon chains are capable of forming membranes or membrane-like structures that separate two aqueous compartments. The natural representatives are the phospholipids (glycerophosphatides, sphingolipids).
- Terpenes and, derived from them, steroids are also predominantly lipophilic natural products, which are biosynthesized from a C5 building block, formally from isoprene (methyl-1,3-butadiene), as recognized by Nobel laureate L. Ružička (1921).

2.5 Heterocyclic Natural Products

In nature, we encounter a great many heterocyclic (Sect. 1.3) structures, which consist of aromatic or (partially) saturated rings. We have already encountered some examples: imidazole (amino acid histidine), indole (amino acid tryptophan), pyrimidine and purine (basic structures of the nucleobases), and the porphyrin structure of heme. Many heterocycles are found in secondary plant metabolites such as alkaloids (Fig. 2.10). Some antibiotics from molds also possess heterocyclic substructures, as do certain animal toxins (Fig. 2.11). Many naturally occurring pigments also contain heterocycles (chlorophyll, among many others)—we have omitted their depiction here. The selection of examples is largely arbitrary and is not intended as material to be memorized. We can only marvel at the diversity of these natural products.

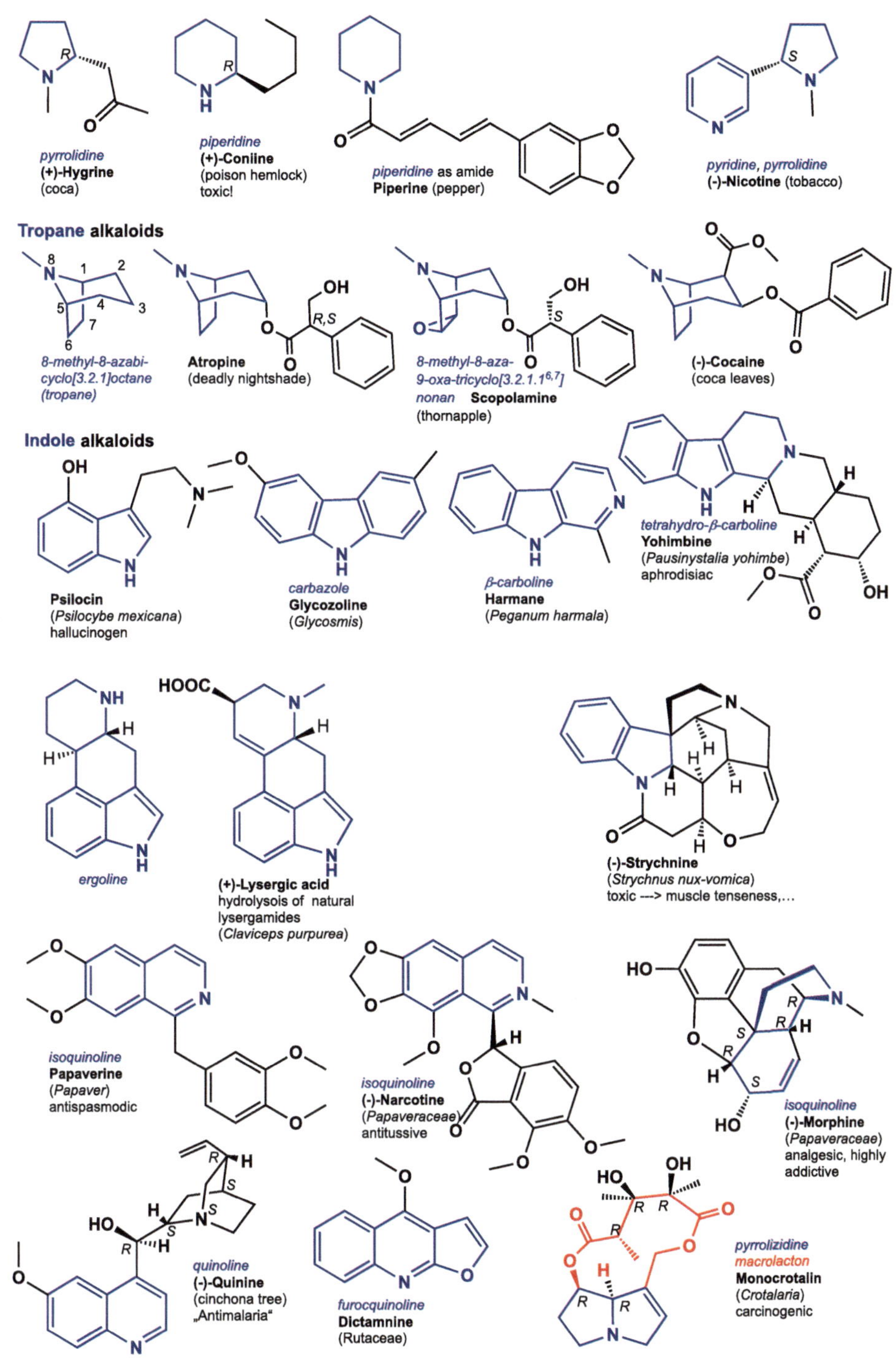

Fig. 2.10 Selection of heterocyclic alkaloid structures

Antibiotics from moulds

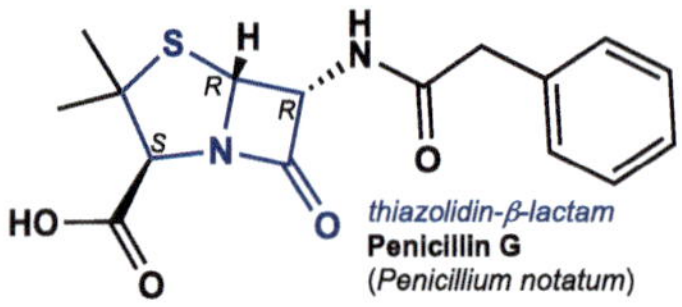

thiazolidin-β-lactam
Penicillin G
(*Penicillium notatum*)

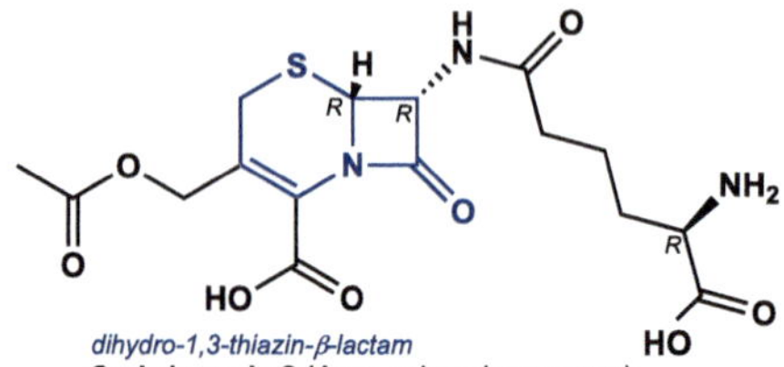

dihydro-1,3-thiazin-β-lactam
Cephalosporin C (*Acremonium chrysogenum*)
less active lead compound of cephalosporin antibiotics

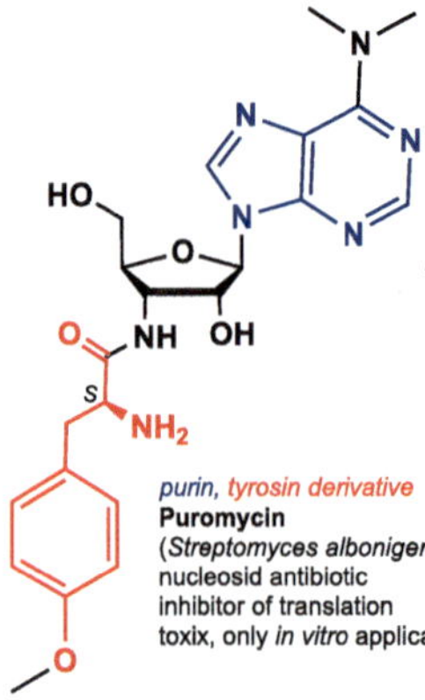

purin, tyrosin derivative
Puromycin
(*Streptomyces alboniger*)
nucleosid antibiotic
inhibitor of translation
toxix, only *in vitro* application

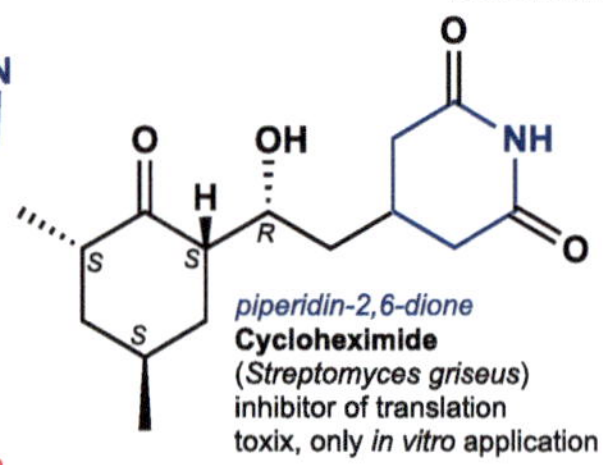

piperidin-2,6-dione
Cycloheximide
(*Streptomyces griseus*)
inhibitor of translation
toxix, only *in vitro* application

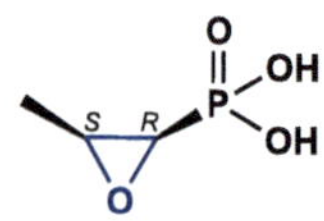

oxirane, epoxide
Fosfomycin
(*Streptomyces*)

Animal toxins

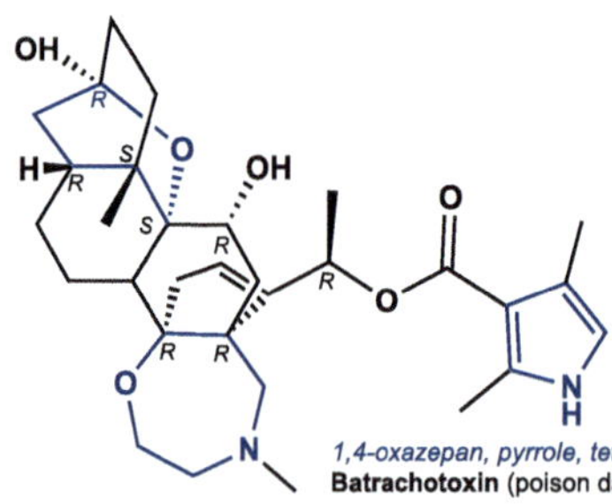

1,4-oxazepan, pyrrole, tetrahydropyran
Batrachotoxin (poison dart frog)

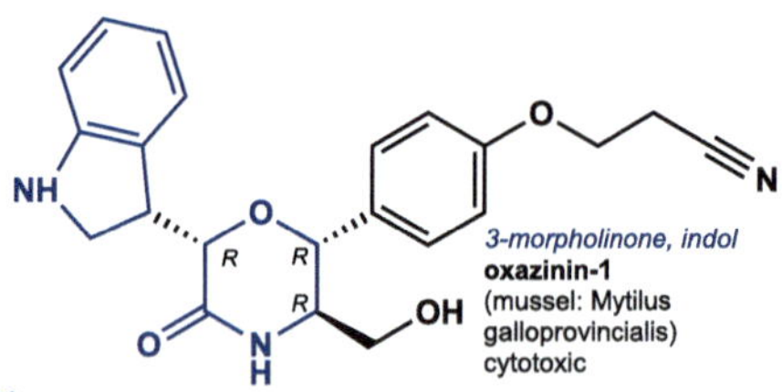

3-morpholinone, indol
oxazinin-1
(mussel: Mytilus galloprovincialis)
cytotoxic

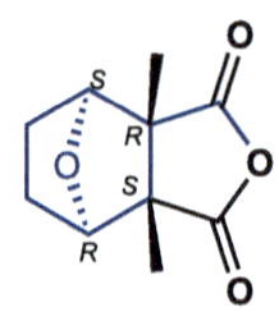

tetrahydrofuran
Cantharidin (meso, not chiral)
(blister beetles)
inhibits protein phosphatases

An extremely complex example:

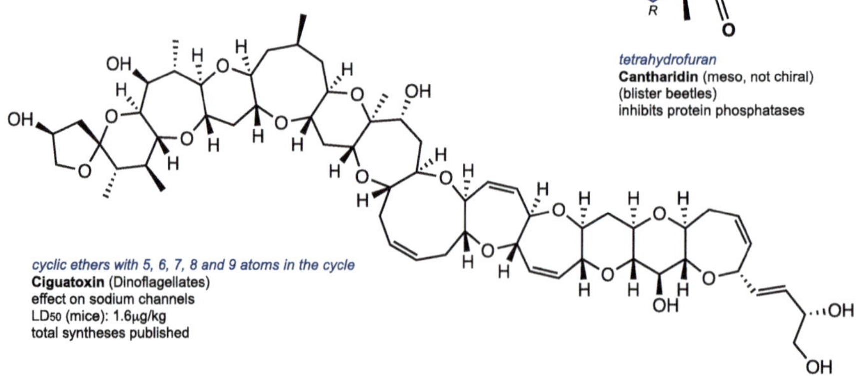

cyclic ethers with 5, 6, 7, 8 and 9 atoms in the cycle
Ciguatoxin (Dinoflagellates)
effect on sodium channels
LD50 (mice): 1.6µg/kg
total syntheses published

Fig. 2.11 Heterocyclic substructures in antibiotics and animal toxins

2.6 Questions

Let's test our understanding of the content of this chapter!

2.1 What is meant by coded proteinogenic amino acids? Try to write them out with names and 3-letter codes.
2.2 Which amino acids have a heterocycle in the side chain? Name these heterocycles and write the formula for these amino acids!
2.3 Using the Fischer projection formula of L-alanine, explain the configuration designation according to the *R*,*S* nomenclature!
2.4 Why is the N atom of a peptide bond not basic?
2.5 Write the Fischer projection formula of D-lyxose and try to derive the Haworth formula for the five-membered ring form! Note that there are two possibilities!
2.6 What is meant by anomers in monosaccharides?
2.7 Why is the glycosidic bond stable under basic conditions?
2.8 What is a nucleoside?
2.9 Write the formula for ATP!
2.10 How are fats structured?
2.11 What is a micelle? Which substances can form micelles?

References

1. Examples from literature concerning our "L configurated protein world": Thiemann, W.: The origin of optical activity. Naturwissenschaften. 61(11), 476–483 (1974). https://doi.org/10.1007/BF00622963; Feringa, B.L., van Delden, R.A.: Absolute asymmetrische Synthese: Ursprung, Beeinflussung und Verstärkung von Chiralität. Angew. Chem. 111(23), 3624–3645 (1999). https://doi.org/10.1002/(SICI)1521-3757(19991203)111:233.0.CO;2-X; Feringa, B.L., van Delden, R.A.: Absolute asymmetric synthesis: the origin, control, and amplification of chirality. Angew. Chem. Int. Ed. 38(23), 3418–3438 (1999). https://doi.org/10.1002/(SICI)1521-3773(19991203)38:23<3418::AID-ANIE3418>3.0.CO;2-V; Maurel, M.-C., Dècout, J.-L.: Origins of life: molecular foundations and new approaches. Tetrahedron. 55(11), 3141–3182 (1999). https://doi.org/10.1002/(SICI)1521-3757(19991203)111:233.0.CO;2-X; Hein, J.E., Blackmond, D.G.: On the origin of single chirality of amino acids and sugars in biogenesis. Acc. Chem. Res. 45(12), 2045–2054 (2012). https://doi.org/10.1021/ar200316n>; Kirschning, A.: On the evolutionary history of the twenty encoded amino acids. Chem. Eur. J. 28(55), e202201419 (2022). https://doi.org/10.1002/chem.202201419; Meinert, C., Cassam-Chenaï, P., Jones, N.C., … Meierhenrich, U.J.: Anisotropy-guided enantiomeric enhancement in alanine using far-UV circularly polarized light. Orig. Life Evol. Biosph. 45(1–2), 149–161 (2015). https://doi.org/10.1007/s11084-015-9413-x
2. Hedges, J.B., Ryan, K.S.: Biosynthetic pathways to nonproteinogenic α-amino acids. Chem. Rev. **120**(6), 3161–3209 (2020). https://doi.org/10.1021/acs.chemrev.9b00408; Armstrong, D.W., Berthod, A.: Occurrence of D-amino acids in natural products. Nat. Prod. Bioprospect. **13**(1), 47 (2023). https://doi.org/10.1007/s13659-023-00412-0

3. Seela, F., Zulauf, M., Sauer, M., … Deimel, M.: 7-substituted 7-deaza-2'deoxyadenosines and 8-Aza-7-deaza-2'deoxyadenosines: fluorescence of DNA-base analogues induced by the 7-alkynyl side chain. Helv. Chim. Acta. 83(5), 910–927 (2000). https://doi.org/10.1002/(SICI)1522-2675(20000510)83:53.0.CO;2-4; Ramzaeva, N., Rosemeyer, H., Leonard, P., … Seela, F.: Oligonucleotides functionalized by fluorescein and rhodamine dyes: Michael addition of methyl acrylate to 2′-deoxypseudouridine. Helv. Chim. Acta. 83(6), 1108–1126 (2000). https://doi.org/10.1002/1522-2675(20000607)83:63.0.CO;2-9

Methods for the Synthesis and Analysis of Peptides and Proteins

3

Abstract

This chapter discusses methods for the synthesis of peptides and provides an overview of analytical techniques for the identification and characterization of peptides and proteins, with a focus on chemical methods—as opposed to purely instrumental methods. The synthetic section covers the most important protecting groups and the main types of coupling methods. Naturally, the solid-phase synthesis method is central, but combinations for the synthesis of larger peptides or small proteins are also mentioned. Methods for the creation of so-called peptide libraries are presented in a separate subsection. In the analytical section, in addition to common instrumental techniques, amino acid analysis and Edman degradation are discussed—methods that, as evidenced by recent literature, remain significant today. The chapter concludes with an example of the combination of solid-phase synthesis and Edman degradation.

3.1 Synthesis of Peptides

▶ **Methodological Reviews** Houben-Weyl, Methods of Organic Chemistry, Vol. 15/1 and 15/2 Synthesis of Peptides, E. Wünsch (Ed.), Thieme, Stuttgart (1974); Houben-Weyl, Methods of Organic Chemistry, vol. E22a—E22e Synthesis of Peptides and Peptidomimetics, M. Goodman (Ed.), Thieme, Stuttgart-New York, (2002) [1].

The synthetic linkage of individual amino acids to form peptides, once a distant goal for organic chemists, today seems to have become an almost trivial routine operation. In this chapter, we will first get to know the most important synthetic methods and also discuss problems and limitations.

D. Krois, *Organic Chemical Methods*, https://doi.org/10.1007/978-3-662-73573-2_3

Let us start at the very basics: The goal is to form a peptide bond between two α-amino carboxylic acids to obtain a *defined* dipeptide. Since the peptide bond is a directional bond—that is, it is not symmetrical—in principle, the formation of two different dipeptides is possible. In Sect. 2.1 we have already established that the peptide bond itself is a normal carboxylic acid amide bond. However, amides cannot be easily synthesized from carboxylic acids and amines. In basic medium, although the proportion of free amine is higher, the carboxylate anion forms, which cannot react with the nucleophile; under acidic conditions, the amine is protonated and thus no longer a nucleophile that could attack the now-activated electrophilic acyl carbon atom. Therefore, activated carboxylic acid derivatives are required. Acid halides of amino acids are not stable or are difficult to prepare, and symmetric anhydrides are rarely used, as half of the valuable amino acid would be lost. So-called "coupling reagents" (Sect. 3.1.2.1) have proven to be very practical, as they form an activated carboxylic acid derivative as an intermediate, which can then react *in situ* with the nucleophile, the amino group of the other amino acid, to form a peptide.

3.1.1 Protecting Groups in Peptide Synthesis

But of course, the coupling reagent does not know which carboxyl group it should activate (Fig. 3.1), and both dipeptides could be formed—but are these really the only two compounds? No, these dipeptides can also further react as amines and/or carboxylic acids, leading to the formation of various mixed oligopeptides. Therefore, "traffic signs" are absolutely necessary to ensure that the correct carboxyl and amino functions react with each other. In chemical synthesis, **protecting groups** are used for this purpose; in biosynthesis within the cell, this is accomplished by enzymes and enzyme complexes. Protecting groups thus serve to temporarily block a functional group of a polyfunctional compound, so that it can later be released again. This leads to the following requirements for protecting groups:

- Easy introduction
- Stability under subsequent reaction conditions
- Selective removal with regeneration of the original functional groups, without side reactions

Two important terms:

> **Key Categories for Protecting Groups**
>
> **Orthogonal** protecting groups: Two protecting groups are orthogonal if each can be selectively removed in the presence of the other.
>
> **Temporary** and **permanent** protecting groups: In the assembly of an oligomer (with repetitive reaction steps), there is a temporary protecting group that must be removed with each extension by a monomer unit, while other functionalities may

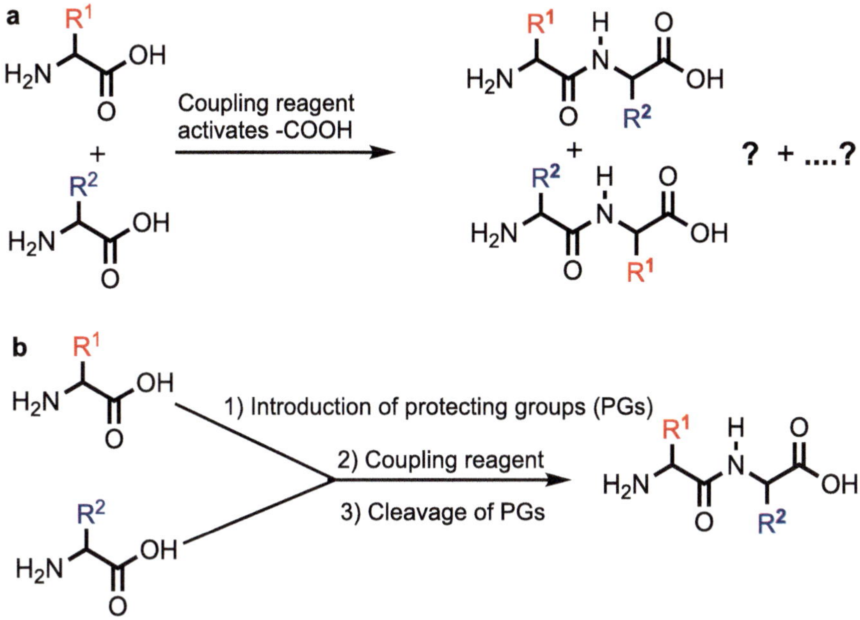

Fig. 3.1 Peptide synthesis: **a** only with coupling reagent, without protecting groups; **b** conditions for obtaining a defined dipeptide product

remain permanently blocked throughout the entire oligomer synthesis. The temporary protecting group must therefore be orthogonal to the permanent protecting groups.

Protecting Groups for the H_2N Function

For the amino function, an amide protecting group is often used in organic synthesis (e.g., the acetyl group for aminobenzenes). This is hardly possible in peptide synthesis, since the peptide bond itself is an amide bond, and thus selective cleavage methods without breaking peptide bonds are difficult to achieve. Therefore, almost exclusively urethane protecting groups are used (Fig. 3.2).

Further advantages of urethane protecting groups include:

- Usually 100% yield upon removal, since one component formed is volatile CO_2 gas.
- Generally, suppression of racemization during activation of the COOH group of urethane-protected amino acids (Sect. 3.1.3).
- Depending on the group R, a wide variety of cleavage conditions are available (Table 3.1).

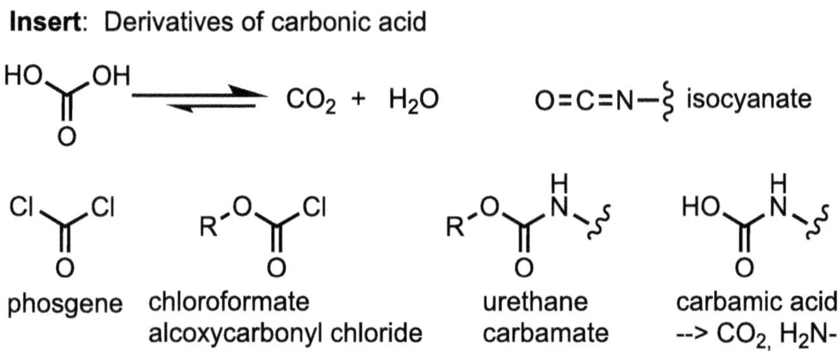

Fig. 3.2 Insert: important derivatives of carbonic acid

Urethane protecting groups are usually introduced using alkoxycarbonyl chlorides, commonly called chloroformates (Fig. 3.3). For the Boc protecting group, the acid chloride does not exist, so the anhydride is used. Alternatively, instead of acid chlorides, activated esters (Sect. 3.1.2.4) can be used, e.g., Fmoc-OSu [1] or Z-OSu.

Removal is carried out depending on the R group and initially yields the carbamic acid, which then spontaneously decarboxylates, thus regenerating the amino function. Table 3.1 lists the three most commonly used protecting groups: Z, Boc, and Fmoc, as well as two more specialized examples that will mainly be discussed in Part 2 of this book. The methods of introduction and the conditions for removal (under which they are labile) are given, as well as the conditions under which they remain stable, i.e., are not removed. From this, one can see, for example, that Fmoc, Boc, and Z protecting groups are orthogonal to each other. Thus, conditions can be found under which only one of the three protecting groups is removed.

For the removal of protecting groups containing *tert*-butyl groups (Boc and tBu protecting groups), trifluoroacetic acid (CF_3–COOH, TFA) is often used as a selective cleavage reagent. Due to the strongly electron-withdrawing CF_3 group, the carboxylate anion is so strongly stabilized that the acid dissociation constant K_a increases by almost five orders of magnitude, i.e., TFA has a pK_a value of about 0 (for comparison, CH_3–COOH: pK_a 4.75), making TFA the strongest carboxylic acid. Nevertheless, it is a much weaker acid than sulfonic acids, sulfuric acid, or HCl and HBr (with strongly negative pK_a values). At the end of Sect. 3.1, the mechanisms of the most common cleavage reactions for the Z, Boc, and Fmoc protecting groups can be found.

[1] –OSu is an abbreviation for *N*-hydroxysuccinimide ester; other abbreviations are –ONSu or –NHS.

Table 3.1 Urethane protecting groups for the amino group in peptide synthesis, with the conditions for their introduction and removal as well as their stability properties. The abbreviations each stand for the following formula: R–O–C(=O)–

Abbr.	R-	Name	Introduction	Stable under	Removal
Z- [1]	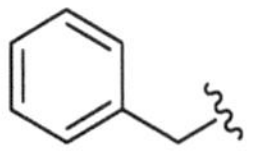	Benzyloxy-carbonyl	Z-Cl, Z-OSu	CF_3COOH (TFA),H_3O^+, bases	H_2/Pd, HBr/AcOH
Boc- [1]	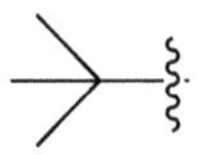	*tert*-Butyl-oxycarbonyl	$(Boc)_2O$	Bases H_2/Pd	TFA, HCl/dioxane, HBr/AcOH,
Fmoc- [2]	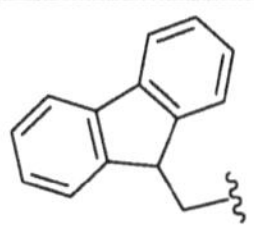	Fluorenyl-methoxycar-bonyl	Fmoc-Cl, Fmoc-OSu	TFA, HBr/AcOH, H_2/Pd	Bases: piperidine, diethylamine
Al(l)oc- [3]	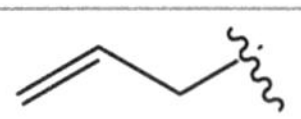	Allyloxycar-bonyl	Al(l)oc-Cl	TFA, bases	Pd(0) + nucleophiles such as mor-pholine or dimedone
AcOZ- [4]	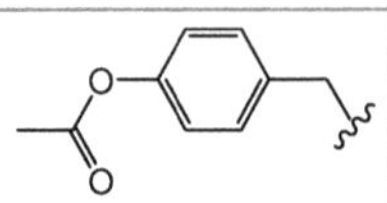	Acetoxyben-zyloxycar-bonyl	AcOZ-Cl	Pd(0), TFA	Acetyl ester-ase (pH = 6)

Protecting Groups for the COOH Function

Carboxyl groups are protected in peptide synthesis, as in organic synthesis in general, as carboxylic acid esters. These protecting groups are introduced as usual either by acidic esterification of the carboxylic acid with gaseous HCl or by an S_N2 reaction of the corresponding alkyl halide with the carboxylate anion of the N-terminally protected amino acid (Fig. 3.4). The specific methods for introduction, removal, and stability conditions are summarized in Table 3.2.

Further Options for Introducing Ester Protecting Groups

- Methyl esters can also be prepared by reaction with acetyl chloride and methanol (HCl_g is formed) or by reaction with methyl chlorosulfinate [CH_3O–S(=O)–Cl], which is produced by cooling from thionyl chloride ($SOCl_2$) and methanol.
- *tert*-Butyl esters are best prepared in larger quantities by reaction with 2-methylpropene (gas) under the influence of strong acids (H_2SO_4 or sulfonic acids) in an autoclave (under pressure).
- Amino acids with protected amino functions can also be converted to esters under activation with coupling reagents in the presence of the corresponding alcohol.

Protecting Groups for Amino Acid Side Chain Functions

Many functional groups in the side chains of amino acids also need to be protected; these must be permanent protecting groups that remain intact throughout the entire peptide

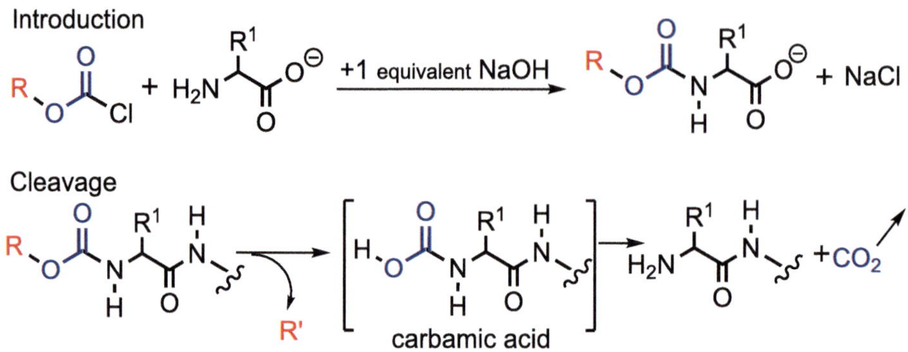

Fig. 3.3 Urethane protecting groups in general: introduction and removal

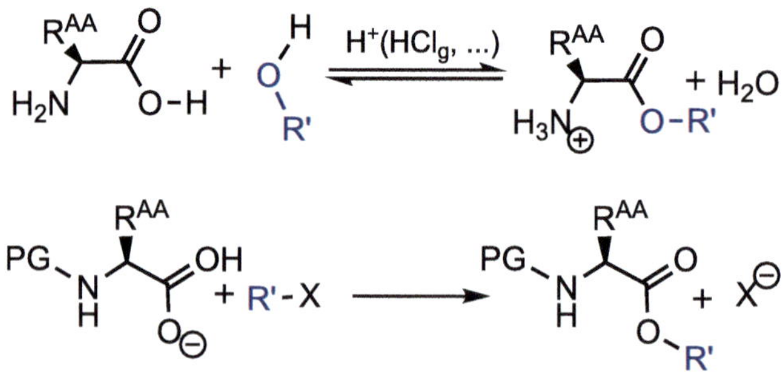

Fig. 3.4 Introduction of ester protecting groups (PG … protecting group, R^{AA}… amino acid side chain)

Table 3.2 Ester protecting groups for the carboxyl group in peptide synthesis, with the conditions for their introduction and removal as well as their stability properties

Abbr.	–C(=O)–	Name	Introduction	Stable under	Removal
–OMe [1]	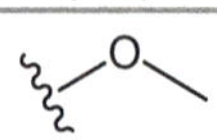	Methyl ester	CH_3OH/ H^+,[CH_3O–S(=O)–Cl]	CF_3COOH (TFA), H_2/Pd	1 equiv. NaOH, H_3O^+
–OBzl [1]	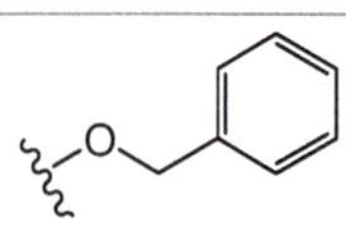	Benzyl ester	Ph–CH_2-OH/ H^+, Ph-CH_2-Br	TFA	H_2/Pd, 1 equiv. NaOH, HF, HBr/ AcOH,
–OtBu [1]	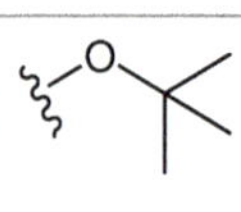	*tert*-butyl ester	2-methylpropene/H_2SO_4 tBuOH + DCC	H_2/Pd, bases, H_3O^+	TFA, HCl/ dioxane,
–O–All [5]	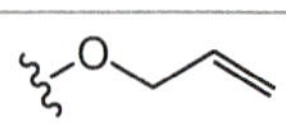	Allyl ester	Via allyl bromide	TFA, amines	Pd(0) + nucleophiles such as morpholine or dimedone

synthesis and are only removed at the end. Their introduction usually requires several reaction steps. We will not discuss these methods in detail here. The following is a brief listing:

- For $-NH_2$ (Lys, Orn) and –COOH (Asp, Glu): protecting groups see tables above
- For –OH (Ser, Thr, Tyr): benzyl (Bzl, Bn) and *tert*-butyl ether (^{t}Bu)
- For –SH (Cys) (many options), e.g.: –S^{t}Bu, –Acm (acetamidomethyl), –Trt (trityl), –THP (tetrahydropyranyl)
- For the guanidino function (Arg): $-NO_2$, $-Z_2$, etc., problematic (removal), or H^+
- For imidazolyl (His): problematic
- Protecting groups are sometimes also needed for Asn, Trp, and "*backbone* NH"

More detailed information can be found, for example, in the volumes of Houben-Weyl [1] as well as in more recent reviews [6–8].

3.1.2 Formation of the Peptide Bond ("Coupling")

To form the peptide bond, the COOH function must be activated. Several methods are available for this purpose—here, we will discuss four fundamentally different types.

3.1.2.1 Use of So-Called Coupling Reagents

In this method, a reagent is added in the presence of both components, the carboxyl and amino components, which activates the COOH group in situ. However, this intermediate is so short-lived that it cannot be isolated and immediately reacts with the nucleophile present, [2] that is, the amino group of the coupling partner. An example is given below:

Activation with Carbodiimides such as DCC (*N,N'*-dicyclohexylcarbodiimide), DIC (*N,N'*-diisopropylcarbodiimide), or the water-soluble EDC (*N*-ethyl-*N'*-dimethylaminopropylcarbodiimide hydrochloride) (Fig. 3.5).

In addition to the H_2N- and COOH-groups to be linked, all other amino and carboxylic acid groups must be protected; hydroxyl groups (as in serine, threonine, and tyrosine) can interfere (→ester formation) and are usually protected as ethers; when reacting in organic solvents (DMF, CH_2Cl_2), proton donors (usually *N*-hydroxy compounds) must be added to suppress the formation of the *N*-acylurea derivative of the carboxylic acid.

The mechanism is schematically explained in Fig. 3.5. As a reactive intermediate, an *O*-acylisourea derivative is postulated, which immediately reacts further in situ. Its

[2] In certain protocols for solid-phase peptide synthesis, a very brief pre-activation is performed, which, however, carries an increased risk of epimerization (racemization of the activated amino acid).

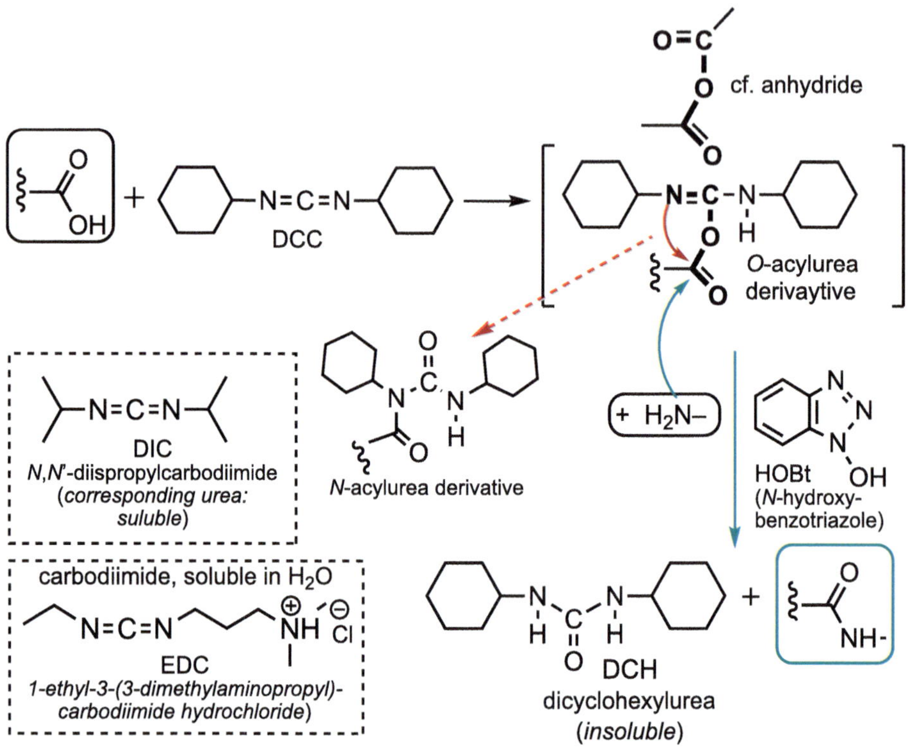

Fig. 3.5 Coupling of the peptide bond with carbodiimides (DCC … *N,N'*-dicyclohexylcarbodiimide)

reactivity can be easily explained by its similarity to an anhydride. As a byproduct, an undesired *N*-acylurea derivative can form, produced by the intramolecular attack of the weakly basic sp^2-hybridized –N=. It has been shown that this side reaction is suppressed in the presence of approximately equimolar amounts of *N*-hydroxybenzotriazole (HOBt) or related *N*-hydroxy compounds. Either an active ester (Sect. 3.1.2.4) is formed as an intermediate, or the weakly acidic *N*-hydroxy compound acts only as a proton carrier in the aprotic solvent (dichloromethane, *N,N*-dimethylformamide … DMF, etc.). In any case, this additive remains unchanged after the reaction—thus, it is essentially a catalyst.

Other Coupling Reagents
Examples include EEDQ (1-ethoxycarbonyl-2-ethoxy-1,2-dihydroquinoline) [9] (mechanism see Sect. 9.1), BOP [benzotriazolyloxytris(dimethylamino)phosphonium hexafluorophosphate] [10], HATU (1-[bis(dimethylamino)methylene-1*H*-1,2,3-triazolo[4,5-b]pyridinium-3-oxide hexafluorophosphate) [11], and COMU (*N*-[1-(cyano-2-ethoxy-2-oxoethylideneaminooxy) dimethylamino(morpholino)uronium hexafluorophosphate) [12] (Fig. 3.6). They are used in a similar way to carbodiimides, namely, activation by the reagent occurs in the presence of both reaction partners: carboxyl and amino components.

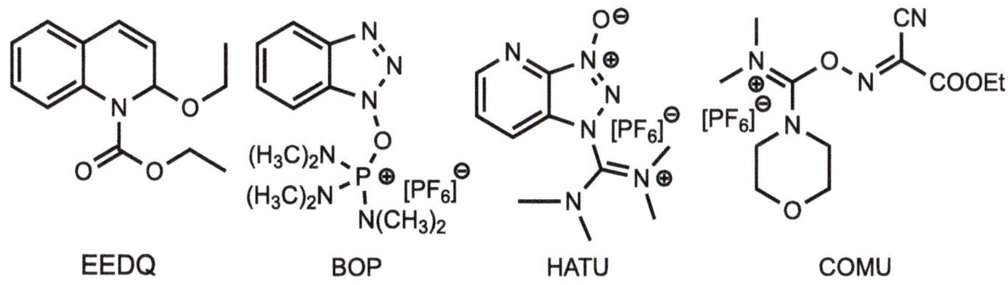

Fig. 3.6 Coupling reagents EEDQ, BOP, HATU, and COMU

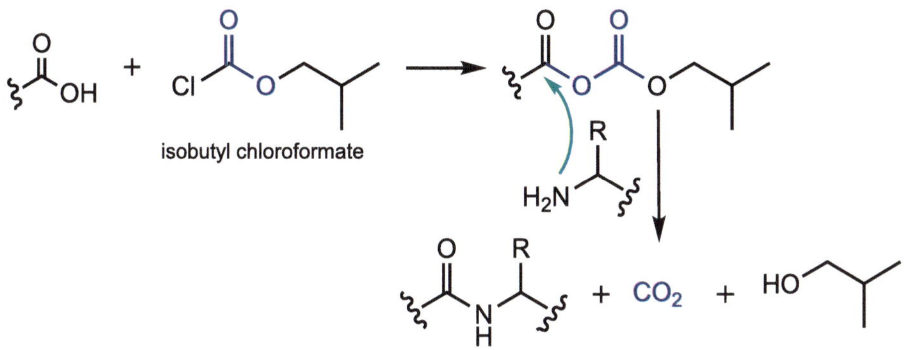

Fig. 3.7 Activation of N-terminally protected amino acids as mixed anhydrides (e.g., with chloroformic acid isobutyl ester)

While a mixed anhydride (Sect. 3.1.2.2) is postulated as an intermediate for EEDQ (mechanism see Sect. 9.1), the other three reagents (BOP, HATU, and COMU) form active esters (Sect. 3.1.2.4) as intermediates. However, since these intermediates are not isolated, activation of the carboxylic acid must occur in the presence of the amino component—as with carbodiimides.

3.1.2.2 Use of Mixed Anhydrides

Activation via **mixed anhydrides** is less commonly used today, as this reaction must be carried out under strong cooling and in complete absence of moisture. First, the COOH group is activated by reaction with chloroformic acid ethyl ester (or isobutyl ester), forming a mixed anhydride that is stable for a certain time at low temperature. The amino component is then added, which may also have unprotected COOH groups (see Fig. 3.7).

3.1.2.3 Activation as *N*-Carboxyanhydrides (NCA)

The cyclic *N*-carboxyanhydrides (NCA) of amino acids were already described by Leuchs in 1906 (also known as Leuchs anhydrides). They have been and are still used today for the production of amino acid polymers, which are important in materials science as well as for biomedical applications [13]. NCAs are prepared by reacting the free amino acid with phosgene [14] (see Fig. 3.8). They are stable and storable under certain conditions, but polymerize very easily and are therefore not suitable for the controlled stepwise assembly of peptides with uniform sequence. Using precisely optimized procedures with *continuous-flow technology*, they can be used for the synthesis of di- and tripeptides in larger quantities [15]. When protected with a urethane protecting group such as Boc, Z, or Fmoc, they are referred to as UNCA, which are indeed suitable for peptide synthesis on solid phase or in solution [16] (see also [8]). The advantage over the previously discussed methods (see Sects. 3.1.2.1 and 3.1.2.2) is that the activated amino acid is stable, yet highly reactive toward nucleophiles (cf. Sect. 3.1.2.4). However, being anhydrides, UNCAs are also easily hydrolyzed—exclusion of water is therefore important, which complicates their use.

3.1.2.4 Use of Active Esters

An active ester is a carboxylic acid ester (CO–O–R) whose acyl C atom is particularly susceptible to nucleophilic attack due to electron-withdrawing groups on the –O–**R** moiety. The following active esters, for example, are used in peptide synthesis: *N*-hydroxysuccinimide ester (–OSu, or –NHS), pentafluorophenyl ester (–OPfp), dinitrophenyl ester (–ODnp), etc. In all these cases, the active ester is prepared separately, e.g., using carbodiimides (Fig. 3.9). The active ester is usually crystalline and stable and can be stored in the refrigerator for extended periods.

To form the peptide bond, only the active ester and the amino component need to be dissolved in a suitable solvent, such as dioxane, acetonitrile, or DMF. The reaction occurs immediately and usually very rapidly. Since in this case the activated carboxylic

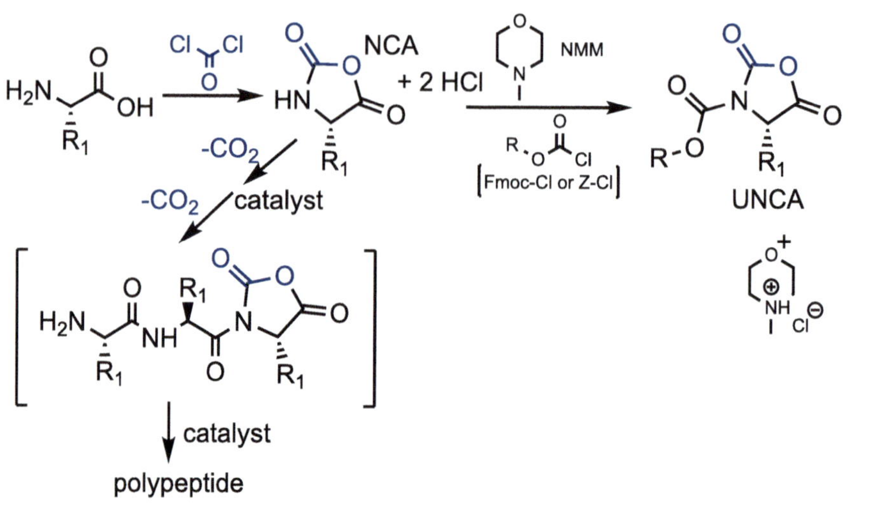

Fig. 3.8 Formation of NCA and their polymerization. The N-terminally protected UNCAs can be used as activated amino acids in targeted peptide synthesis

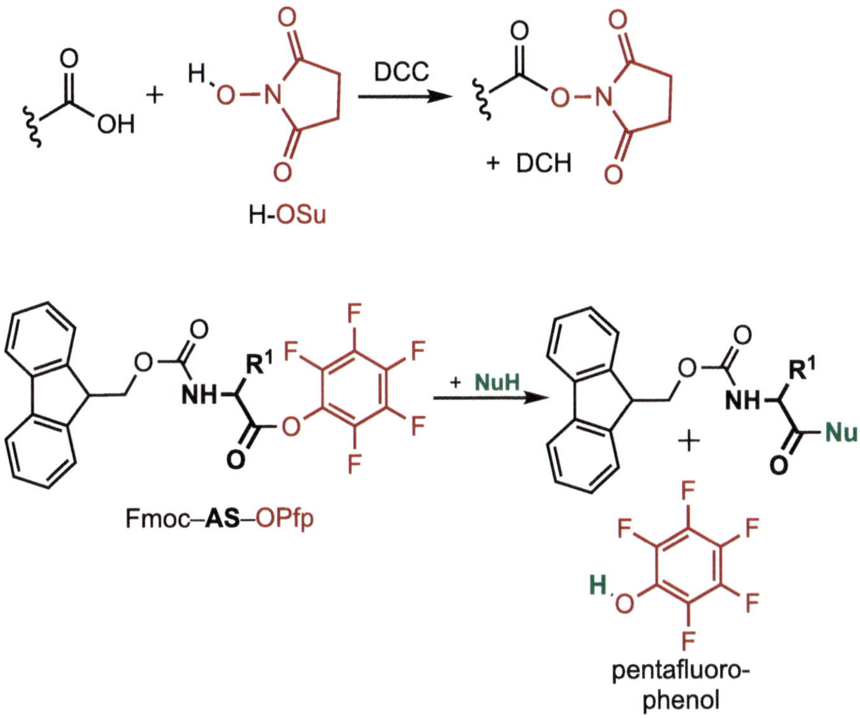

Fig. 3.9 Preparation and use of active esters

acid is defined from the outset, COOH groups on the amino component do not need to be protected. At least in the case of *N*-hydroxysuccinimide esters, any hydroxyl groups present in the side chains do not interfere either; these active esters react very selectively with primary and secondary amines. Pentafluorophenyl esters are particularly reactive and are used in solid-phase synthesis; however, the pentafluorophenol released is highly toxic, whereas *N*-hydroxysuccinimide (HOSu) is even biocompatible. *N*-hydroxysuccinimide esters are therefore widely used for labeling proteins or other natural products, even *in vivo,* (see Chap. 7).

3.1.3 Strategies for the Synthesis of Longer Peptides

With the methods discussed above for the use of protecting groups and the formation of the peptide bond, peptide synthesis can now be approached. As follows from the previous section (Sect. 3.1.2), the carboxyl group is almost always activated and reacted with the amino group as nucleophile, which, of course, must not be fully protonated. However, a problem can sometimes arise: partial racemization at the α-C atom of the activated amino acid may occur. This leads to diastereomers in a peptide, as this amino acid will then be present in part in the opposite D-configuration.

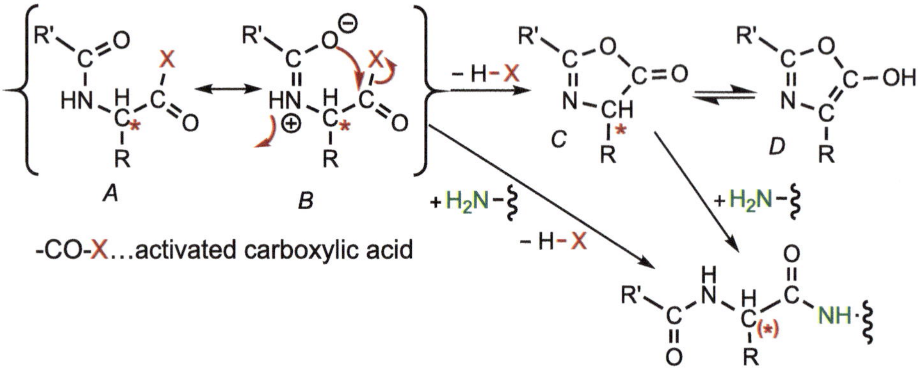

Fig. 3.10 Mechanism of racemization of activated amino acids

Racemization of Activated Amino Acids

The mechanism of this racemization is shown in Fig. 3.10. The resonance structure *B* of the activated amino acid explains the tendency to form the 1,3-oxazolin-5-one derivative *C*. However, a tautomeric equilibrium with form *D* then leads to the loss of chiral information at the α-C atom. Since both the activated carboxylic acid derivative (*A*, *B*) and the oxazolinone *C* can react with the nucleophile, the configuration at the corresponding α-C atom will be partially inverted in the resulting peptide. However, if the activated amino acid is protected at the α-amino group as a urethane, i.e., the R′ group is an alkoxy group, this racemization practically does not occur; the influence of this O atom apparently reduces the significance of resonance structure *B* and the corresponding conformation, so that ring closure to the oxazolinone does not take place.

Through precise measurements of the proportion of D-configured amino acids (Sect. 3.2.3) after peptide couplings, it has been proven that this racemization pathway is suppressed in the case of activation of amino acids protected as urethanes. For this reason, urethane protecting groups such as those in Table 3.1 are used almost exclusively for the α-amino group, and synthesis is started at the C-terminus, so that only N-terminally protected amino acids need to be activated and not entire peptide fragments. A synthesis proceeding from the N- to the C-terminus would also be fundamentally unreasonable, since increasingly larger peptides would have to be activated; this would certainly lead to very different reactivities. However, since even small amounts of racemized amino acid residues in peptides cause problems in purification, very detailed studies have been conducted on this topic. The proportion of oppositely configured amino acids depends on the amino acid, the coupling method, the basicity of the base used, and the temperature [6, 7].

3.1.3.1 Convergent Synthesis in Solution

This is the method of peptide synthesis originally developed (from about 1950), which is still used today—especially for the production of larger quantities. The usual procedure is as follows:

I. Preparation of peptide fragments (about 2–10 amino acids long), each by successive coupling of protected amino acids, starting from the C-terminus (solvent: usually dichloromethane or DMF), followed by removal of the temporary α-amino protecting group (e.g., Z or Boc, possibly also Fmoc) including extractive purification after each coupling step or precipitation.
II. Removal of the protecting group from the COOH or NH_2 groups to be linked.
III. Coupling of the individual peptide fragments (so-called fragment coupling), avoiding racemization (see Sect. 3.1.3 above). For this, glycine, but also proline, which cannot racemize by this mechanism (Fig. 3.10), are suitable at the C-terminus. Several methods may need to be tested for this step, and insufficient solubility can also be a problem.
IV. Removal of the remaining permanent protecting groups (release of the peptide), followed by purification by gel chromatography and/or RP-HPLC.
V. Characterization by AAA (Sect. 3.2.3) (→peptide content), determination of configurational homogeneity, MS.

↗In this way, peptides with up to 40 amino acids have been and are produced in large quantities (several grams to kilograms) (including as pharmaceuticals, such as somatostatin). Of interest in this context is also the review by L. Moroder on the production of insulin and its variants, comparing molecular biological and chemical synthetic methods [17].

3.1.3.2 Solid Phase Peptide Synthesis (SPPS)

This method was introduced by **R. B. Merrifield** (Nobel Prize 1984). What is special about it? The oligomer (in our case, the peptide) grows on a polymeric insoluble support, often referred to as resin. This allows the growing peptide to be separated at any time from all reagents and by-products, which must all be highly soluble, simply by filtration (washing cycles!). Time-consuming extraction purifications after each reaction can thus be avoided. The process is also very easy to automate. Commercial synthesizers are quite simple machines. They essentially consist of pumps and valves that introduce reagents and solvents into the reaction vessel and then remove them again (Fig. 3.11). There are two variants: either the solid phase (on which the peptide is synthesized) is in a vessel closed by a fritted glass seal (*in batch*), or it is in a kind of column (*in column, continuous-flow*, as in chromatography), and the reagents are cyclically pumped over the column. This can achieve very high local concentrations, which has a beneficial effect on the reaction kinetics [7, 18]. The principle of solid phase synthesis is, of course, no longer used only in peptide synthesis today, but is particularly suitable for all repetitive syntheses (Sects. 6.2 and 4.2).

Regarding the synthesis strategy, Merrifield worked with Boc as a temporary protecting group for the α-amino group and Z, benzyl esters and ethers as permanent protection

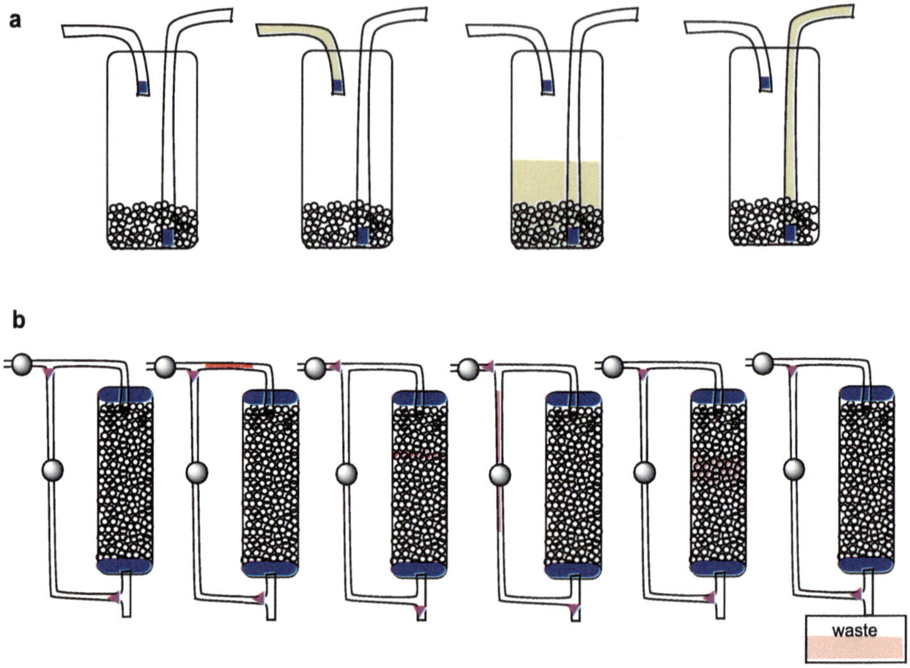

Fig. 3.11 Principle of solid phase peptide synthesis, timeline of a treatment with a reagent including washing: **a** *in batch* method; **b** *in column* method

for side chain functionalities. As resin, he used cross-linked polystyrene with a Cl–CH_2–Ph anchor for the C-terminal amino acid. The cleavage of the side chain protecting groups was carried out together with the cleavage from the support (polymer-bound benzyl ester) using anhydrous HF.

The method was improved by **R. C. Sheppard** [19]: He used Fmoc as a temporary protecting group for the α-amino function, and the protection of functionalities in the side chains is essentially based on *tert*-butyl protecting groups (i.e., Boc, *tert*-butyl esters, and *tert*-butyl ethers), so that the cleavage of these protecting groups along with cleavage from the resin (e.g., polyacrylamide, via an acid-labile linker) can be achieved with 95% TFA. Today, specialists have access to many different polymeric supports and anchors. By the way: polymer + anchor can be considered as a C-terminal protecting group for the peptide, and accordingly, there are various variants with different cleavage conditions. A typical synthesis scheme is shown in Fig. 3.12. At the end of the synthesis cycles, consisting of peptide coupling (with coupling reagents such as DIC or activated esters such as pentafluorophenyl ester), washing cycle, cleavage of the temporary amino protecting group, and washing cycle, depending on the anchor used, the peptide can be simultaneously freed from the side chain protecting groups **and** cleaved from the resin (→ free peptide), or **only** the side chain protecting groups can be cleaved, resulting in a peptide bound to a solid phase, which is advantageous for certain screening methods.

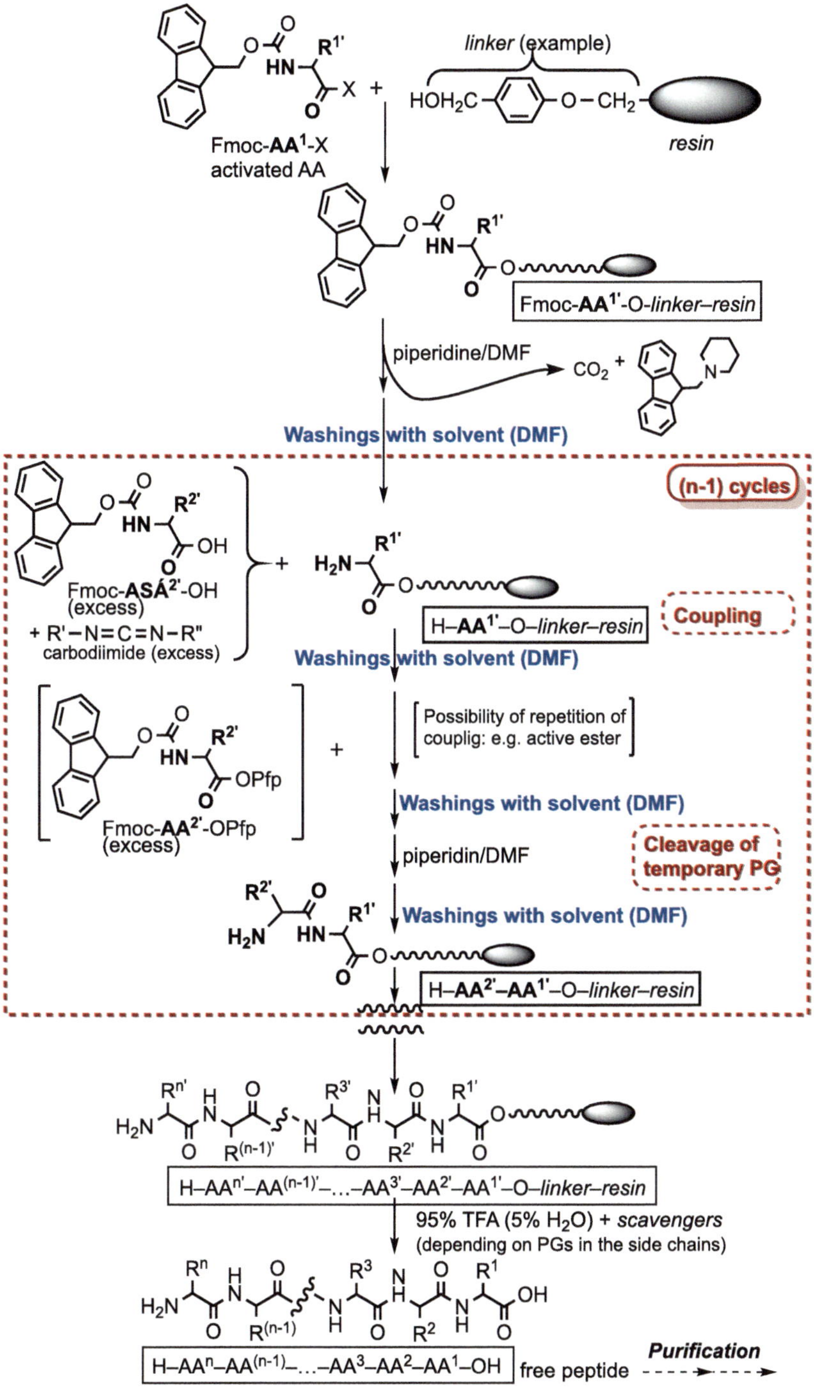

Fig. 3.12 Synthesis scheme of solid phase peptide synthesis (SPPS); AA … amino acid, DMF … *N*,*N*-dimethylformamide; TFA … trifluoroacetic acid

The advantages of the method are obvious: reagents and soluble by-products are removed simply by washing with solvent. Automation saves a lot of time and technician hours. Syntheses of medium-sized peptides are completed in one to two days, or today even much faster [18], whereas solution-phase synthesis often required weeks. Especially for the production of small amounts of peptide (milligram scale), as often needed in biological research, SPPS is now the only viable alternative.

But of course, there are also disadvantages: large amounts of solvents are consumed, and typically about fivefold excesses of reagents must be used. This results in very high material costs when producing larger quantities. Why must such large excesses be used?—One must strive to achieve yields of 99–100% per reaction step, otherwise products with gaps (missing amino acids) accumulate on the solid support. Table 3.3 lists the percentage of peptides with one, two, or more gaps depending on the yield achieved per cycle (extension by one monomer unit). For longer peptides, yields of 99% per cycle are indeed necessary to ultimately ensure usable peptide yields and practical purification. Just as an example: For a 30-mer peptide, with "only" 95% yield per chain extension, 30 peptides with **one** missing amino acid are expected, totaling 34% yield, as well as 435 peptides with **two** gaps, which still account for 26% of the total.

If one wants to obtain a highly pure peptide, chromatographic purification must, of course, be performed after cleavage from the solid phase. This is usually done by RP-HPLC. However, given the large number of "gap peptides" (each in small amounts), it is likely that several of them will have very similar retention to the desired peptide and will therefore remain "hidden" under the main peak. For many purposes, this will be irrelevant. However, if the goal is to identify very specific biological activities, even a peptide present at a few percent or even less can mimic an activity of the target peptide. For peptides approved as pharmaceuticals, proven high purity of each batch is absolutely essential. One must therefore be aware of these possibilities and, if necessary, perform more detailed analyses. In solution-phase synthesis, larger amounts remain unreacted at each individual step (since usually only equimolar amounts of reagents are used)—but purification is performed at each step, and thus these "gap peptides" do not accumulate over the entire synthesis. In SPPS, so-called *capping* can be used to facilitate the purification of "gap peptides" by adding a reagent after each coupling step that blocks the unreacted free NH_2 groups, so that these "gap peptides" are then much shorter and easier to separate.

An important parameter in SPPS is, understandably, the ***loading*** of the polymer with anchor groups. This determines how many peptide chains grow on a polymer particle, which in turn determines the distance between individual peptide chains. This is illustrated schematically in Fig. 3.13. If this distance is too small, stronger interactions between the peptide chains can occur (depending on the sequence and the side chain protecting groups used), leading to the formation of β-sheet structures. In this case, the peptide chains are no longer properly solvated, and the reagents can no longer reach the

Table 3.3 Dependence of the amounts (% n/n) of peptide with the desired sequence and the "gap peptides" formed on the effective coupling yield (= coupling yield, yield of cleavage of the temporary amino protecting group and formation of the peptide bond) in SPPS

Peptides with *r* gaps	Number of peptides	At 99% coupling yield		At 97% coupling yield		At 95% coupling yield	
		Each at % *n*/*n*	% total	Each at % *n*/*n*	% total	Each at % *n*/*n*	% total
For a peptide with ten amino acid residues:							
0	1	90.44	90.44	73.74	73.74	59.87	59.87
1	10	0.91	9.1	2.28	22.81	3.15	31.51
2	45	0.0092	0.42	0.071	3.17	0.166	7.46
3	120	0.00009	0.01	0.0022	0.26	0.0087	1.05
4	210	0.000001	0.002	0.000067	0.014	0.00046	0.097
For a peptide with 30 amino acid residues:							
0	1	73.97	73.97	40.1	40.10	21.46	21.46
1	30	0.747	22.42	1.24	37.21	1.13	33.89
2	435	0.00755	3.28	0.0384	16.69	0.0595	25.86
3	4060	0.000076	0.31	0.00119	4.82	0.00313	12.70
4	27,405	0.00000077	0.02	0.000037	1.01	0.000165	4.51
5	142,506	0	0.00	0.0000011	0.16	0.0000087	1.24
For a peptide with 100 amino acid residues:							
0	1	36.6	36.60	4.76	4.76	0.592	0.59
1	100	0.3697	36.97	0.147	14.71	0.0312	3.12
2	4,950	0.00373	18.49	0.00455	22.52	0.00164	8.12
3	161,700	0.000038	6.10	0.000141	22.75	0.000086	13.96
4	3,921,225	0.00000038	1.49	0.0000044	17.06	0.0000045	17.81
5	75,287,520	0.0000000038	0.29	0.0000001	10.13	0.00000024	18.00

For calculation:
Number of possible combinations of *r* "gaps" within a peptide of *n* amino acids: $\binom{n}{r} = \frac{n!}{(n-r)!r!}$

% yield for a peptide with *r* "gaps" at an expected chain length of *n* amino acids in the case of a coupling yield of *a* %: $\% = 100\left(\frac{a}{100}\right)^{(n-r)} \cdot \left(1 - \frac{a}{100}\right)^{r}$

reaction sites ⇛ the synthesis more or less stops, and only severely truncated peptides are isolated. In modern synthesizers, the yield of the cleavage of the temporary protecting group (especially Fmoc) can be monitored online spectroscopically, so that such a synthesis interruption can be detected at an early stage.

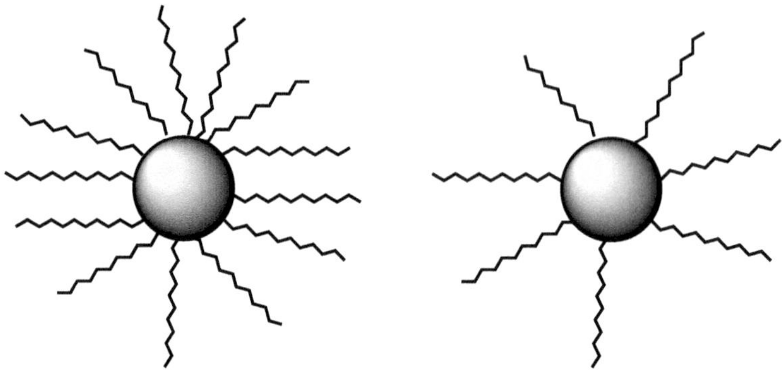

Fig. 3.13 Schematic representation of a polymer particle with higher or lower loading of growing peptide chains

The reactions in SPPS are therefore not actually reactions in the solid phase, but rather the reactive parts of the peptide chain must be "dissolved" in the solvent; they are only "suspended" on the surface of a solid particle. The "insolubility" of the peptide chains on the solid support also depends strongly on the amino acid sequence. In some cases, a special tactic must be used for such "difficult peptides": special protecting groups for the backbone itself or the introduction of modified amino acids that can later be converted into the desired one [20].

Due to the environmental impact factors, such as high consumption of hazardous or toxic solvents and large excesses of reagents, which themselves must be produced in environmentally harmful ways, as well as overall high disposal costs, there are serious efforts to solve these problems of SPPS. Discussing this in detail would go beyond the scope of this book, as it also involves technological details. Some references on this topic can be found in [21].

Advantages and Disadvantages of Solid Phase Synthesis

Advantages of the method:

- No complicated purification operations during synthesis (extraction, recrystallization, etc.), instead washing cycles
- High yields per coupling step
- Automation $\Rrightarrow$ low personnel costs
- Significant time savings
- Well suited for all types of peptide screening

Disadvantages of the method:

- High solvent consumption
- High consumption of reagents due to large excesses ⇛ resulting in high material and disposal costs
- Accumulation of "gap peptides" ⇛ more elaborate final purification for highly pure peptides
- Aggregation of peptide chains is not as readily apparent as in solution-phase synthesis, where the peptide may no longer dissolve, ⇛ risk of synthesis failure

3.1.3.4 Combination of Multiple Methods for the Synthesis of Large Peptides or Proteins

When larger peptides or even proteins are to be synthesized, a combination of several methods is often used today.

(a) Fragment couplings: Synthesis of fragments using SPPS, ensuring that the protecting groups for the side chain functionalities (permanent protecting groups) and the temporary protecting group remain intact when the fragments are cleaved from the resin, followed by coupling of these fragments to the resin-bound C-terminal fragment [22].
(b) Synthesis of fragments using SPPS—as above—but fragment couplings are then performed in solution as in (a) (coupling by chemical methods).
(c) Synthesis of fragments using SPPS, but cleavage of all or most protecting groups and subsequent fragment couplings in solution using enzymatic methods [23].
(d) Protein synthesis: Parts of the protein are produced by molecular biology (by expression), and are linked to fully synthetically produced (modified) peptide segments by *native chemical ligation* (NCL) or other ligation methods (see Chap. 8).

3.1.4 Cleavage of Protecting Groups and Purification

After any type of chemical peptide synthesis, the permanent protecting groups on the side chains (usually based on benzyl or *tert*-butyl groups) must be removed at the end of the synthesis. This general cleavage is usually carried out under acidic catalysis. In the process, carbocations (electrophiles), R^+, are formed, which can react with aromatics via substitution (especially Tyr, Trp) ⇛ undesirably modified amino acids. Therefore, so-called *scavengers* (traps for electrophiles, i.e., nucleophiles themselves) such as H_2O, ethanedithiol, anisole (methoxybenzene), or thioanisole often need to be added during the cleavage of protecting groups. In any case, one must always expect the formation of modified peptides (easily recognized in RP-HPLC by longer retention times, i.e., they elute only at a higher proportion of acetonitrile in the eluent). The release of the peptide in the final step of a synthesis is therefore often first tested with small amounts to determine the optimal conditions, and only then is the main batch released.

3.1.5 Multiple Peptide Syntheses and Production of "Peptide Libraries"

To use peptides for screening-methods, many varied peptides must be produced simultaneously. This can be achieved by multiple peptide syntheses. More or less systematically constructed peptide libraries are generated. These methods were first developed in the 1980s [24–27].

Some methods for producing such libraries:

- *Tea-bag* method: labeled polypropylene nets, each containing about 100 mg of resin sealed inside; couplings are performed separately, cleavage and washing are done together; → peptides can be analyzed, approx. 50 mg [28]
- Synthesis on PE sticks (8 × 12/block, fits into microtiter plates), peptide is not isolated, for ELISA (PEPSCAN), approx. 300 nmol/stick (*pin*) [29]
- Synthesis using synthesis robots with a variety of polymers and vessels
- Peptide libraries can also be generated molecular genetically, usually expressed within a fusion protein, e.g., by a phage (*phage-display*). For a complete peptide library of, for example, a pentadecapeptide (15-mer peptide), 20^{15} = approx. 3.3×10^{19} clones would be required. If only a small fraction of these is available, repeated mutagenesis and selection can be used to find an optimal peptide sequence, which can then be identified by sequencing. This process is called "affinity maturation," in analogy to the development of optimal antibodies against antigens by the immune system [30]. New studies using rapid sequencing methods (NGS … *next generation sequencing*) show that there is not a statistically equal distribution of all possible peptide sequences, since the incorporation of certain nucleotides (in this example, guanine) at variable positions is preferred. Therefore, certain amino acids are overrepresented in the peptides [31].
- The discovery of biologically active *cyclic* peptides has recently become particularly important for drug development (*drug design*) [32]. An innovative chemical-synthetic method for producing a corresponding library of cyclic peptides was recently introduced [33]. Of course, there are also several genetically encoded, i.e., molecular biological, approaches for this [34].

In the following, we will look at the general procedure for "fishing" peptides that show a defined interaction with a biologically relevant molecule (receptor, antibody, enzyme), as described in reference [26].

Within a peptide, different types of positions are assigned. The so-called fixed positions are those for which a specific amino acid has already been identified as essential;

defined positions are those for which certain amino acids are to be tested. This results in so-called sublibraries; (if only coded proteinogenic amino acids are used: 1 def. pos. → 20 sublibraries, 2 → 400, $m \rightarrow 20^m$). The remaining positions are "mixed positions". At these sites, an equimolar mixture of all amino acids should be present. For n mixed positions, each sublibrary contains 19^n (cysteine is usually excluded here due to dimer formation via disulfide bridges) different peptides in the mixture. The defined position of the *sublibrary* that shows the highest activity (affinity) in the test is now assigned as the new fixed position. Table 3.4 shows the procedure for the so-*called* iterative deconvolution process, and Table 3.5 shows the *positional scanning deconvolution process*. To ensure accuracy, in the second method, the peptide of the identified sequence must be synthesized in a defined manner at the end.

But how are such *mixed positions* synthesized? There are basically two possibilities for this.

First, the more precise method, called *divide—couple—recombine* (Fig. 3.14). For each mixed position X, the polymer is divided into as many portions as amino acids are to be tested (so, for only coded proteinogenic amino acids without cysteine: 19), then each is coupled separately with the respective activated N-terminally protected amino acid, and then all are recombined and mixed. If another mixed position X is to follow, the process is repeated, and so on. One must consider whether there are enough growing peptide chains to allow all desired combinations. A special variant for peptide libraries is the OBOC (*one-bead-one-compound*) method, which uses *mixed positions* of this type—an example of which we will discuss in Sect. 3.3.

The simpler, but less precise method is to attempt to couple all desired amino acids simultaneously in ***one*** coupling step in the same ratio. Table 3.6 lists the relative amounts of amino acid derivatives used to achieve the most even distribution possible:

These figures also provide information about the relative coupling rates of amino acid derivatives during peptide bond formation. Most striking among these numbers are the two β-branched amino acids isoleucine and valine, which react significantly more slowly than all others due to steric reasons.

APPENDIX (On the cleavage of the amino protecting group)Protecting group

The most selective cleavage of the Z-protecting group as well as other benzyl protecting groups is based on hydrogenolysis (Fig. 3.15), a reaction typical for benzyl derivatives. Hydrogen dissolved in a finely dispersed metal, namely Pd(0), is transferred to the benzylic bond, causing it to split homolytically. Toluene and the corresponding hydroxylated compound are formed (for Z → carbamic acid, for benzyl esters → carboxylic acid, for benzyl ethers → alcohol).

Cleavage of *tert*-butyl protecting groups occurs via protonation of the corresponding C=O oxygen atom. This automatically leads, through resonance, to a positive charge on the O atom adjacent to the tertiary C atom. Since this is known to form a highly stabilized carbocation (carbenium ion), the protonated compound consequently decomposes into the *tert*-butyl cation, which then reacts further either by elimination to methylpropene or by addition to a nucleophile (e.g., H_2O) (Fig. 3.16).

Table 3.4 ***Iterative deconvolution process*** for identifying defined sequences using a hexapeptide library (X corresponds to an equimolar mixture of the 19 coded proteinogenic amino acids, without cysteine) (→ 480 samples to be tested). Each row corresponds to a cycle of the iterative process; during selection, the sublibrary with the strongest interaction is identified in a test

Sublibraries											
$20^2 = 400$ OOXXXX	AAXXXX	ACXXXX	…	AYXXXX	CAXXXX	…	**RRXXXX**	RSXXXX	…	YWXXXX	YYXXXX
Selection and synthesis											
$20^1 = 20$ FFOXXX	RRAXXX	RRCXXX	…	RRKXXX	RRLXXX	…	RRTXXX	RRVXXX		**RRWXXX**	RRYXXX
Selection and synthesis											
$20^1 = 20$ FFFOXX	RRWAXX	**RRWCXX**	…	RRWKXX	RRWLXX	…	RRWTXX	RRWVXX		RRWWXX	RRWYXX
Selection and synthesis											
$20^1 = 20$ FFFFOX	RRWCAX	RRWCCX	…	**RRWCKX**	RRWCLX	…	RRWCTX	RRWCVX		RRWCWX	RRWCYX
Selection and synthesis											
$20^1 = 20$ FFFFFO	RRWCKA	RRWCKC	…	RRWCKQ	**RRWCKR**	…	RRWCKT	RRWCKV		RRWCKW	RRWCKY
Selection											
RRWCKR = H–Arg–Arg–Trp–Cys–Lys–Arg–OH											
F … fixed position; **O …** defined position → *sublibraries*; **X** … mixed position											

Table 3.5 ***Positional scanning deconvolution process*** for identifying defined sequences using a hexapeptide library (X corresponds to an equimolar mixture of the 19 coded proteinogenic amino acids without cysteine) (→ 120 samples), the hit sublibrary is highlighted in bold

AXXXXX	XAXXXX	XXAXXX	XXXAXX	XXXXAX	XXXXXA
CXXXXX	XCXXXX	XXCXXX	**XXXCXX**	XXXXCX	XXXXXC
DXXXXX	XDXXXX	XXDXXX	XXXDXX	XXXXDX	XXXXXD
....	....	....	....	....	....
KXXXXX	XKXXXX	XXKXXX	XXXKXX	**XXXXKX**	XXXXXK
....	....	....	....	....	....
RXXXXX	**XRXXXX**	XXRXXX	XXXRXX	XXXXRX	**XXXXXR**
....	....	....	....	....	....
WXXXXX	XWXXXX	**XXWXXX**	XXXWXX	XXXXWX	XXXXXW
YXXXXX	XYXXXX	XXYXXX	XXXYXX	XXXXYX	XXXXXY
Selection	Selection	Selection	Selection	Selection	Selection
Synthesis					
RRWCKR = HArg–Arg–Trp–Cys–Lys–Arg–OH					

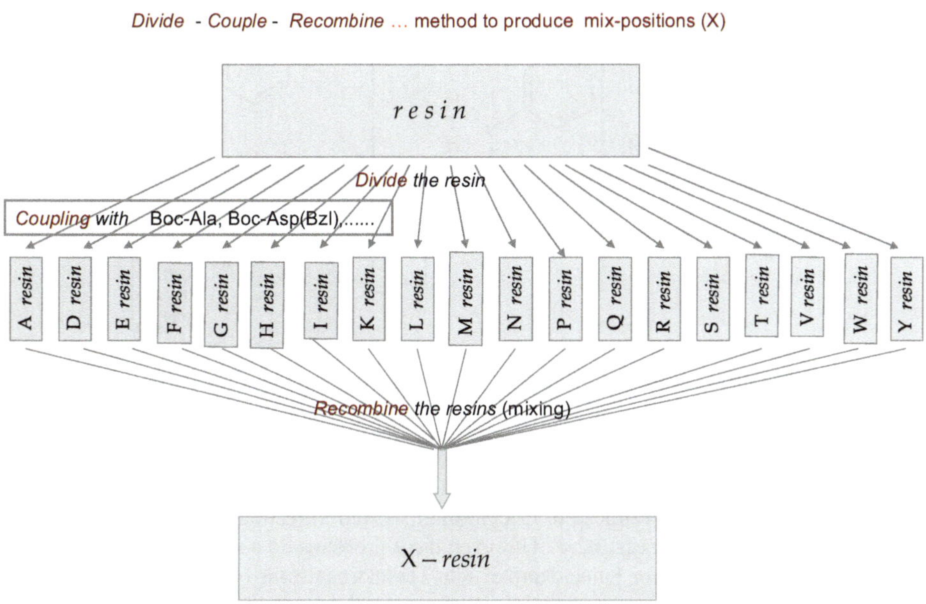

Fig. 3.14 Production of a mixed position on a polymer resin

Table 3.6 Molar ratio, in mole percent (% *n*/*n*), of protected amino acid for the most equimolar peptide coupling ⇛ ***X-Resin***

	Amino acid derivative	% *n*/*n*		Amino acid derivative	% *n*/*n*
A	Boc-Ala-OH	3.44	N	Boc-Asn-OH	5.42
D	Boc-Asp(Bzl)-OH	3.55	P	Boc-Pro-OH	4.38
E	Boc-Glu(Bzl)-OH	3.69	Q	Boc-Gln-OH	5.39
F	Boc-Phe-OH	2.56	R	Boc-Arg(Tos)-OH	6.61
G	Boc-Gly-OH	2.92	S	Boc-Ser(Bzl)-OH	2.82
H	Boc-His(DNP)-OH	3.61	T	Boc-Thr(Bzl)-OH	4.85
I	Boc-Ile-OH	17.6	V	Boc-Val-OH	11.41
K	Boc-Lys(ClZ)-OH	6.31	W	Boc-Trp(Formyl)-OH	3.84
L	Boc-Leu-OH	5.03	Y	Boc-Tyr(BrZ)-OH	4.20
M	Boc-Met(→O)-OH	2.33			

Side chain protecting groups: Bzl ... benzyl; DNP ... imidazole-NH with dinitrophenyl group; Tos ... guanidino-NH_2 as *p*-toluenesulfonamide; Formyl ... indole-NH as "formamide"; ClZ (BrZ) ... 2-chloro(2-bromo)benzyloxycarbonyl; methionine is used as the sulfoxide (→O) due to its sensitivity to oxidation

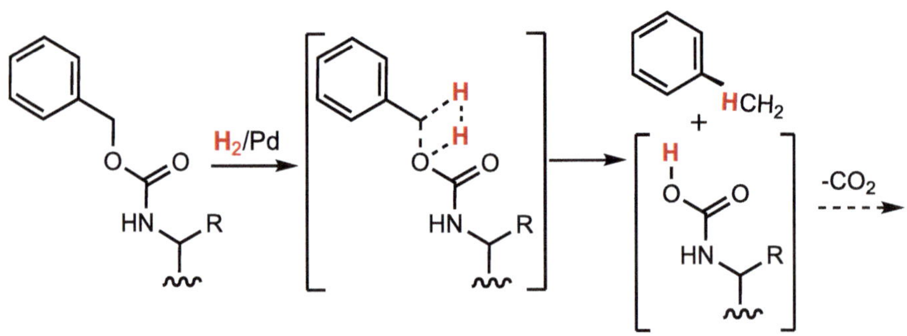

Fig. 3.15 Mechanism of hydrogenolytic cleavage of benzyl protecting groups

The Fmoc protecting group is one of the few protecting groups used in peptide synthesis that can be cleaved under basic conditions. Basic conditions are generally to be avoided for peptides, as abstraction of the α-H atom can lead to racemization, as well as rearrangements (e.g., in peptides containing Asp or Asn, see Sect. 3.2.4). However, these problems do not occur in the presence of the secondary amines used for Fmoc deprotection. The mechanism is based on the acidity of the H atom in the central five-membered ring, since the anion has aromatic character (↗cyclopentadienyl anion) (Fig. 3.17). This is followed by a β-elimination, forming a fulvene derivative (which would otherwise polymerize). In the presence of piperidine, an addition product is formed that is soluble.

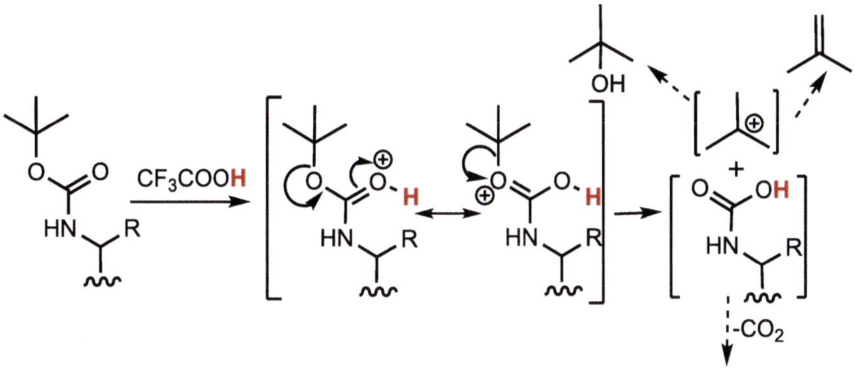

Fig. 3.16 Mechanism of cleavage of *tert*-butyl protecting groups by acid (e.g., TFA)

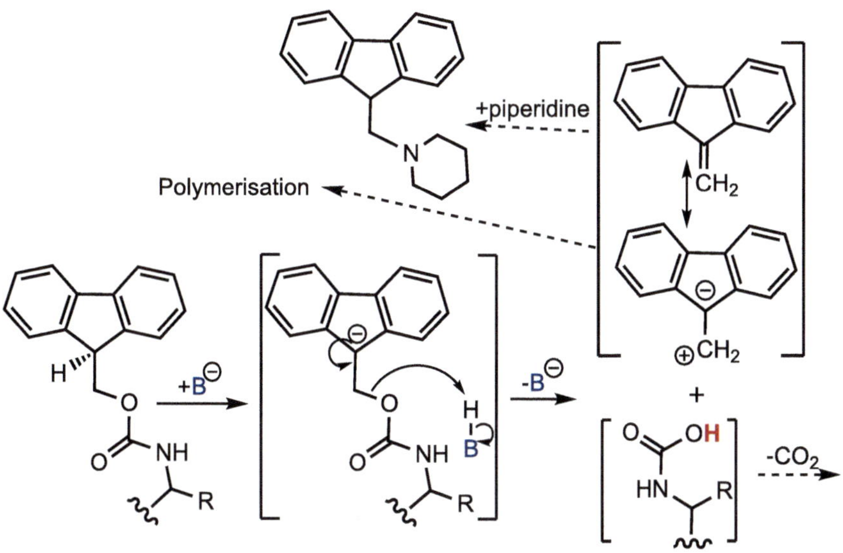

Fig. 3.17 Mechanism of cleavage of Fmoc protecting groups

Summary and Outlook

In the now at least 75 years of peptide synthesis, a wealth of methodology has been developed, so that the synthesis of peptides, which at first seemed almost a titanic feat in competition with nature, has become practically routine. Nevertheless, each individual peptide sequence, especially when it comes to producing larger quantities of highly pure peptides, still presents a certain challenge for a synthetic chemist. The synthesis of proteins (Chap. 8), glycopeptides or glycoproteins (Sects. 5.2 and 8.2.2) or lipopeptides and lipoproteins (Chap. 9) is then even more complex.

Of course, development is constantly progressing: new amino acids are being developed that have very special properties and can decisively influence the structure, reactivity, and imaging properties of peptides and proteins. In Part 2 (Chap. 12), where examples from recent literature are discussed, we will learn more about this.

3.2 Analysis of Peptides and Proteins

The analysis of peptides (proteins) [35] is, of course, crucial for all synthetically produced peptides, but also for peptides isolated from natural sources. In the broadest sense, this also includes the characterization of proteins present in the cell, the so-called **proteome** (analogous to the **genome**).

3.2.1 Methods for Checking Homogeneity (Purity)

First, a brief listing of general analytical methods for checking purity (homogeneity) with references in relation to peptides.

HPLC (high performance/pressure liquid chromatography)
For molecules such as peptides and carbohydrates, *reversed-phase high performance (or high pressure) liquid chromatography* (RP-HPLC, RPLC) [36, 37] is used with C_{18}- or C_8-modified silica gel as the stationary phase (hydrophobicity as the separation parameter), since these polar molecules would remain too strongly bound to unmodified silica gel. As the mobile phase, a gradient of water↗acetonitrile (each with 0.1% TFA) is usually used (sometimes buffer↗acetonitrile).

Detection is primarily by UV absorption ($\lambda = 214$ nm for peptide bond, $\lambda = 254$ nm or 280 nm for aromatic residues; sensitivity depends on ε!). LC-MS instruments (mass spectrometric detection in addition to UV) are increasingly common, with electrospray ionization (ESI-MS) becoming established (Sect. 3.2.2).

RP-HPLC is an essential technique for checking the purity of synthetic peptides and, on a preparative scale, for purifying synthetic peptides. It is also used to check the homogeneity or heterogeneity of isolated proteins or their fragments in the trace range (10 pmol to 10 nmol), and by LC/MS coupling, sequence information can also be obtained.

SFC (supercritical fluid chromatography)
Recently, a chromatographic method that was first introduced in 1962 has gained increasing importance [38]. A supercritical fluid is used as the mobile phase. CO_2 has become established as the supercritical phase due to its favorably located critical point, good availability, and unproblematic disposal. Supercritical CO_2 has a polarity comparable to hexane and cannot dissolve polar substances such as amino acids, peptides, etc. Therefore, a cosolvent (usually an alcohol) is added for SFC of most biomolecules.

The method is very fast (due to the low viscosity of the mobile phase) and also has very good resolution. Similar materials as in HPLC are used as the stationary phase. There are purely analytical instruments, but SFC is also very well suited for preparative separations, since the eluent is very easily and cheaply removed; one can speak of a "green" separation method.

Supercritical Fluid Using Carbon Dioxide as an Example

If pressure and temperature rise above values beyond the critical point, the boundary between liquid (L) and gaseous (G) phase becomes blurred. This is called the supercritical region and the supercritical fluid. Figure 3.18 schematically shows the pressure/temperature diagram of CO_2. CO_2 is particularly well suited as a supercritical fluid because its critical point is at a low temperature (31.1 °C) and relatively low pressure (only about 73 atm). Therefore, the technical requirements for SFC instruments are not so high.

Capillary Electrophoresis (CE)

In CE [39, 40], separation occurs primarily by charge and only secondarily by size (a complementary technique to RP-HPLC!). Due to the electroosmotic flow, all molecules migrate toward the cathode (negative pole). The separation principle is electrophoretic migration: positively charged molecules migrate faster, negatively charged ones more

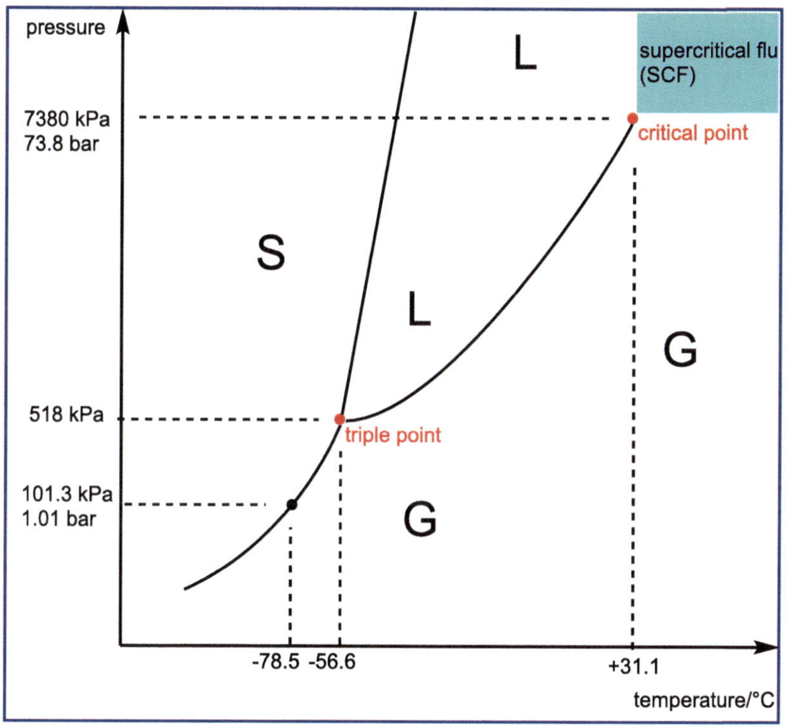

Fig. 3.18 Phase diagram of carbon dioxide

slowly toward the cathode. CE is superior to gel electrophoresis in terms of resolution and separation speed, as high voltages can be applied because the resulting heat can be dissipated very effectively.

On Electroosmotic Flow in CE

This effect is explained—perhaps not entirely physically accurate, but illustrative—in Ref. [39] as follows: The inner silicate surface of the glass capillary is negatively charged (the cations partially enter the solution), so that in the adjacent solution layer there is an increased cation concentration. These cations migrate toward the cathode, which is at the end of the capillary (i.e., toward the detector). This leads to a general flow, including of neutral H_2O molecules, toward the cathode. Accordingly, all sample molecules also migrate toward the cathode—the cationically charged ones particularly quickly, the anionic ones as well, just much more slowly.

Some technical data on CE

Voltage: 20–30 kV, capillary diameter: 50–75 μm, length: 20–50 cm (corresponding to 1–3 μL capillary volume), usually cooled, UV detection; 5–50 nL injection volume of solutions of 10–20 μg/mL peptide or protein in water (possibly +0.1% TFA); buffer solution: 50 mM Na-phosphate buffer pH = 2–3 (at low pH, lower probability of irreversible adsorption of peptides to the capillary surface).

There are various possible configurations: CE in uncoated capillaries, in coated capillaries, isoelectric focusing (IEF), capillary gel electrophoresis (CGE), and micellar electrokinetic chromatography (MEC) for separating uncharged molecules using SDS micelles.

Today, mass spectrometric detection is also very often used (especially for proteome analysis) [41]. CE is applied to peptides, proteins, nucleic acids, but also smaller charged molecules.

Gel Chromatography and Gel Electrophoresis

These methods, with somewhat lower resolution, are still used. They are assumed to be known.

3.2.2 Mass Spectrometric Methods

Mass spectrometry (MS) has increasingly become one of the most important analytical methods in biological research. Here, only a brief summary of the most important ionization methods suitable for large biomolecules will be given. Of course, these are used not only in the analysis of proteins, but also of carbohydrates and nucleic acids.

FAB-MS (fast atom bombardment mass spectrometry) [42]

This ionization method was the first in MS to be suitable for larger polar molecules.

Principle: In high vacuum in the mass spectrometer, the—preferably homogeneous—sample, dissolved in a matrix (e.g., glycerol, possibly with addition of TFA or HCl), is bombarded with a particle beam of neutral atoms (Xe) of high energy (e.g., 5–30 keV) and thereby ionized. The analysis of the ions is carried out as in any MS technique by electric and possibly magnetic fields. There is usually only slight fragmentation, so that the molecular mass can be determined accurately. Typically, in positive FAB-MS, the $(M + H)^+$ ions are generated and determined [in some cases also $(M + Na)^+$]. Not all peptides give equally strong signals → no quantitative information in mixtures. The molecular masses of polypeptides up to the size of insulin (5750 Da) can be determined very accurately.

FAB-MS/MS (tandem MS) allows the determination of the sequence of a selected molecular ion: The ion selected in the first mass spectrometer is fragmented by collisions with gas molecules, and the fragments are analyzed in the second mass spectrometer. Since fragmentation occurs along the peptide chain, the sequence of amino acids can usually be determined from the pattern of fragment masses.

MALDI-MS (matrix-assisted laser desorption-ionization mass spectrometry) [36, 43]

Principle: A pulsed laser beam (usually N_2, 337 nm) is directed onto the sample dissolved in the matrix [e.g., α-cyano-4-hydroxycinnamic acid (CHCA) or 2-(4-hydroxyphenylazo)benzoic acid (HABA)] and causes an "explosion" of matrix + ions into the vacuum of the ion source. The ions are accelerated and focused in the pulsed electric field and then released into the field-free region of the time-of-flight (*time of flight*, TOF) MS. Since all ions have received the same kinetic energy, the flight time (departure time results from the pulse of the laser and electric field) is proportional to the mass (m/z).

Mass range: from 100 up to 200,000 and more; sample amounts in the range of 5–500 pmol.

In mixtures, the relative peak heights (areas) do not necessarily correlate with the amount (→ no quantitative information).

Plasma-Based Ionization Methods in MS

Review [44] provides a very detailed comparison of various mass spectrometric methods in which ionization is achieved by a plasma [3].

Historically, *plasma desorption MS* was probably the first such method [45]. Principle: A radioactive Cf-252 source is used. During spontaneous atomic fission, two particles are generated simultaneously. One triggers the "start time," the other hits the sample and ionizes it. The ion is accelerated in the electric field and the flight time is measured (since the start time is known). This method is thus used like MALDI in combination with TOF-MS, and here too, very large molecular masses can be analyzed. However, it has been far surpassed in importance by the MALDI method.

[3] Just as a reminder: A plasma in the physical sense is a mixture of particles consisting of ions and electrons, but also neutral particles. It is an electrically conductive "gas," but is often also referred to as the fourth state of matter.

ICP-MS (inductively-coupled plasma ionization mass spectrometry) [46, 47]. In this method, an aqueous solution is finely sprayed into a chamber of an inductively coupled hot argon plasma. Positive ions are formed. In addition to ions derived from Ar and H_2O, mainly higher elements are generated as M^+ or MO^+ ions. A small portion of these ions enters the high vacuum system of the MS instrument and is analyzed there. This method has recently become established especially for the highly sensitive quantitative determination of heteroatoms (such as P, S, I, Br), metalloids (such as As and Se), and metals (such as Fe, Cu, Zn) in biomolecules. Due to this high sensitivity, this MS method is also suitable for determining labels of biomolecules with stable isotopes [47].

ESI-MS (electrospray ionization mass spectrometry) [48, 49]
Principle: Sample molecules dissolved in a relatively volatile solvent (as free as possible from Na and other ions, no detergents) are ionized when sprayed from a fine nozzle in a strong electric field (at atmospheric pressure). Singly or multiply positively charged ions are formed, in the case of neutral molecules usually $[M + Na]^+$, $[M + H]^+$ or $[M + K]^+$—for a peptide with a basic amino acid and free N-terminus, for example, $[M + 2H]^{2+}$ and less of the $[M + H]^+$ ion. These ions are then transferred via a system of capillaries and restrictors into the high vacuum of the MS and analyzed there (m/z!). The method is particularly suitable for LC-MS or CE-MS coupling.

▶ The following section discusses the two most important organic-chemical methods for the analysis of peptides and proteins. These are procedures based on chemical reactions, i.e., not purely physical measurement or separation methods.

3.2.3 Amino Acid Analysis (AAA)

An automated method for the analysis of amino acid mixtures after total hydrolysis of peptides or proteins was introduced by Moore and Stein in the 1960s [50]. A more recent summary of this method was published in 1997 [51].

The analysis of a peptide or protein is carried out in three steps:

I. Complete hydrolysis of the peptide bonds: 6 M HCl, 110 °C, 24 h or 48 h (sealed in ampoules, Ar) or 12 M HCl/propionic acid (1/1 v/v) at 150 °C, 90 min, or enzymatically (e.g., aminopeptidase, which stops at D-amino acids, but cleaves sterically hindered peptide bonds between Val or Ile residues more completely than acid-catalyzed hydrolysis). The necessity of such rather drastic hydrolysis conditions underscores for us the strength of the peptide bond (Sect. 2.1).
Sample amount: a few micrograms to a few milligrams (for quantitative AAA: precisely weighed)

II. Evaporation of the solution (vacuum). Then alternatively III or IV.
III. In the so-called ***postcolumn** AAA*: Derivatization *after* separation by ion exchange chromatography on sulfonated divinylbenzene/polystyrene resin. Elution by stepwise salt and pH gradient (e.g., Beckman Instruments); average separation time 60 min; then "online" reaction with ninhydrin reagent at over 100 °C and UV/Vis detection (primary amino derivatives at $\lambda = 570$ nm, secondary at $\lambda = 440$ nm). Improved and more recent implementations of this method were summarized in a 2013 review [52].
IV. In ***precolumn** AAA*: derivatization is performed **before** separation, e.g.,

- with *o*-phthalaldehyde and detection after RP-HPLC by fluorescence at $\lambda = 455$ nm (excitation at $\lambda = 360$ nm) (commercial instrument HP),
- with PITC (phenylisothiocyanate) (PTC chemistry) and UV detection ($\lambda = 360$ nm) [53],
- with aminoquinolylhydroxysuccinimidylcarbamate (AQC), detection by fluorescence at $\lambda = 395$ nm (excitation at $\lambda = 245$ nm) [54].
  Advantage: higher sensitivity, especially with fluorescence detection [*postcolumn deriv.* →dilution]
  Disadvantage: interference from the matrix, since derivatization must occur before separation

Norleucine (Nle, 2-aminohexanoic acid) is usually added as an internal standard to the solvent in which the hydrolysates evaporated to dryness are dissolved. Calibration is performed using a standard amino acid mixture. Accuracy: repeated analyses of the same hydrolysate yield 1%—different hydrolysates 5%—deviations.

Interferences:

- Asn and Gln are, of course, measured as the carboxylic acids Asp and Glu (referred to as Asx and Glx).
- During hydrolysis, Cys and Trp are completely destroyed; Cys can only be determined as cysteic acid (by oxidation with performic acid, HCOOOH, before hydrolysis), Trp can be approximately determined after hydrolysis with 4 M methanesulfonic acid.
- Ser and Thr can be partially destroyed (up to 10–15%).
- Some peptide bonds such as Ile-Ile or Val-Val are incompletely cleaved.

Result: relative abundance of amino acids in proteins and peptides and **peptide content** (protein content). If the amino acid content in a solution is to be determined, steps I and II are of course omitted.

Today, the analysis of amino acids is not only of interest after total hydrolysis of a peptide or protein—the distribution of various proteinogenic and non-proteinogenic amino acids in plants, foods, or even in the human body is of increasing importance, also for the diagnosis of metabolic disorders (such as diabetes). Therefore, miniaturized determination methods are also being developed, which are mostly based on electrophoretic separation [55].

The **configurational homogeneity** of the amino acid after total hydrolysis can be determined by derivatization and subsequent GC on capillary columns with chiral stationary phase (Chirasil-Val®). Application: optimization of fragment couplings in peptide synthesis or determination of the content of D-amino acids in peptides that do not originate from ribosomal protein biosynthesis. Averaged values are obtained if an amino acid occurs more than once in the peptide [56]! Chromatographic methods for configuration determination, either by pre-column derivatization with a chiral reagent or on chiral column material, have also been described [57].

3.2.4 Peptide Sequencing by Edman Degradation

F. Sanger elucidated the first sequence of a (small) protein, insulin, through painstaking work by labeling the respective N-terminal amino acid of many isolated fragments with Sanger's reagent (2,4-dinitrofluorobenzene) and subsequent total hydrolysis + AAA, for which he was awarded the Nobel Prize in 1958. The degradation of protein fragments in an automated process developed by **P. Edman** then greatly simplified protein sequencing.

Important chemical characteristics of the Edman degradation (Fig. 3.19) [58, 59] are **three reaction steps**:

I. Under distinctly **basic** conditions, the N-terminal amino group is reacted with **phenylisothiocyanate** (PITC) to form a thiohydantoin derivative (*phenylthiocarbamoyl peptide, PTC-peptide*), then the excess reagent (PITC) is washed out. These basic conditions are the main reason for the interferences and errors, which will be briefly discussed later.
II. Then, by the action of an **anhydrous acid** (usually trifluoroacetic acid, TFA), the N-terminal amino acid is cleaved off as an anilinothiazolinone amino acid (ATZ-AA), which is separated by extraction, while the now shortened peptide remains in the reaction vessel. This step proceeds rapidly and with good yield only for α-amino acid residues; ATZ-AAs are extracted with organic solvents and thus separated from the remaining peptide, which is then subjected to step I again in the next cycle.
III. The unstable ATZ–AA is rearranged by the action of **aqueous acid** at slightly elevated temperature to the more stable phenylthiohydantoin amino acid (PTH-AA), which is subsequently identified by HPLC (UV absorption as detection). Evaluation is performed by comparison with a standard mixture of PTH-AAs. If aqueous acid were already used in step II, the PTH-AA would be obtained immediately, but other peptide bonds would also be partially cleaved, and satisfactory sequencing results would not be obtained.

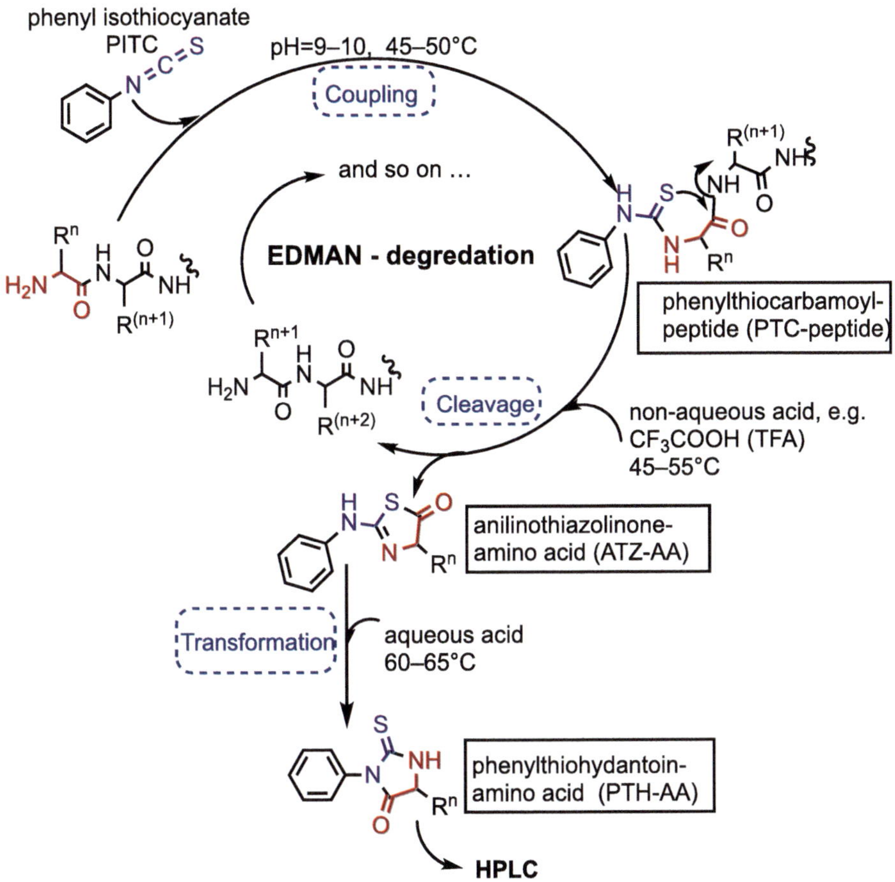

Fig. 3.19 Edman degradation

In the first cycle, all amino groups in side chains (Lys, Orn, etc.) are also derivatized during the reaction with phenylisothiocyanate (PITC), but no cleavage occurs there. These amino acids are then only cleaved in subsequent cycles and are detected as PTH derivatives with PTC-protected side chains.

When analyzing protected peptides (e.g., from SP peptide synthesis), protecting groups in side chains are partially retained (corresponding standards required), and the reaction conditions must be slightly modified compared to free peptides.

Interferences and Errors in Edman Degradation [59]

- Peptide washout: If not bound to a solid support, the peptide is washed out of the reaction vessel by repeated extractions with organic solvents—especially as the remaining peptide becomes shorter or is particularly lipophilic. Nevertheless, the penultimate and sometimes even the C-terminal amino acid can usually be identified if the peptide is not longer than 30 amino acids.

- Degradation of certain amino acids: Stable are Asp, Glu, Gly, Ala, Tyr, Pro, Val, Ile, Leu, Phe, and Lys; Thr, Ser, and Cys, on the other hand, are degraded by β-elimination. Thr and Ser are usually still partially visible intact. Dithiothreitol (DTT) is often added to the sequencing solvents, and some β-elimination products can then be detected as DTT derivatives (Ser and Cys give identical products); Met and Trp are easily oxidized. The presence of DTT, however, reduces any Met (sulfoxide) formed; Asn and Gln can be partially deamidated; for His and Arg, the PTH derivatives still have charge character, so that due to poor extraction yield, weaker signals are observed.
- Background signals: due to impurities from reagents, etc., and as a result of minor cleavage of internal peptide bonds (Asp-Pro bonds are particularly labile, to a lesser extent also Asp-X).
- *Carryover*: due to less than 100% cleavage of the N-terminal amino acid, so that PITC-coupled peptide remains. Most pronounced after Pro residues (cleavage of the imino acid is less efficient).
- *Preview*: appearance of the amino acid already in the previous cycle, usually due to partial deletion of an amino acid (gap, especially in synthetic peptides).
- Cysteine is only detectable after derivatization (e.g., with acrylamide or bromopropylamine). Cystine, on the other hand, is not degraded. The Edman degradation proceeds and gives a blank cycle at the position of the first part of the cystine (*blank*), since the formed derivative is still attached to the second cysteine residue; at the position of the second cysteine, PTH-bis(cystine) is finally eluted (albeit with reduced yield).
- Formation of cyclic imides (Fig. 3.20) mainly from Asn, to a lesser extent from Asp (Gln) under basic conditions; they usually reopen (path b in Fig. 3.20), resulting either in *iso*-Asp peptides ⇛ sequencing stops, or partial epimerization occurs, i.e., the appearance of D-Asp in the sequence; in rare cases (path a), imide formation can also lead to chain cleavage (practically only relevant for Asn-Pro). H-Gln– (less so H-Glu–) at the N-terminus can cyclize to pyroglutamic acid under basic conditions ⇛ no Edman degradation.

▶ **Formation of cyclic imides within the peptide chain as an undesirable side reaction (Fig. 3.20)** This undesirable side reaction of Asn and Asp during Edman degradation is also relevant in peptide synthesis or, more generally, whenever peptides or proteins are exposed to strongly basic conditions. In some specific sequences, epimerizations at Asp (L → D) or continuation of the peptide chain in the side chain (*iso*-Asp) can also occur frequently and lead to larger amounts of unwanted byproducts.

Edman degradation is generally not suitable for quantitative determinations and for excluding heterogeneities. Therefore, peptides or protein fragments should be highly purified beforehand and their homogeneity ensured (HPLC, CE, MS). However, according to a 2013 publication [60], quantitative conclusions can be drawn by combining Edman degradation with MS methods and internal isotope-labeled standards. At the same time, the sample requirement has been reduced to 1–100 pmol.

For sequence analysis of unknown proteins, the following procedure should be followed:

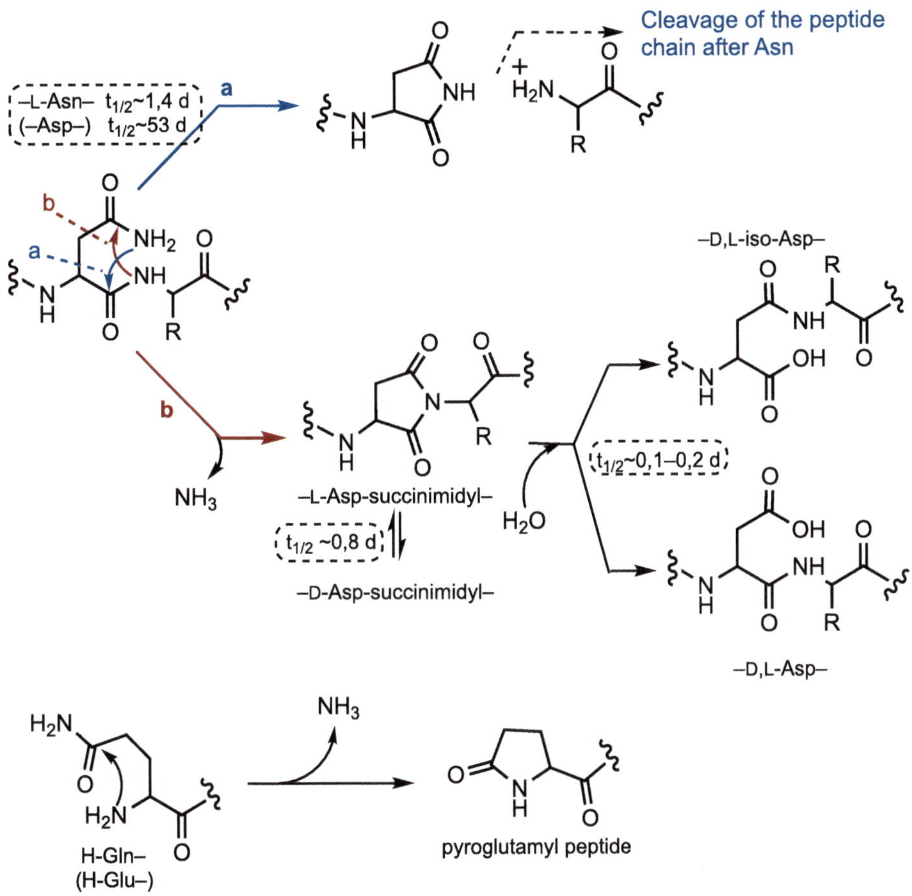

Fig. 3.20 Formation of cyclic imides under basic conditions

I. Purification of the protein
II. Fragmentation (chemically, e.g., BrCN, or enzymatically)
III. Purification of the individual fragments by several HPLC purification steps
IV. Analysis of the fragments by AAA and Edman degradation.

Previously, about 1–10 nmol of protein were required for this. By using micro-HPLC systems, only a few picomoles of protein would be needed, but for sequencing (after Edman) larger amounts are still required (e.g., 10 nmol for up to 90 amino acids, 10 pmol → 15–30 amino acids using gas-phase sequenators with adsorption of the peptide on glass fiber filters using polybrene).

Today, a number of methods are in development as alternatives to pure MS/MS methods [61], which also use the principle of sequencing by chemical degradation (as in Edman), but should be faster and able to analyze mixtures of proteins in parallel.

Reviews of these *next-generation sequencing methods* are listed in Ref. [62]. Why is complete protein sequencing still needed in times of rapid and reliable DNA sequencing methods (Sect. 6.1)? It would suffice to determine only the first N-terminal amino acids and the molecular mass of a protein and then derive the primary structure from the DNA sequence. However, from a medical and biological perspective, the entire proteome in an isolate is of interest, and these new methods should be able to achieve this, with post-translational modifications also becoming visible. There is therefore still much need for development in this area, and it appears that purely instrumental analytical methods alone do not yet achieve the desired goal.

Determination of Sequence and Configuration of Individual Amino Acids in Peptides
If synthetic peptides are to be examined for the homogeneity of the configuration of the individual amino acids, or if the configurations of the individual amino acids of a non-proteinogenic peptide (originating from a non-ribosomal peptide synthesis, such as peptide antibiotics) are to be determined, then sequence information including configuration is of interest. A method was developed for this purpose [63]. First, the peptide is reacted with *tert*-butyl isocyanate, forming the corresponding *tert*-butyl isocarbamoyl peptide, which upon reaction with 1 N HCl in 2-propanol yields the *tert*-butyl isocarbamoyl "amino acid" isopropyl ester and the peptide shortened by one amino acid. The amino acid ester can be analyzed by GC on a capillary column with chiral coating (Chirasil-Val®). Starting from 0.5 mg peptide, the first 8–10 amino acids can be examined for their enantiomeric purity. The amino acids Arg, Lys, His, and Trp cannot be analyzed by this method.

3.2.5 Analysis of the Structure of Peptides and Proteins

Amino acid composition and sequence of a protein are important for characterizing a protein, but provide little information about its function. Unfortunately, computational methods in theoretical chemistry are not yet sufficiently developed to derive the native 3D structure from the sequence—that was the state of research at the time of the first edition in 2016/17. However, in the last three to four years, there have been tremendous advances, also with the help of *artificial intelligence* (AI) [64]. With *deep learning* methods such as AlphaFold2 and RoseTTaFold, it is now possible in many cases to predict protein structures almost flawlessly (compared to experimentally determined ones), despite the large number of possible conformations and the influence of protein folding on the native conformation, which does not necessarily have to be the thermodynamically most stable one.

Nevertheless, experimental methods have not lost their importance. For the structure of proteins in the solid state, X-ray crystallography of single crystals is still used, for a long time the only way to determine protein structures. However, the methodology has also greatly improved and changed here. Electron scattering and cryo-electron microscopy have now also achieved high atomic resolution. [65].

For structure elucidation in solution, NMR methods have now also become routine. However, specially isotope-labeled variants of the protein must be produced by molecular biology: Isotope labeling with ^{13}C and ^{15}N is practically always necessary, but often certain amino acids must also be specifically labeled. The specific synthesis of the labeled amino acid or its precursors from the side of organic chemistry is in any case necessary. The introduction of fluorine atoms can also provide additional conformational information by ^{19}F-NMR. [66]. Through these labels and the various NMR techniques, it should be possible to assign most NMR signals to the individual amino acids in the sequence. Based on distance information from the observed so-called N.O.E.s, structures of the protein in solution can be derived using computational programs. Very specific new NMR methods can provide information about non-bonding interactions and dynamic behavior [67].

The method of **circular dichroism** (CD) can be used especially to observe conformational changes or to compare the conformations of different mutants. CD, a UV-Vis absorption spectrum with chiral light (left- and right-circularly polarized light), shows more or less strong positive or negative absorption bands. Polymers of enantiomerically pure monomer building blocks (such as amino acids or nucleotides) sometimes show very different circular dichroism depending on their spatial structure. In proteins, the proportion of the main secondary structure elements (β-sheet and α-helix) can be estimated from the CD spectrum. However, it is of course only an effect averaged over the entire protein.

Specific conformational information can also be obtained by **fluorescence**. Tryptophan in particular shows pronounced fluorescence, which—like any fluorescence—is strongly influenced by the environment (aqueous milieu or hydrophobic region). However, conformational effects can also be studied using specifically introduced fluorophores.

3.3 An Illustrative Application Example for SPPS and Edman Degradation

A publication from 2006 [Aggarwal, S., Harden, J.L., Denmeade, S.R.: Synthesis and Screening of a Random Dimeric Peptide Library Using the One-Bead-One-Dimer Combinatorial Approach. *Bioconjugate Chem.*, **17** (2), 335–340 (2006)] [68] provides a very illustrative example for the statements made in Chap. 3. The biological context is not so essential here, but the goal is to find peptides that can specifically recognize a certain type of cancer cell. For this purpose, a peptide library is screened, specifically, the authors want to test dimeric dodecapeptides. They use the so-called *One-bead-one-compound* method, often abbreviated as OBOC. How does this work? The procedure is schematically depicted in Fig. 3.21 . A library with twelve mix positions is constructed on an H–Lys–Gly-*resin* using the *divide—couple—recombine* (or *split—couple—mix*) method, and because lysine can also react via its side chain, "dimeric" peptides are obtained.

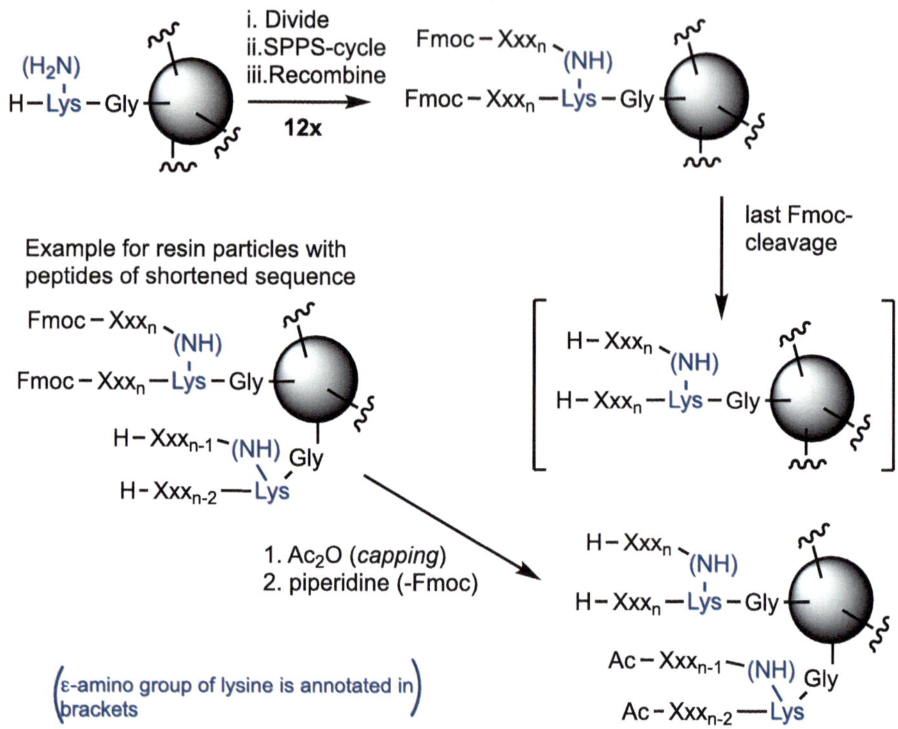

Fig. 3.21 Synthesis of a dimeric one-bead-one-compound peptide library

This has proven advantageous for enhanced interactions. The key point of this method (OBOC) is that all peptide chains growing on ***one*** polymer bead must have the ***same sequence*** (apart from possible deletion peptides).

During screening, all polymer beads carrying a strongly interacting peptide were then individually— in this case, manually under the microscope—isolated, and each was subjected to Edman degradation to determine the sequence. Hit peptides were then selectively synthesized and their interactions verified.

This publication specifically addresses some problems of the method and introduces improvements, which is very enlightening for us as readers. First, an exemplary "dimeric" dodecapeptide was synthesized on a polymer resin with a loading of 0.4–0.5 mmol/g and compared to a resin with 0.2 mmol/g. The RP-HPLC of the two products demonstrates that at higher peptide chain density on the polymer bead, much less of the desired peptide and even predominantly by-products are obtained (see Fig. 3 in [68]). The Edman degradation of peptides on a selected polymer bead (*bead*) yielded unclear results in some cycles, so that no sequence could be derived. This was attributed to premature termination of synthesis on some chains. As a remedy, *capping*, i.e., saturating free, non-reactive aminogroups, could be used. Ideally, this should be done after each coupling cycle, but even the reaction with acetic anhydride before the last Fmoc deprotection achieved the desired effect. The Edman sequencing then became unambiguous. Examples of such HPLC diagrams are shown in Fig. 4 in Ref. [68].

A fundamental problem of this type of library is, of course, also addressed: How many different dimeric dodecapeptides are there in total? Since only the coded proteinogenic amino acids (excluding cysteine) were used, there are a total of $12^{19} = 2.2 \times 10^{15}$ different peptides. However, only about 10^5 polymer beads were used. Thus, it was accepted that by no means all possible sequences were present. This problem also exists for all peptide libraries generated not synthetically, but by molecular biology methods such as *phage display*. In this publication, the authors intended in a second round to introduce certain fixed positions to possibly obtain even better hits.

3.4 Synthetic Peptides—Current Significance for Medicine and Materials Science

3.4.1 Peptides in Drug Research and Production (peptide drugs)

The first peptide to be approved as a drug was insulin (1923). Originally isolated from porcine pancreas, there are now numerous variants produced recombinantly or semi-synthetically [17]. Other hormones followed, e.g., somatostatin (14 aa, one –S–S– bridge), which was produced on a kilogram scale by solution-phase peptide synthesis, now replaced by octreotide (8 aa, including one D-Phe and one D-Tyr, one –S–S– bridge, C-terminal threoninol). These modifications, among other things, reduce enzymatic degradation and increase the residence time in the body.

The Major Challenges for the Use of Peptides as Drugs

- Peptides are rapidly degraded by peptidases and proteases ➔ short residence time in the body, necessitating high dosages.
- Very poor oral bioavailability due to degradation in gastric juice and poor permeability in the intestine.
- Short peptides are usually flexible and therefore have lower bioactivity.
- Peptides are generally quite hydrophilic ➔ poor cell permeability.
- Longer peptides may be immunogenic.
- Although SPPS is a very good synthetic method, it is problematic for large-scale production due to the high consumption of reagents and solvents.

Despite the problems listed above, the number of approved peptide drugs, as well as those in clinical trials, has increased significantly in recent years. Major advances are expected from peptide-based pharmaceuticals. By varying amino acids (including the use of non-proteinogenic amino acids), improved variants can be developed in a targeted and relatively simple manner—always using the same peptide synthesis methods. Cyclic peptides, in particular, are considered especially suitable for this purpose, as they are more resistant to degradation and less flexible [32, 69]. Further improvements are expected through conjugation (e.g., with lipophilic chains, polyethylene glycol, …), N-methylation of the chain, or incorporation of isopeptide building blocks (esters instead of peptide bonds) [70]. The lack of oral bioavailability can probably best be compensated by nasal administration. There are already approved formulations of this type, e.g., for insulin and calcitonin [71].

3.4.2 Polypeptides in Materials Science

Polypeptides with long homologous sequences are readily accessible via *ring-opening polymerisation* (ROP) of NCAs (Sect. 3.1.2.3). They have found a wide range of applications in materials science today, either alone or as block copolymers with other polymers, especially for pharmacological and medical purposes. To obtain polymers that are as uniform and application-specific as possible, intensive research is still being conducted on the various polymerization methods (initiator, catalyst, physical conditions) [13, 72]. What can we imagine these materials to be? The range is very broad:

- Polymer brushes on a solid surface, which on the one hand exhibit special electrical properties (e.g., due to the oriented dipoles of α-helical polypeptides), but can also undergo pH- and/or temperature-induced conformational changes, or serve as chiral column materials.
- Hydrogels that can continuously release a drug into the body in the form of implants
- Polymersomes that can be used as *carriers* for drugs (*drug delivery systems,* DDS)
- Biocompatible polymer materials for implants, including in orthopedics
- Production of nanostructures for a wide variety of purposes
- Surface protection films
- And many more.

As we can see, there is still much synthetic-chemical and technological work to be done in this area. Apart from the very specific applications mentioned above, the development of plastics based on renewable raw materials has become a very important topic overall in terms of sustainability.

3.5 Questions

Let's test our understanding of the content of this chapter!

3.1 How would you proceed with the synthesis of a defined dipeptide such as H–Ala–Gly–OH?
3.2 Explain the mechanism of activation of carboxyl groups by carbodiimides (with formulas).
3.3 Explain the difference between peptide bond formation via active esters or coupling reagents!
3.4 List the advantages and disadvantages of SPPS!
3.5 Why is it so important to achieve a very high yield (>95 %) in each coupling step during SPPS?
3.6 What are urethanes, and why are they particularly suitable for protecting the amino group in peptide synthesis?
3.7 What are peptide libraries, and what are they used for?
3.8 Which methods are used to determine the purity of peptides (and proteins)?
3.9 What is meant by amino acid analysis?
3.10 Describe the Edman degradation of peptides!
3.11 Which experimental methods exist for determining the spatial structure of proteins?
3.12 By which method can polymers of a single amino acid be easily produced?
3.13 Choose one of the references listed in the bibliography, read it, and write a brief summary!

References

1. Houben-Weyl. Methoden der Organischen Chemie, Bd. 15/1, 15/2, Synthese von Peptiden, Wünsch, E. (Hrsg.), Thieme, Stuttgart (1974), https://doi.org/10.1055/b-003-115789; Houben-Weyl, Methods of Organic Chemistry, vol. E22a–E22e, Synthesis of Peptides and Petidomimetics, Goodman, M. (Hrsg.), Thieme, Stuttgart/New York (2002), https://doi.org/10.1055/b-003-111719
2. Carpino, L.A., Han, G.Y.: 9-Fluorenylmethoxycarbonyl amino-protecting group. J. Org. Chem. **37**(22), 3404–3409 (1972). https://doi.org/10.1021/jo00795a005
3. Kunz, H., Unverzagt, C.: The allyloxycarbonyl (Aloc) moiety—conversion of an unsuitable into a valuable amino protecting group for peptide synthesis. Angew. Chem. Int. Ed. **23**(6), 436–437 (1984). https://doi.org/10.1002/anie.198404361; Kunz, H., Unverzagt, C.: Der Allyloxycarbonyl(aloc)-Rest—die Verwandlung einer untauglichen in eine wertvolle Aminoschutzgruppe für die Peptidsynthese. Angew. Chem. **96**(6), 426–427 (1984). https://doi.org/10.1002/ange.19840960612
4. Waldmann, H., Schelhaas, M., Nägele, E., …, Silvius, R.J.: Chemoenzymatic synthesis of fluorescent N-Ras lipopeptides and their use in membrane localization studies in vivo. Angew. Chem. Int. Ed., **36** (20), 2238–2241 (1997). https://doi.org/10.1002/anie.199722381;

Waldmann, H., Schelhaas, M., Nügele, E., …, Silvius, J. R.: Chemoenzymatische Synthese fluoreszierender N-Ras-Lipopeptide und ihre Verwendung bei In-vivo-Studien zur Membranlokalisierung. Angew. Chem. **109**(20), 2334–2337 (1997). https://doi.org/10.1002/ange.19971092028

5. Friedrich-Bochnitschek, S., Waldmann, H., Kunz, H.: Allyl esters as carboxy protecting groups in the synthesis of *O*-glycopeptides. J. Org. Chem. **54**(4), 751–756 (1989). https://doi.org/10.1021/jo00265a007
6. Behrendt, R., White, P., Offer, J.: Advances in Fmoc solid-phase peptide synthesis. J. Pept. Sci. **22**(1), 4–27 (2016). https://doi.org/10.1002/psc.2836
7. Gordon, C.P.: The renascence of continuous-flow peptide synthesis—an abridged account of solid and solution-based approaches. Org. Biomol. Chem. **16**(2), 180–196 (2018). https://doi.org/10.1039/C7OB02759A
8. Masui, H., Fuse, S.: Recent advances in the solid- and solution-phase synthesis of peptides and proteins using microflow technology. Org. Process Res. Dev. **26**(6), 1751–1765 (2022). https://doi.org/10.1021/acs.oprd.2c00074
9. Belleau, B., Malek, G.: A new convenient reagent for peptide syntheses. J. Am. Chem. Soc. **90**(6), 1651–1652 (1968). https://doi.org/10.1021/ja01008a045
10. Le-Nguyen, D., Heitz, A., Castro, B.: Renin substrates. Part 2. Rapid solid phase synthesis of the ratine sequence tetradecapeptide using BOP reagent. J. Chem. Soc. Perkin Trans. **1**, 1915–1919 (1987). https://doi.org/10.1039/P19870001915; Coste, J., Le-Nguyen, D., Castro, B.: PyBOP®: a new peptide coupling reagent devoid of toxic by-product. Tetrahedron Lett. **31**(2), 205–208 (1990). http://dx.doi.org/10.1016/S0040-4039(00)94371-5
11. Carpino, L.A., Imazumi, H., Foxman, B.M., …, Bienert, M.: Comparison of the effects of 5- and 6-HOAt on model peptide coupling reactions relative to the cases for the 4- and 7-Isomers. Org. Lett. **2**(15), 2253–2256 (2000). https://doi.org/10.1021/ol006013z; Carpino, L.A., Imazumi, H., El-Faham, A., …, Bienert, M.: The uronium/guanidinium peptide coupling reagents: finally the true uronium salts. Angew. Chem. Int. Ed. **41**(3), 441–445 (2002). https://doi.org/10.1002/1521-3773(20020201)41:3<441::AID-ANIE441>3.0.CO;2-N
12. El-Faham, A., Funosas, R. S., Prohens, R., Albericio F.: COMU: a safer and more effective replacement for benzotriazole-based uronium coupling reagents. Chem. Eur. J., 15 (37), 9404–9416 (2009). https://doi.org/10.1002/chem.200900615; Manne, S.R., de la Torre, B.G., El-Faham, A., Albericio, F.: OxymaPure coupling reagents: beyond solid-phase peptide synthesis. Synthesis. **52**(21), 3189–3210 (2020). https://doi.org/10.1055/s-0040-1706296
13. Rasines Mazo, A., Allison-Logan, S., Karimi, F., …, Quiao, G.G.: Ring opening polymerization of α-amino acids: advances in synthesis, architecture and applications of polypeptides and their hybrids. Chem. Soc. Rev. **49**(14), 4737–4834 (2020). https://doi.org/10.1039/C9CS00738E; Song, Z., Han, Z., Lv, S., …, Cheng, J.: Synthetic polypeptides: from polymer design to supramolecular assembly and biomedical application. Chem. Soc. Rev. **46**(21), 6570–6599 (2017). https://doi.org/10.1039/C7CS00460E; Lv, W., Wang, Y., Li, M., …, Tao, Y.: Precision synthesis of polypeptides via living anionic ring-opening polymerization of N-carboxyanhydrides by tri-thiourea catalysts. J. Am. Chem. Soc. **144**(51), 23622–23632 (2022). https:/doi.org/10.1021/jacs.2c10950; Wang, W., Fu, H., Lin, Y., …, Song, Z.: Cooperative covalent polymerization of N-carboxyanhydrides: from kinetic studies to efficient synthesis of polypeptide materials. Acc. Mat. Res. **4**(7), 604–615 (2023). https://doi.org/10.1021/accountsmr.3c00046
14. Semple, J.E., Sullivan, B., Sill, K.N.: Large-scale synthesis of α-amino acid-N-carboxyanhydrides. Synth. Commun. **47**(1), 53–61 (2017). https://doi.org/10.1080/00397911.2016.1249289

15. Jolley, K.E., Nye, W., González Niño, C., …, Blacker A.J.: Highly productive continuous flow synthesis of di- and tripeptides in water. Org. Process Res. Dev. **21**(10), 1557–1565 (2017). https://doi.org/10.1021/acs.oprd.7b00214
16. Fuller, W.D., Goodman, M., Naider, F.R., Zhu, Y.-F.: Urethane-protected α-amino acid N-carboxyanhydrides and peptide synthesis. Pept. Sci. **40**(2), 183–205 (1996). https://doi.org/10.1002/(SICI)1097-0282(1996)40:2<183::AID-BIP1>3.0.CO;2-S
17. Moroder, L., Musiol, H.-J.: Insulin—from its discovery to the industrial synthesis of modern insulin analogues. Angew. Chem. Int. Ed. **56**(36), 10656–10669 (2017). https://doi.org/10.1002/anie.201702493
18. Hartrampf, N., Saebi, A., Poskus, M., …, Pentelute, B.L.: Synthesis of proteins by automated flow chemistry. Science. **368**(6494), 980–987 (2020). https://doi.org/10.1126/science.abb2491
19. Atherton, E., Sheppard, R.C.: Solid Phase Peptide Synthesis. IRC Press, Oxford (1989)
20. Paradís-Bas, M., Tulla-Puche, J., Albericio, F.: The road to the synthesis of "difficult peptides". Chem. Soc. Rev. **45**(3), 631–654 (2016). https://doi.org/10.1039/C5CS00680E
21. Declerck, V., Nun, P., Martinez, J., Lamaty, F.: Solvent-free synthesis of peptides. Angew. Chem. Int. Ed. **48**(49), 9318–9321 (2009). https://doi.org/10.1002/anie.200903510; Martin, V., Egelund, P.H.G., Johansson, H., …, Pedersen, D.S.: Greening the synthesis of peptide therapeutics: an industrial perspective. RSC Adv. **10**(69), 42457–42492 (2020). https://doi.org/10.1039/D0RA07204D; Al Musaimi, O., de la Torre, B.G., Albericio, F.: Greening Fmoc/tBu solid-phase peptide synthesis. Green Chem. **22**(4), 996–1018 (2020). https://doi.org/10.1039/C9GC03982A; Jaradat, D.M.M., Al Musaimi, O., Albericio, F.: Advances in solid-phase peptide synthesis in aqueous media (ASPPS). Green Chem. **24**(17), 6360–6372 (2022). https://doi.org/10.1039/D2GC02319A; Ferrazzano, L., Catani, M., Cavazzini, A., …, Tolomelli, A.: Sustainability in peptide chemistry: current synthesis and purification technologies and future challenges. Green Chem. **24**(3), 975–1020 (2022). https://doi.org/10.1039/D1GC04387K
22. Benz, H.: The role of solid-phase fragment condensation (SPFC) in peptide synthesis. Synthesis. **1994**(4), 337–358 (1994). https://doi.org/10.1055/s-1994-25472
23. Bordusa, F.: Substrate mimetics in protease catalysis: characteristics, kinetics, and synthetic utility. Curr. Protein Pept. Sci. **3**(2), 159–180 (2002). https://doi.org/10.2174/1389203024605322
24. Jung, G., Beck-Sickinger, A.G.: Multiple peptide synthesis methods and their applications. new synthetic methods (87). Angew. Chem. Int. Ed. **31**(4), 367–383 (1992). https://doi.org/10.1002/anie.199203673; Jung, G., Beck-Sickinger, A. G.: Methoden der multiplen Peptidsynthese und ihre Anwendungen. Angew. Chem. **104**(4), 375–391 (1992). https://doi.org/10.1002/ange.19921040404
25. Houghten, R.A., Wilson, D.B., Pinilla, C.: Drug discovery and vaccine development using mixture-based synthetic combinatorial libraries. Drug Discov. Today. **5**(7), 276–285 (2000). https://doi-org/10.1016/S1359-6446(00)01513-0; Eichler, J., Houghten, R.A.: Generation and utilization of synthetic combinatorial libraries. Mol. Med. Today. **1**(4), 174–180 (1995). https://doi.org/10.1016/S1357-4310(95)91867-1
26. Ostresh, J.M., Blondelle, S.E., Dörner, B., Houghten, R.A.: Generation and use of nonsupport-bound peptide and peptidomimetic combinatorial libraries. Methods Enzymol. **267**, 220–234 (1996). https://doi-org/10.1016/S0076-6879(96)67015-3
27. Chen, C.L., Strop, P., Lebl, M., Lam, K.S.: One bead-one compound combinatorial peptide library: different types of screening. Methods Enzymol. **267**, 211–219 (1996). https://doi.org/10.1016/S0076-6879(96)67014-1

28. Houghten, R.A.: General method for the rapid solid-phase synthesis of large numbers of peptides: specificity of antigen-antibody interaction at the level of individual amino acids. Proc. Natl. Acad. Sci. USA. **82**(15), 5131–5135 (1985). https://doi.org/10.1073/pnas.82.15.513
29. Geysen, H.M., Meloen, R.H., Barteling, S.J.: Use of peptide synthesis to probe viral antigens for epitopes to a resolution of a single amino acid. Proc. Natl. Acad. Sci. (USA). **81**(13), 3998–4002 (1984). https://doi.org/10.1073/pnas.81.13.3998; Geysen, H.M., Rodda, S.J., Mason, T.J.: A priori delineation of a peptide which mimics a discontinuous antigenic determinant. Mol. Immunol. **23**(7), 709–715 (1986). https://doi.org/10.1016/0161-5890(86)90081-7
30. Yu, J., Smith, G.P.: Affinity maturation of phage-displayed peptide ligands. Methods Enzymol. **267**, 3–27 (1996). https://doi.org/10.1016/S0076-6879(96)67003-7
31. Ryvkin, A., Ashkenazy, H., Weiss-Ottolenghi, Y., …, Gershoni, J.M.: Phage display peptide libraries: deviations from randomness and correctives. Nucl. Acid Res. **46**(9), e52–e52 (2018). https://doi.org/10.1093/nar/gky077
32. Ji, X., Nielsen, A.L., Heinis, C.: Cyclic peptides for drug development. Angew. Chem. Int. Ed. **63**(3), e202308251 (2024). https://doi.org/10.1002/anie.202308251
33. Bruce, A., Adebomi, V., Czabala, P., …, Raj, M.: A tag-free platform for synthesis and screening of cyclic peptide libraries. Angew. Chem. Int. Ed. **63**(21), e202320045 (2024). https://doi.org/10.1002/anie.202320045
34. Sohrabi, C., Foster, A., Tavassoli, A.: Methods for generating and screening libraries of genetically encoded cyclic peptides in drug discovery. Nat. Rev. Chem. **4**(2), 90–101 (2020). https://doi.org/10.1038/s41570-019-0159-2; Valentine, J., Tavassoli, A.: Chapter six—genetically encoded cyclic peptide libraries: from hit to lead and beyond. In: Lesburg, C.A. (Hrsg.) Methods in Enzymology, Bd. 610, S. 117–134. Academic Press (2018). https://doi.org/10.1016/bs.mie.2018.09.020
35. Barrett, G.C., Elmore, D.T.: Amino Acids and Peptides. Cambridge University Press (1998), 9780521468275. http://www.cambridge.org/cr/academic/subjects/chemistry/chemistry-general-interest/amino-acids-and-peptides. Zugegriffen am 25.12.2025
36. Swiderek, K.M., Lee, T.D., Shively, J.E.: Trace structural analysis of proteins. Methods Enzymol. **271**, 68–86 (1996). https://doi.org/10.1016/S0076-6879(96)71005-4
37. Fekete, S., Veuthey, J.-L., Guillarme, D.: New trends in reversed-phase liquid chromatographic separations of therapeutic peptides and proteins: theory and applications. J. Pharm. Biomed. Anal. **69**, 9–27 (2012). https://doi.org/10.1016/j.jpba.2012.03.024; Mitulović, G.: New HPLC techniques for proteomics analysis: a short overview of latest developments. J. Liq. Chromatogr. Relat. Technol. **38**(3), 390–403 (2015). https://doi.org/10.1080/10826076.2014.941266
38. Molineau, J., Hideux, M., West, C.: Chromatographic analysis of biomolecules with pressurized carbon dioxide mobile phases—a review. J. Pharm. Biomed. Anal. **193,** 113736 (2021). https://doi.org/10.1016/j.jpba.2020.113736; Kupnik, K., Knez, Ž., Primožič, M., Leitgeb M.: Separation of amino acids and peptides with supercritical fluids chromatography. Sep. Purif. Rev. **52**(1), 58–74 (2023). https://doi.org/10.1080/15422119.2022.2038625
39. Engelhardt, H., Beck, W., Kohr, J., Schmitt, T.: Capillary electrophoresis: methods and scope. Angew. Chem. Int. Ed. **32**(5), 629–649 (1993). https://doi.org/10.1002/anie.199306293; Engelhardt, H., Beck, W., Kohr, J., Schmitt. T.: Kapillarelektrophorese: Methoden und Möglichkeiten. Angew. Chem. **105**, 659–680 (1993); Sanchez, A., Smith, A.J.: Capillary electrophoresis. Methods Enzymol. **289**, 469–478 (1997). https://doi.org/10.1016/S0076-6879(97)89059-3
40. Harstad, R.K., Johnson, A.C., Weisenberger, M.M., Bowser, M.T.: Capillary electrophoresis. Analyt. Chem. **88**(1), 299–319 (2016). https://doi.org/10.1021/acs.analchem.5b04125; Kristoff, C.J., Bwanali, L., Veltri, L.M., …, Holland, L.A.: Challenging bioanalyses with

capillary electrophoresis. Analyt. Chem., **92**(1), 49–66 (2020). https://doi.org/10.1021/acs.analchem.9b04718

41. Herrero, M., Ibanez, E., Cifuentes, A.: Capillary electrophoresis-electrospray-mass spectrometry in peptide analysis and peptidomics. Electrophoresis, **29**(10), 2148–2160 (2008). https://doi.org/10.1002/elps.200700404; Seyfinejad, B., Jouyban, A.: Capillary electrophoresis-mass spectrometry in pharmaceutical and biomedical analyses. J. Pharm. Biomed. Anal. **221**, 115059 (2022). https://doi.org/10.1016/j.jpba.2022.115059
42. Beranovà-Giorgianni, S., Desiderio, D.M.: Fast atom bombardment mass spectrometry of synthetic peptides. Methods Enzymol. **289**, 478–499 (1997). https://doi.org/10.1016/S0076-6879(97)89060-X; Claeys, M., Claereboudt, J.: Fast atom bombardment ionization in mass spectrometry. In: Lindon, J.C., Tranter, G.E., Koppenaal, D.W. (Hrsg.) Encyclopedia of Spectroscopy and Spectrometry (Third Edition), S. 581–587. Academic Press (2017). https://doi.org/10.1016/B978-0-12-803224-4.00320-4
43. Karas, M., Hillenkamp, F.: Laser desorption ionization of proteins with molecular masses exceeding 10,000 daltons. Anal. Chem. **60**(20), 2299–2301 (1988). https://doi.org/10.1021/ac00171a028; Moore, W.T.: Laser desorption mass spectrometry. Methods Enzymol. **289**, 520–542 (1997). https://doi.org/10.1016/S0076-6879(97)89062-3; Dreisewerd, K.: Recent methodological advances in MALDI mass spectrometry. Anal. Bioanal. Chem. **406** (9–10), 2261–2278 (2014). https://doi.org/10.1007/s00216-014-7646-6; Heeren, R.M.A.: MALDI techniques in mass spectrometry imaging. In: Lindon, J.C., Tranter, G.E., Koppenaal, D.W. (Hrsg.) Encyclopedia of Spectroscopy and Spectrometry (Third Edition), S. 711–718. Academic Press (2017). https://doi.org/10.1016/B978-0-12-803224-4.00008-X; *eine interessante Anwendung im Zusammenhang mit Peptidbibliotheken*: Schey, G.L., Buttery, P.R., Hildebrandt, E.R., …, Distefano, M.D.: MALDI-MS analysis of peptide libraries expands the scope of substrates for farnesyltransferase. Int. J. Mol. Sci. **22**(21), 12042 (2021). https://doi.org/10.3390/ijms222112042
44. Molnar, B.T., Shelley, J.T.: Modern plasma-based desorption/ionization from atoms and molecules to chemical synthesis. Mass Spectrom. Rev. **40**(5), 609–627 (2021). https://doi.org/10.1002/mas.21645
45. Roepstorff, P.: Plasma desorption mass spectrometry of peptides and proteins. Acc. Chem. Res. **22**(12), 421–427 (1989). https://doi.org/10.1021/ar00168a003
46. Houk, R.S., Fassel, V.A., Flesch, G.D., …, Taylor, C.E.: Inductively coupled argon plasma as an ion source for mass spectrometric determination of trace elements. Anal. Chem. **52**(14), 2283–2289 (1980). https://doi.org/10.1021/ac50064a012
47. Sanz-Medel, A.: ICP-MS for multiplex absolute determinations of proteins. Anal. Bioanal. Chem. **398**(5), 1853–1859 (2010). https://doi.org/10.1007/s00216-010-4091-z; Liu, R., Zhang, S., Wei, C., …, Zhang, X.: Metal stable isotope tagging: renaissance of radioimmunoassay for multiplex and absolute quantification of biomolecules. Acc. Chem. Res. **49**(5), 775–783 (2016). https://doi.org/10.1021/acs.accounts.5b00509
48. Burdick, D.J., Stults, J.T.: Analysis of peptide synthesis products by electrospray ionization mass spectrometry. Methods Enzymol. **289**, 499–519 (1997). https://doi.org/10.1016/S0076-6879(97)89061-1; Loo, J.A., Kilby, G.W.: Electrospray ionization mass spectrometry of peptides and proteins methodologies and applications. Pract. Spectrosc., **32**(Applied Electrospray Mass Spectrometry), 251–281 (2002)
49. Konopka, A., Wild, C., Boehm, M.E., Lehmann, W.D.: Proteomics standards with controllable trueness- absolute quantification of peptides, phosphopeptides and proteins using ICP- and ESI-MS. New Dev. Mass Spectrom. **1**(Quantitative Proteomics), 110–128 (2014). https://doi.org/10.1039/9781782626985-00110

50. Moore, S., Stein, W.H.: Chromatographic determination of amino acids by the use of automatic recording equipment. Methods Enzymol. **6**, 819–831 (1963). https://doi.org/10.1016/0076j6879(63)06257-1
51. Smith, A.J.: [18] Amino acid analysis. Methods Enzymol. **289**, 419–426 (1997). https://doi.org/10.1016/S0076-6879(97)89057-X; Ferré, S., González-Ruiz, V., Guillarme, D., Rudaz, S.: Analytical strategies for the determination of amino acids: past, present and future trends. J. Chrom, B. **1132**, 121819 (2019). https://doi.org/10.1016/j.jchromb.2019.121819
52. Rigas, P.G.: Post-column labeling techniques in amino acid analysis by liquid chromatography. Anal. Bioanal. Chem. **405**(25), 7957–7992 (2013). https://doi.org/10.1007/s00216-013-7127-3
53. Molnár-Perl, I.: Advances in the high-performance liquid chromatographic determination of phenylthiocarbamyl amino acids. J. Chromatogr. A. **661**(1–2), 43–50 (1994). https://doi.org/10.1016/0021-9673(94)85175-1
54. Strydom, D.J., Cohen, S.A.: Comparison of amino acid analyses by phenylisothiocyanate and 6-aminoquinolyl-N-hydroxysuccinimidyl carbamate precolumn derivatization. Anal. Biochem. **222**(1), 19–28 (1994). https://doi.org/10.1006/abio.1994.1448; Masuda, A., Dohmae, N.: Amino acid analysis of sub-picomolar amounts of proteins by precolumn fluorescence derivatization with 6-aminoquinolyl-N- hydroxysuccinimidyl carbamate. BioSci. Trends, **5**(6), 231–238 (2011). https://doi.org/10.5582/bst.2011.v5.6.231
55. Ou, G., Feng, X., Du, W., …, Liu, B.-F.: Recent advances in microchip electrophoresis for amino acid analysis. Anal. Bioanal. Chem. **405**(25), 7907–7918 (2013). https://doi.org/10.1007/s00216-013-6830-4; Denoroy, L., Parrot, S.: Analysis of amino acids and related compounds by capillary electrophoresis. Sep. Purif. Rev. **46**(2), 108–151 (2017). https://doi.org/10.1080/15422119.2016.1212378; Cai, D., Hou, B., Xie, S.L.: Amino acid analysis as a method of discovering biomarkers for diagnosis of diabetes and its complications. Amino Acids, **55**(5), 563–578 (2023). https://doi.org/10.1007/s00726-023-03255-8
56. Frank, H., Nicholson, G.J., Bayer, E.: Chiral polysiloxanes for resolution of optical antipodes. Angew. Chem. Int. Ed. **17**(5), 363–365 (1978). https://doi.org/10.1002/anie.197803631; Frank, H., Nicholson, G.J., Bayer, E.: Chirale Polysiloxane zur Trennung von optischen Antipoden. Angew. Chem. **90**(5), 396–398 (1978). https://doi.org/10.1002/ange.19780900525
57. Todoroki, K., Ishii, Y., Toyoda, K., …, Toyo'oka, T.: Dress-up chiral columns for the enantioseparation of amino acids based on fluorous separation. Anal. Bioanal. Chem. **405**(25), 8121–8129 (2013). https://doi.org/10.1007/s00216-013-7207-4; Bhushan, R., Brückner, H.: Use of Marfey's reagent and analogs for chiral amino acid analysis: assessment and applications to natural products and biological systems. J. Chrom. B. **879**(29), 3148–3161 (2011). https://doi.org/10.1016/j.jchromb.2011.05.058; Wernisch, S., Bisi, F., Cazzato, A.S., …, Lindner, W.: 2-Acyl-dimedones as UV-active protective agents for chiral amino acids: enantiomer separations of the derivatives on chiral anion exchangers. Anal. Bioanal. Chem. **405**(25), 8011–8026 (2013). https://doi.org/10.1007/s00216-013-6932-z; Hellinger, R., Horak, J., Lindner, W.: Enantioseparation of 6-aminoquinolyl-N-hydroxysuccinimidyl carbamate tagged amino acids and other zwitterionic compounds on cinchona-based chiral stationary phases. Anal. Bioanal. Chem. **405**(25), 8105–8120 (2013). https://doi.org/10.1007/s00216-013-7121-9
58. Edman, P.: Method for determination of the amino acid sequence in peptides. Acta Chem. Scand. **4**, 283–293 (1950). https://doi.org/10.3891/acta.chem.scand.04-0283; Edman, P.: Phenylthiohydantoins in protein analysis. Ann. N. Y. Acad. Sci. **88**, 602–610 (1960). https://doi.org/10.1111/j.1749-6632.1960.tb20056.x; Edman, P., Begg, G.: A protein sequenator. Eur. J. Biochem. **1**(1), 80–91 (1967). https://doi.org/10.1111/j.1432-1033.1967.tb00047.x; Edman, P.: Sequence determination. In: Needleman, S.B. (Hrsg.) Molecular Biology, Biochemistry and

Biophysics, Bd. 8, Protein sequence determination, S. 211–255, Springer, Berlin/Heidelberg (1970). https://doi.org/10.1007/978-3-662-12834-3_8

59. Grant, G.A., Crankshaw, M.W., Gorka, J.: [17] Edman sequencing as tool for characterization of synthetic peptides. Methods Enzymol. **289**, 395419 (1997). http://dx.doi.org/10.1016/S0076-6879(97)89056-8; Hunkapiller, M., Kent, S., Caruthers, M., …, Hood, L.: A microchemical facility for the analysis and synthesis of genes and proteins. Nature. **310**(5973), 105111 (1984). https://doi.org/10.1038/310105a0; Hunkapiller, M.W., Hood, L.E.: [43] Analysis of phenylthiohydantoins by ultrasensitive gradient high-performance liquid chromatography. Methods Enzymol. **91**, 486–493 (1983). https://doi.org/10.1016/S0076-6879(83)91045-5
60. Satoh, R., Goto, T., Lee, S.H., Oe, T.: Can Edman degradation be used for quantification? Isotope-dilution liquid chromatography-electrospray ionization tandem mass spectrometry and the long-term stability of 20 phenylthiohydantoin-amino acids. Anal. Bioanal. Chem. **405**(25), 8001–8010 (2013). https://doi.org/10.1007/s00216-013-6901-6
61. Liu, X., Dekker, L.J.M., Wu, S., …, Pevzner, P.A.: De novo protein sequencing by combining top-down and bottom-up tandem mass spectra. J. Proteome Res. **13**(7), 3241–3248 (2014). https://doi.org/10.1021/pr401300m
62. Tullman, J., Marino, J.P., Kelman, Z.: Leveraging nature's biomolecular designs in next-generation protein sequencing reagent development. Appl. Microbiol. Biotechnol. **104**(17), 7261–7271 (2020). https://doi.org/10.1007/s00253-020-10745-2 (2020); Floyd, B.M., Marcotte, E.M.: Protein sequencing, one molecule at a time. Annu. Rev. Biophy. **51**(2022), 181–200 (2022). https://doi.org/10.1146/annurev-biophys-102121-103615; He, H., Wu, C., Saqib, M., Hao, R.: Single-molecule fluorescence methods for protein biomarker analysis. Anal. Bioanal. Chem. **415**(18), 3655–3669 (2023). https://doi.org/10.1007/s00216-022-04502-9 (2023); Dorey, A., Howorka, S.: Nanopore DNA sequencing technologies and their applications towards single-molecule proteomics. Nat. Chem. **16**(3), 314–334 (2024). https://doi.org/10.1038/s41557-023-01322-x
63. Bolte, T., Yu, D., Stüwe, H.-T., König, W.A.: Sequential configuration analysis of peptides. Angew. Chem. Int. Ed. **26**(4), 331–332 (1987). https://doi.org/10.1002/anie.198703311; Bolte, T., Yu, D., Stüwe, H.-T., König, W.A.: Sequentielle Konfigurationsanalyse von Peptiden. Angew. Chem. **99**(4), 362–363 (1987). https://doi.org/10.1002/ange.19870990425
64. Versini, R., Sritharan, S., Aykac Fas, B., …, Taly, A.: A perspective on the prospective use of AI in protein structure prediction. J. Chem. Inform. Modeling, **64**(1), 26–41 (2024). https://doi.org/10.1021/acs.jcim.3c01361; Saharkhiz, S., Mostafavi, M., Birashk, A., …, Akhavan-Sigari, R.: The state-of-the-art overview to application of deep learning in accurate protein design and structure prediction. Topics Curr. Chem. **382**(3), 23 (2024). https://doi.org/10.1007/s41061-024-00469-6; Wuyun, Q., Chen, Y., Shen, …, Zheng, W.: Recent progress of protein tertiary structure prediction. Molecules. **29**(4), 832 (2024). https://doi.org/10.3390/molecules29040832; Lupas, A.N., Pereira, J., Alva, V., …, Hartmann, M.D.: The breakthrough in protein structure prediction. Biochem. J. **478**(10), 1885–1890 (2021). https://doi.org/10.1042/bcj20200963; Marx, V.: Method of the year: protein structure prediction. Nat. Methods. **19**(1), 5–10 (2022). https://doi.org/10.1038/s41592-021-01359-1
65. Schaefer, A., Naser, D., Siebeneichler, B., …, Meiering, E.M.: Methodological advances and strategies for high resolution structure determination of cellular protein aggregates. J. Biol. Chem. **298**(9), 102197 (2022). https://doi.org/10.1016/j.jbc.2022.102197; Liu, W., Wacker, D., Gati, C.,…, Cherezov V.: Serial femtosecond crystallography of G protein–coupled receptors. Science, **342**(6165), 1521–1524 (2013). https://doi.org/10.1126/science.1244142; Kurpiewska, K., Sławek, J., Klonecka, A., Kozak, M.: The promise and challenge of high pressure macromolecular crystallography. Crystals, **13**(4), 560 (2023). https://doi.org/10.3390/cryst13040560; Clabbers, M.T.B., Xu, H.: Macromolecular crystallography using microcrys-

tal electron diffraction. Acta Cryst. Sect. D. **77**(3), 313–324 (2021). https://doi.org/10.1107/S2059798320016368

66. Lichtenecker, R.J., Schörghuber, J., Bisaccia, M.: Synthesis of metabolic amino acid precursors: tools for selective isotope labeling in cell-based protein overexpression. Synlett. **26**(19), 2611–2616 (2015). https://doi.org/10.1055/s-0034-1378878; Schörghuber, J., Geist, L., Platzer, G., …, Lichtenecker, R.J.: Late metabolic precursors for selective aromatic residue labeling. J. Biomol. NMR. **71**(3), 129–140 (2018). https://doi.org/10.1007/s10858-018-0188-z; Jaipuria, G., Krishnarjuna, B., Mondal, S., Dubey, A., Atreya, H.S.: Amino acid selective labeling and unlabeling for protein resonance assignments. In: Atreya, H.S. (Hrsg.) Advances in Experimental Medicine and Biology, Bd. 992, Isotope Labeling in Biomolecular NMR, S. 95–118, Springer Netherlands (2012). https://doi.org/10.1007/978-94-007-4954-2_6; Verardi, R., Traaseth, N.J., Masterson, L.R., Vostrikov, V.V., Veglia, G.: Isotope labeling for solution and solid-state NMR spectroscopy of membrane proteins, In: Atreya, H.S. (Hrsg.) Advances in Experimental Medicine and Biology, Bd. 992, Isotope Labeling in Biomolecular NMR, S. 35–62. Springer Netherlands (2012). https://doi.org/10.1007/978-94-007-4954-2_3; Urbanek, A., Elena-Real, C.A., Popovic, M., …, Bernardo, P.: Site-Specific Isotopic Labeling (SSIL): access to high-resolution structural and dynamic information in low-complexity proteins. ChemBioChem. **21**(6), 769–775 (202). https://doi.org/10.1002/cbic.201900583; Sengupta, I.: Insights into the structure and dynamics of proteins from 19F solution NMR spectroscopy. Biochemistry, **63**(22), 2958–2968 (2024). https://doi.org/10.1021/acs.biochem.4c00534
67. Negroni, M., Kurzbach, D.: Missing pieces in structure puzzles: how hyperpolarized NMR spectroscopy can complement structural biology and biochemistry. ChemBioChem. **24**(6), e202200703 (2023). https://doi.org/10.1002/cbic.202200703; Tugarinov, V., Ceccon, A., Clore, G.M.: NMR methods for exploring ‚dark' states in ligand binding and protein-protein interactions. Progr. Nucl. Magn. Res. Spectr., **128**, 1–24 (2022). https://doi.org/10.1016/j.pnmrs.2021.10.001
68. Aggarwal, S., Harden, J.L., Denmeade, S.R.: Synthesis and screening of a random dimeric peptide library using the one-bead-one-dimer combinatorial approach. Bioconjug. Chem. **17**(2), 335–340 (2006). https://doi.org/10.1021/bc0502659
69. Buckton, L.K., Rahimi, M.N., McAlpine, S.R.: Cyclic peptides as drugs for intracellular targets: the next frontier in peptide therapeutic development. Chem. Eur. J. **27**(5), 1487–1513 (2021). https://doi.org/10.1002/chem.201905385
70. Muttenthaler, M., King, G.F., Adams, D.J., Alewood, P.F.: Trends in peptide drug discovery. Nat. Rev. Drug Discov. **20**(4), 309–325 (2021). https://doi.org/10.1038/s41573-020-00135-8; Apostolopoulos, V., Bojarska, J., Chai, T.-T., …, Toth, I.: A global review on short peptides: frontiers and perspectives. Molecules, **26**(2), 430 (2021). https://doi.org/10.3390/molecules26020430; Wang, L., Wang, N., Zhang, W., …, Fu, C.: Therapeutic peptides: current applications and future directions. Signal Transduct. Target. Ther. **7**(1), 48 (2022). https://doi.org/10.1038/s41392-022-00904-4; Luo, X., Chen, H., Song, Y., …, Dessie, W.: Advancements, challenges and future perspectives on peptide-based drugs: focus on antimicrobial peptides. Eur. J. Pharm. Sci. **181**, 106363 (2023). https://doi.org/10.1016/j.ejps.2022.106363; Pereira, A. J., de Campos, L.J., Xing, H., Conda-Sheridan, M.: Peptide-based therapeutics: challenges and solutions. Med. Chem. Res. **33**(8), 1275–1280 (2024). https://doi.org/10.1007/s00044-024-03269-1
71. Luo, D., Ni, X., Yang, H., …, Bai, L.: A comprehensive review of advanced nasal delivery: specially insulin and calcitonin. Eur. J. Pharm. Sci. **192**, 106630 (2024). https://doi.org/10.1016/j.ejps.2023.106630

72. Shen, Y., Li, Z., Klok, H.-A.: Polypeptide brushes grown via surface-initiated ring-opening polymerization of α-amino acid N-carboxyanhydrides. Chin. J. Polym. Sci. **33**(7), 931–946 (2015). https://doi.org/10.1007/s10118-015-1654-7; Liarou, E., Varlas, S., Skoulas, D., …, Iatrou, H.: Smart polymersomes and hydrogels from polypeptide-based polymer systems through α-amino acid N-carboxyanhydride ring-opening polymerization. From chemistry to biomedical applications. Progr. Polym. Sci. **83**, 28–78 (2018). https://doi.org/10.1016/j.progpolymsci.2018.05.001; Varlas, S., Maitland, G.L., Derry, M.J.: Protein-, (poly)peptide-, and amino acid-based nanostructures prepared via polymerization-induced self-assembly. Polymers. **13**(16), 2603 (2021). https://doi.org/10.3390/polym13162603

Oligosaccharides (Carbohydrates)

4

Abstract

This chapter deals with the synthesis and analysis of oligosaccharides. Due to the structure of monosaccharides, these biopolymers are of significantly higher complexity than peptides or nucleic acids. Therefore, it will not be possible within the scope of this book to explain all synthetic methods in detail. Classical synthesis and automated solid-phase synthesis are presented. The use of enzymes is also important in oligosaccharide synthesis. The analytical section primarily covers chemical methodology, but also addresses the possibilities offered by mass spectrometry and NMR spectroscopy. Selected examples of oligo- and polysaccharides are presented. The topic of glycoconjugates is covered in a separate chapter (Chap. 5).

4.1 Properties and Significance

The number of possible structures (= complexity) of oligosaccharides is much greater (even without considering D- and L-forms) [1, 2] than for peptides and oligonucleotides (Table 4.1).

For clarity, the calculation of the number of disaccharides according to Ref. [1] in the right column of the table lists the various possible linkages. The numbers taken from Ref. [2] are not as easy to follow and may not be entirely chemically accurate, and are intended here only as an example of the order of magnitude. In any case, these numbers result in greater demands on analytical and synthetic methodology compared to the other biopolymers (proteins and nucleic acids).

Precisely for this reason, interest in the structure and interactions of carbohydrate chains is particularly pronounced today, as the methodology is still under development

D. Krois, *Organic Chemical Methods*, https://doi.org/10.1007/978-3-662-73573-2_4

Table 4.1 Number of all possible linkage isomers of biopolymers in comparison [1, 2]

Zahl der Monomere	Zusammensetzung	Zahl der Isomere		
		Peptide, Oligonucleotide	Oligosaccharide	
			aus Lit. [1][a]	aus Lit. [2][b]
1	Z	1	1	
2	Z, Z	1	11[c]	
3	Z, Z, Z	1	120	
4	Z, Z, Z, Z	1	1424	
5	Z, Z, Z, Z, Z	1	17872	
1	Z	1	1	2
2	Y, Z	2	20[d]	256
3	X, Y, Z	6	720	38016
4	W, X, Y, Z	24	34560	7602176
5	V, W, X, Y, Z	120	2144640	2633600000

c) Dimere einer Aldohexose:
α (1→2), β(1→2)
α (1→3), β(1→3)
α (1→4), β(1→4)
α (1→6), β(1→6)
(α1→ α1)
(α1→β1), (β1→β1)

d) Dimere z. B. von. Glc u. Man
Glc*p*→ Man | Man*p*→Glc
α(1→2), β(1→2) | α(1→2), β(1→2)
α(1→3), β(1→3) | α(1→3), β(1→3)
α(1→4), β(1→4) | α(1→4), β(1→4)
α(1→6), β(1→6) | α(1→6), β(1→6)
Glc*p*↔Man*p*
(α1→α1), (β1→α1)
(α1→β1), (β1→β1)

[a]Number of linear isomers, without considering branching and different ring sizes; [b]Number of possible isomers, considering branching and different ring sizes. For references [c] and [d] see right column of the table

and many questions remain open. *Glycobiology* has thus become a field of intensive research [3].

4.2 Oligosaccharide Syntheses

Because of the complexity of carbohydrate structures, the chemical synthesis of oligosaccharide chains is very demanding. So, could one not try to produce them molecular biologically—like proteins? The answer is simple: there is no template-driven oligosaccharide biosynthesis comparable to protein biosynthesis. Carbohydrate chains are produced by multi-enzyme complexes and therefore always exhibit significant microheterogeneity, i.e., even within a single cell, for example, glycoprotein molecules of one type do not all have the same carbohydrate content. However, if one wishes to study their influence, uniform compounds would be desirable. Consequently, the chemical synthesis of oligosaccharides has gained great importance [4, 5]. In recent years, automated oligosaccharide synthesis (analogous to SPPS, Sect. 3.1.3.2) has also been continuously developed (Sect. 4.2.3) [6–8].

As in peptide synthesis, protection group strategy and the unambiguous formation of the glycosidic bond (Sect. 2.2, keyword: anomers) are central.

4.2.1 Protecting Groups for the Hydroxyl Functions of Aldoses and Ketoses

Due to the numerous possible linkages, even the introduction of permanent and temporary protecting groups is a challenge. Why a plurality of temporary protecting groups?—Carbohydrates are often branched, so one must first deprotect one OH group and at a later stage of the synthesis another. Table 4.2 lists many of the common protecting groups, including abbreviations and at least one cleavage method. Mainly ester and ether protecting groups are used. Acetal protecting groups direct the introduction of other protecting groups to specific positions—they are thus used in the preparation of specifically protected monosaccharides. Acetone reacts mainly with 1,2-*cis*-hydroxyl groups, benzaldehyde more with 1,3-*cis*- or 1,2-*trans*-hydroxyl groups. The NH_2 group of amino sugars can be introduced and protected as phthalimide or as azide. For some of the less common protecting groups, the reaction schemes for introduction and cleavage are given in the appendix Sect. 4.2.4.

It is striking that practically no protecting groups are used that must be cleaved under acidic conditions—in stark contrast to peptide synthesis. But this is also logical. While peptides and amino acids often undergo transformations under basic conditions (cf. Sect. 3.2.4), the glycosidic bond, as an acetal bond, is stable under basic conditions but is already relatively easily cleaved under acidic conditions.

The choice of protecting groups is crucial not only for selective deprotection but also for the reactivity of the glycosyl donor during glycosidic bond formation, known as the neighboring group effect. This mainly concerns the 2-position of aldoses. But cyclic protecting groups can also influence the conformation of the pyranose ring and thus control the formation of a specific anomer during glycosylation. A more in-depth discussion of protecting group strategy and glycosylation control would go beyond the scope of this book; for this, see the review articles [1, 4] and [6].

A monograph on the topic of protecting groups in carbohydrate chemistry was published in 2019: [17]

4.2.2 Formation of the Glycosidic Bond

While the 1986 review by Richard R. Schmidt still posed the question, "Are there alternatives to the Koenigs-Knorr method?" [1], today numerous other methods for forming the glycosidic bond have been established, most of which are also based on activation of the glycosyl donor, i.e., the 1-position in aldoses. The most important routes for linking two monosaccharides are summarized in Fig. 4.1 (from Refs. [1, 4, 5] and [18]).

While the classic Koenigs-Knorr synthesis, published in 1901 (Fig. 4.1, **A**), was long the method of choice, today the trichloroacetimidate method (**E**), activation via phosphates (**F**),

Table 4.2 Selection of important protecting groups (PG) for oligosaccharide synthesis, (**temporary PG** in bold):

PG	Abbr.	Name	Formula	Most common cleavage conditions (see Ref. [9])
Ester	Ac-	Acetyl-	CH_3–CO–	K_2CO_3/MeOH; MeOH/NaOMe (*cat.*); NH_3/MeOH
	Bz-	Benzoyl-	Ph–CO–	1 % NaOH/MeOH
	Piv-, Pv-	Pivaloyl-	$(CH_3)_3C$–CO–	MeOH/NaOMe; $MeNH_2/H_2O$,
	Fmoc-	Fluorenylmethoxycarbonyl-	(see Tab. 3.1) Fmoc–	Et_3N/pyridine; piperidine/DMF
	Lev-	Levulinoyl-	CH_3–CO–CH_2–CH_2–CO–	H_2N–NH_2/pyridine/AcOH, [10]
	ClAc-	2-Chloroacetyl-	$ClCH_2$–CO–	H_2N–C(=S)–NH_2; H_2N–NH_2·AcOH [11, 12]
	AMB-, **AZMB**-	2-Azidomethylbenzoyl-	(2-CH_2N_3)Ph–CO–	Bu_3P, [13]
Ether	Me-	Methyl-	CH_3–	BBr_3/CH_2Cl_2
	(Bzl-)Bn-	Benzyl-	Ph–CH_2–	H_2/Pd; Na/NH_3; Na/EtOH
	All-	Allyl-	CH_2=CH–CH_2–	$Pd(PPh_3)_4$ (*cat.*)/p-TolSO_2H/DME [14]
	NAP-	2-Naphthylmethyl-	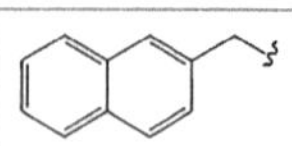	DDQ/MeOH, 60 °C; DDQ/β-pinene [11, 12, 15, 16]
	TMS-	Trimethylsilyl-	$(CH_3)_3Si$–	$Bu_4N^+F^-$(TBAF)/THF; K_2CO_3/MeOH; DDQ; $BF_3 \cdot Et_2O$,
Acetal PGPG		Isopropylidene-	$(CH_3)_2$C<	H^+(… TFA, p-Tol-SO_3H, …)/H_2O, …
		Benzylidene-	Ph–CH<	H_2/Pd/AcOH; Na/NH_3; Et_3SiH/TFA; BH_3/THF; $NaCNBH_3$/HCl
Amino-PG	PhthN-	Phthalimido-	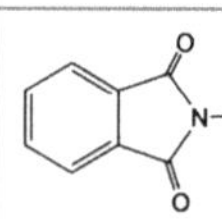	H_2N–NH_2/MeOH; H_2N–NH_2·AcOH
	N_3-	Azido-	N^-=N^+=N–	H_2/Pd

Abbreviations (Abbr.): cat. … catalytic, DME … dimethoxyethane, THF … tetrahydrofuran, DDQ … 2,3-dichloro-5,6-dicyano-*p*-benzoquinone, *p*-TolSO_2H … *p*-toluenesulfinic acid

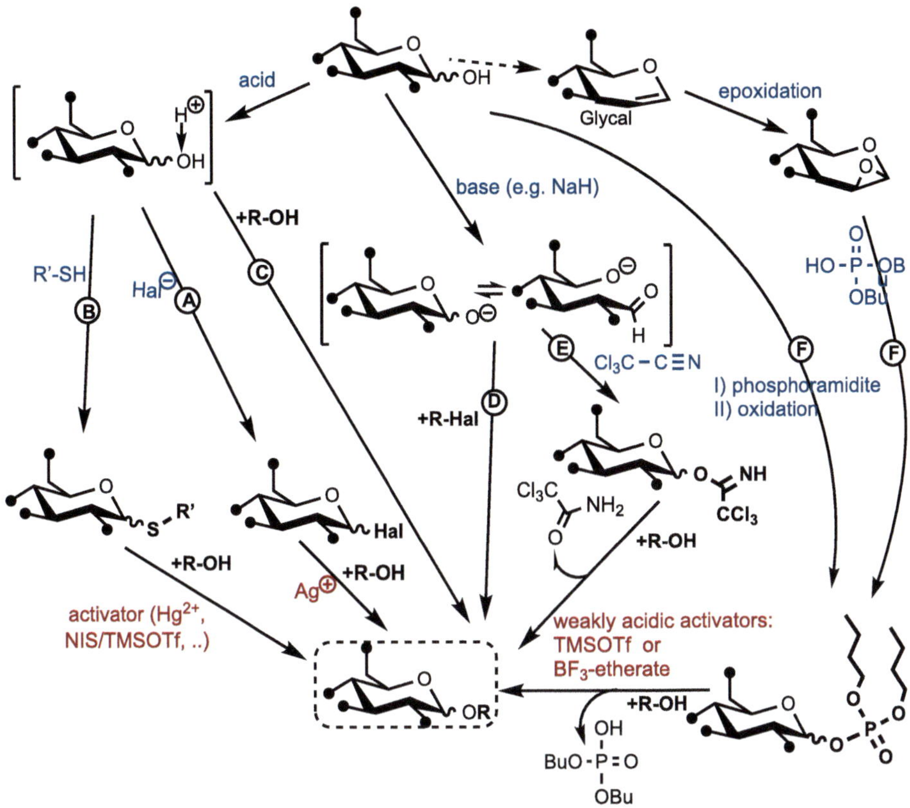

Fig. 4.1 Overview of methods for forming the glycosidic bond: **A** Koenigs-Knorr method (halogenose activation), **B** activation via thioglycosides, **C** Fischer-Helferich method (acid activation), **D** 1-O-alkylation (base activation), **E** trichloroacetimidate method (base-acid activation), **F** activation via phosphate esters (introduction via glycals or with phosphoramidites)—according to Refs. [1, 4]. and [18]. NIS … *N*-iodosuccinimide; TMSOTf … trimethylsilyl triflate; phosphoramidite … Sect. 6.2.2

and via thioglycosides (**B**) are perhaps the most common methods. Instead of trichloroacetonitrile, other compounds with electron-deficient triple bonds or cumulated double bonds (such as ketenimines) can also be used. These activation methods are also applied in the increasingly important solid-phase synthesis (see below). Glycals can also be used directly as glycosyl donors after activation with an electrophile such as iodine or by epoxidation [19, 20]. A variant of the Koenigs-Knorr method, which uses glycosyl iodides previously considered too reactive, has shown interesting potential for the formation of glycosidic bonds [21].

4.2.3 Automated Oligosaccharide Syntheses

Analogous to the successful solid-phase peptide synthesis (SPPS) (Sect. 3.1.3) or oligonucleotide synthesis (Sect. 6.2), solid-phase oligosaccharide synthesis (SPOS or *automated glycan assembly* AGA) was also developed and continuously refined [6–8, 12, 20, 22], albeit later due to the complexity of carbohydrate synthesis.

As an example, we would like to mention one of the first more complex **automated** oligosaccharide syntheses, namely that of the fully protected Lewis^X–Lewis^Y nonasaccharide by Peter Seeberger from 2004, which was carried out on the synthesizer developed by his group [23]. Figure 4.2 sketches the assembly of this branched oligosaccharide from the individual, appropriately protected building blocks. The anchor with the isolated double bond enables subsequent coupling with other olefins via metathesis reaction. Fmoc and the orthogonal levulinoyl protecting group are used as temporary protecting groups. Table 4.3 lists the conditions for one cycle (formation of the glycosidic bond and removal of the temporary protecting group). Two coupling steps are included to ensure the highest possible yield. As a reminder: In SPPS as well (Sect. 3.1.3.2), we have seen how important an almost quantitative conversion is in solid-phase synthesis.

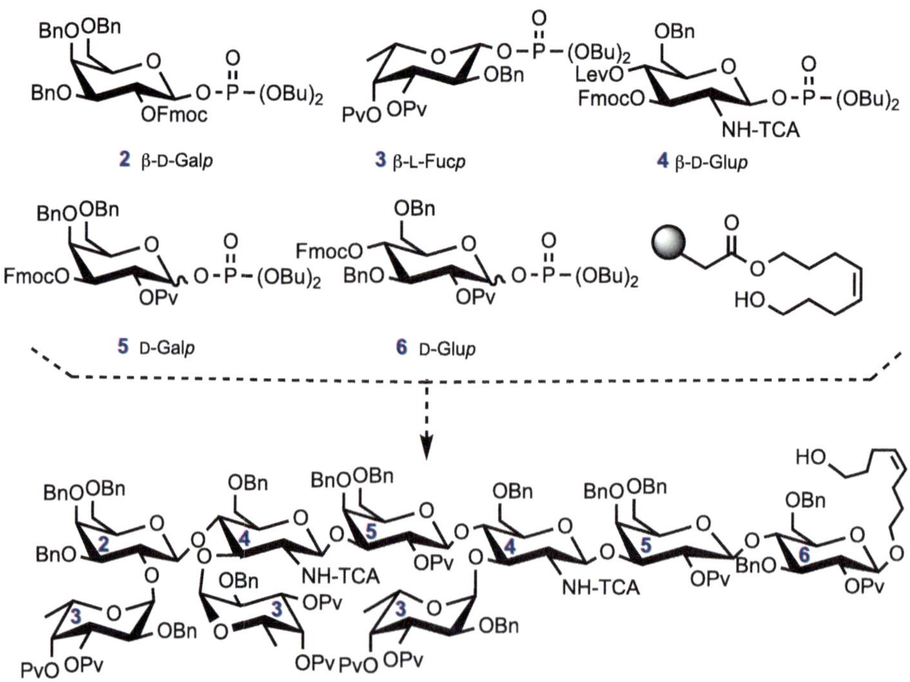

Fig. 4.2 Synthesis of the fully protected Lewis^X-Lewis^Y nonasaccharide from appropriately protected building blocks of glucose (Glc), galactose (Gal), and 6-deoxygalactose = fucose (Fuc) [for abbreviations see Table 4.2, TCA … trichloroacetyl-]. (Adapted from [23])

Table 4.3 General cycle of SPOS for glycosyl phosphates as donor group according to [23]

No.	Process type	Reagents	Time/min
1	Coupling	5 equivalents building block + 5 equivalents TMSOTf	21
2	Washing	CH_2Cl_2 (dichloromethane)	9
3	Coupling	5 equivalents building block + 5 equivalents TMSOTf	21
4	Washing	Dimethylformamide (DMF)	9
5	Removal of temporary protecting group	3 × 175 equivalents piperidine/DMF (cleavage of Fmoc ester) or 5 × 10 equivalents $H_2N–NH_2$/DMF (cleavage of Lev ester)	34 or 80
6	Washing	DMF	9
7	Washing	0.2 M AcOH/THF	9
8	Washing	THF (tetrahydrofuran)	9
9	Washing	CH_2Cl_2	9
	Total time per cycle		130 or 176

TMSOTf … trimethylsilyl triflate $(CH_3)_3Si–O–SO_2–CF_3$

As already mentioned, the methodology has been continuously improved. Certain glycosidic linkages were initially not possible, but are now accessible through appropriate choice of protecting groups and coupling methods; by introducing up to four temporary protecting groups (e.g., [7, 11, 12, 16]), it is possible to create more complex branches. Since a 100% conversion in glycosylation is usually not achieved within a reasonable time, a cycle blocking any remaining free HO groups—a so-called *capping*—has generally been introduced, which was missing in the above example. We have already encountered this in SPPS (Sect. 3.1.3.2). Temperature control has also proven to be particularly important. On the one hand, glycosylation should be carried out at the lowest possible temperature [24], but removal of the protecting groups requires significantly higher temperatures if the cycle time is to be reduced. Therefore, a specially designed synthesizer with a combination of constant cooling via a thermostat and brief heating by microwave is proposed as a solution [7, 11].

So far, we have discussed chemical synthesis using elaborate protecting group strategies—the specific formation of glycosidic bonds with specialized enzymes could dispense with protecting groups. This is sometimes applied in manual synthesis, but automation has not yet been developed very far, as described in the review [25]. For specific problems, the enzymatic introduction and subsequent removal of phosphate groups as special protecting groups has also been introduced in the SPOS discussed above [26].

As an alternative to solid-phase synthesis, some groups have also developed the synthesis of oligosaccharides in solution using fluorocarbon tags (*fluorous tags* [27, 28]. In this approach, the growing chain is attached not to a polymer but to a soluble anchor, which usually carries several C_nF_{2n+1} residues. For purification after coupling and removal of protecting groups, it is exploited that a solvent such as C_6F_{14} (FC72™) is immiscible with both water and most "normal" organic solvents, and the growing oligosaccharide, due to its tag (F-tag), is the only compound soluble in the fluorocarbon-containing solvent (an optimal mixture is FC72 and CH_3–O–C_4F_9). Similar to synthesis on an insoluble polymer, this greatly facilitates purification.

4.2.4 Appendix

Here, for those interested, we compile the reactions for the introduction and removal of some protecting groups (Table 4.2 and references therein). It should be noted that the relevant literature often does not specify the exact product of the cleavage. However, plausible structures are shown in the schemes in Fig. 4.3.

4.3 Analysis of Oligosaccharides

Due to the diversity of possible isomers, the analysis of oligosaccharides is also far more complex than that of peptides and oligonucleotides, even if one restricts oneself to the sequence (naturally including the linkage and type of glycosidic bond)—that is, apart from the spatial structure. Historically, chemical degradation reactions in combination with instrumental methods were initially used here as well. Gas chromatography probably initially provided the best resolution for separating individual monomer building blocks, but prior derivatization was necessary to achieve the required volatility [29]. The following describes the degradation reactions for determining composition, ring size, and type of linkage.

4.3.1 Composition of Carbohydrates (glycosyl composition)

Procedure: Fig. 4.4

I. Total hydrolysis (e.g., 2 M TFA) → mixture of monosaccharides
II. Reduction ($NaBH_4$) and acetylation to alditol acetates
III. Identification of the monosaccharide derivatives by GC or GC-MS (alternatively: h.p.a.e. = *high pH anion exchange chromatography* with amperometric detection); the absolute configuration can be determined after derivatization with enantiomerically pure 2-butanol or 2-octanol by GC analysis.

Errors and interferences can occur due to incomplete hydrolysis or decomposition of a product. It should also be noted that an alditol can sometimes arise from two different aldoses (e.g., D-altrose and D-talose both yield the same alditol).

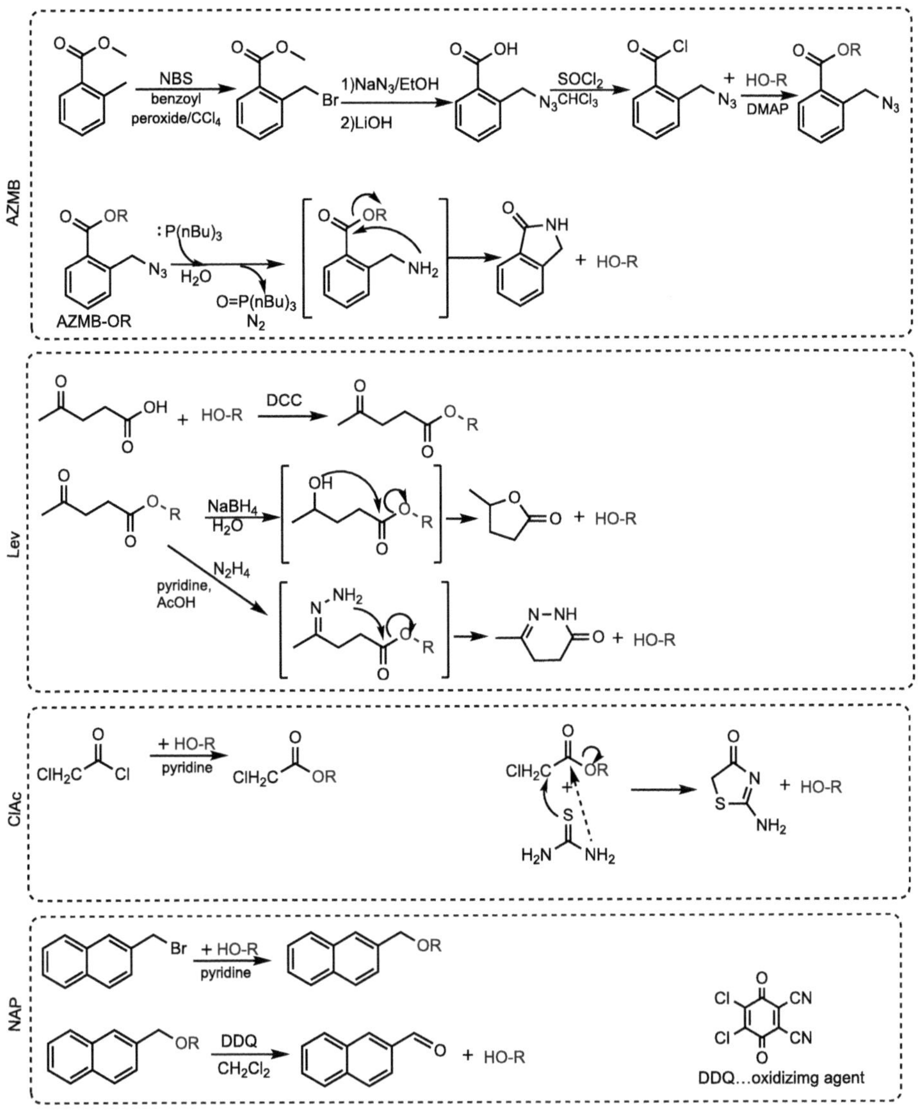

Fig. 4.3 Introduction and removal of some temporary protecting groups in oligosaccharide synthesis; carbohydrate residue … blue

Chromatographic separation of aldoses or ketoses is in principle not meaningful due to the constant interconversion of the anomeric forms. Therefore, these are always reduced to the corresponding alditols. For GC analysis, acetylation is performed to increase volatility.

4.3.2 Linkage Analysis

The next most important information is the ring size (pyranose or furanose form) and the linkage position, i.e., the hydroxyl group involved in the glycosidic bond with the preceding monosaccharide (Fig. 4.5). Here too, derivatizations are performed prior to analytical separation.

I. Complete methylation of free –OH: Methylation is usually carried out under strongly basic conditions with methyl iodide (CD_3I/NaOH/DMSO). By introducing deuterated methyl ethers, pre-existing methyl ethers can be identified. Under basic methylation, *O*-acyl substituents would be cleaved beforehand. To identify such conjugates, a different approach is used: methylation under non-basic conditions (e.g., with methyl triflate), followed by basic cleavage of the esters (*O*-deacylation) and labeling of these HO groups by ethylation [30].

II. Total hydrolysis with TFA [following steps II, III, and IV as in Sect. 4.3.1]

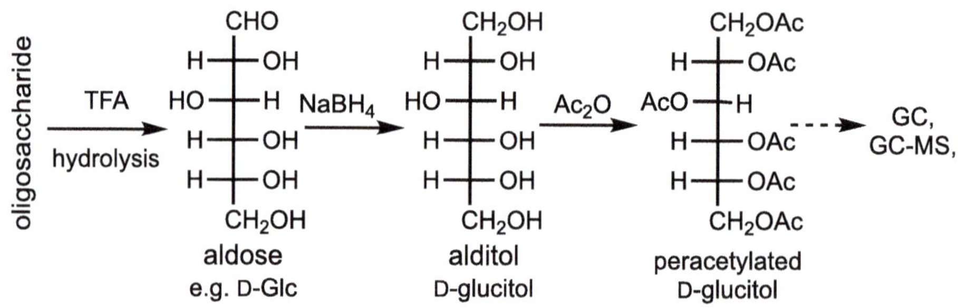

Fig. 4.4 Analysis of the composition of carbohydrates

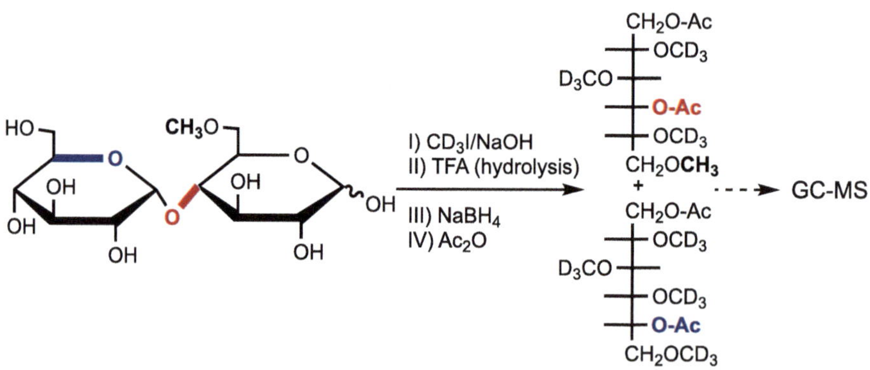

Fig. 4.5 Scheme of linkage analysis using the example of a methylated maltose: α-D-Glc*p*-(1 → 4)-6-*O*-methyl-D-Glc. Note: As a reducing disaccharide, the ring size of the second glucose molecule is, of course, not fixed, since it is in equilibrium with the open-chain form

III. Reduction with $NaBH_4$ to alditols.
IV. Acetylation of all remaining free HO groups. The position of the acetyl groups now indicates the linkage sites and respective ring size. Of course, the hemiacetalic hydroxyl group is also acetylated in each case.
V. GC-MS.

4.3.3 Chromatographic and Electrophoretic Methods in Carbohydrate Analysis

In addition to the already mentioned gas chromatography (GC), the methods already discussed for peptides, such as HPLC and capillary electrophoresis (CE), are of course also used for the separation of oligosaccharides or monosaccharides (Sect. 3.2.1). In addition, ion chromatography (IC) and *high performance anion exchange chromatography* (HPAEC) should be mentioned, where, in contrast to HPLC, NaOH or KOH (i.e., strongly basic eluents) are used as the mobile phase, which promote the formation of ions [31]. A review in the journal "Food Chemistry" [32] also describes sample preparation, further chromatographic, as well as colorimetric and spectroscopic methods. It is not surprising that a great deal is published on carbohydrate analytics in nutrition science journals, since our foods—especially those of plant origin—primarily contain these biopolymers or mono- and disaccharides. Verifying the authenticity and purity of these foods is therefore an important task for analysts.

4.3.4 Mass Spectrometry (MS)

Today, the composition and sequence analysis of oligo- and polysaccharides is mostly performed using MS in sometimes very specialized variants. ESI is primarily used as the ionization method. For sequencing, highly specific fragmentation of the oligosaccharide chains is required, which is induced in so-called MS/MS (or tandem MS) experiments on individual isolated ions. There are a number of methods in the mass spectrometer for this, in which energy is transferred to isolated ions in the form of collisions with neutral atoms or electrons, or by interaction with photons (IR or UV) or electric fields. Which bonds dissociate then depends strongly on the method used, so that different and complementary results are obtained. A very detailed and comprehensible presentation can be found in [33]. Sometimes, additional information can be obtained by H/D exchange [34].

4.3.5 NMR Spectroscopy

A major problem in the analysis of carbohydrate chains, but also monosaccharides, is the large number of possible isomers (with the same mass). With certain methods, MS can still achieve some differentiation, but NMR spectroscopy is certainly capable of making much more definitive assignments. In particular, the determination of the anomeric configuration (α or β) is only directly possible with NMR. The H–H coupling constant of the proton at the anomeric carbon atom (to the neighboring proton), as well as the ^{13}C–H coupling, provide relevant information. However, with more complex oligosaccharides with possibly several similar branches, strong signal overlap is a problem. Here, too, certain experiments can help. To determine the number and type of monomer building blocks, it is often necessary to perform total hydrolysis (Sect. 4.3.1) beforehand. A major difference from MS, however, is the required sample amount, and the sample must also be homogeneous for NMR studies, whereas MS experiments can yield results even with mixtures.

There is also a very comprehensive and informative review on this topic with all possible NMR techniques [35]. A somewhat more illustrative and practice-oriented guide is provided in this *hitchhiker's guide*: [36].

4.3.6 Examples of Other Methods

As already mentioned above, there is great interest in rapid and low-effort analysis of naturally occurring carbohydrates in plant foods, as well as processed foods, etc. *Nanopore based methods* are, however, still under development. Compared to protein analysis and especially DNA sequencing, the requirements for oligo- and polysaccharides are much higher. The methodology is based on the fact that, when a single molecule is forced through a nanopore in a membrane in an electric field, the recorded ion current provides information about the molecular segments in the pore. Due to the much higher complexity of carbohydrates, it is more difficult to obtain unambiguous sequencing information [37]. A little-known spectroscopic method that provides material information with little effort is THz spectroscopy, in the frequency range between microwaves and far-IR. It is already used today in various fields, from astronomy to dentistry, including in the area of food quality control. The obtained absorption spectra allow, for example, the distinction between the isomers glucose and fructose [38].

4.3.7 Structural Analysis of Oligosaccharide Chains

As with proteins, the spatial structure of carbohydrates is of great importance, especially when it comes to understanding biological activity and interactions with other biomolecules. A prerequisite for such studies is a homogeneous, uniform sample substance. However, in the field of carbohydrates, microheterogeneities from natural sources are always present. Today, very pure oligosaccharides can be produced by chemical synthesis, enabling corresponding structural studies. The methodology would in principle be the same as for proteins (Sect. 3.2.5). However, crystals are usually difficult to obtain, and therefore NMR is the method of choice. The specific techniques in combination with appropriate computational methods are described, among others, in [39].

4.4 Mono-, Oligo- and Polysaccharides

Here we will mention only a few significant representatives as examples.

▶ A Small Selection of Commonly Occurring Monosaccharide Building Blocks

The importance of oligosaccharides, especially in glycoconjugates with lipids and peptides and proteins, is increasingly being recognized today. A small selection of the most important building blocks is shown in Fig. 4.6 . They will be particularly important for the following chapter on glycoconjugates and glycoproteins (Chap. 5). As can be seen, fucose is a 6-deoxygalactose and rhamnose is a 6-deoxymannose. Neuraminic acid (or sialic acid) is probably the most important C9 sugar and is usually transferred to the free end of oligosaccharide chains by sialyltransferases. The proportion of neuraminic acid on the cell surface often also provides clues to cancer cells. In Sect. 7.9.1 a method for labeling such neuraminic acid residues is presented. The alduronic acids, here exemplified by glucose and galactose, are formed by enzymatic oxidation of the primary alcohol of aldoses and, like aldoses and ketoses, can be incorporated into oligosaccharides via glycosidic bonds.

Oligosaccharides in Human Milk
There is a wealth of publications on this topic. Why especially nowadays?—It has been understood that these rather indigestible components of breast milk make an important contribution to infant health—apparently through their effect on the human gut flora. Therefore, both the analysis (see also [33]) of these oligosaccharides (*human milk oligosaccharides* HMO) has become very important, as well as their possible production for infant formula substitutes [40].

Heparin, One of the Most Important Drugs for the Prevention of Thrombosis
Heparin is a polysaccharide with a molecular mass of about 16 kDa, consisting of alternating glucuronic acid (or iduronic acid) and glucosamine disaccharide units with about 2–3 sulfate residues per disaccharide. It is largely obtained from the intestinal mucosa of pigs, but most often a partially degraded (enzymatically or chemically) shorter polymer (*low molecular weight heparin* LMWH) heparin with a molecular mass between 3000 and 6000 Da is used. A major problem is quality assurance, as it is an extremely heterogeneous mixture [41]. It would be desirable to produce it independently of animal

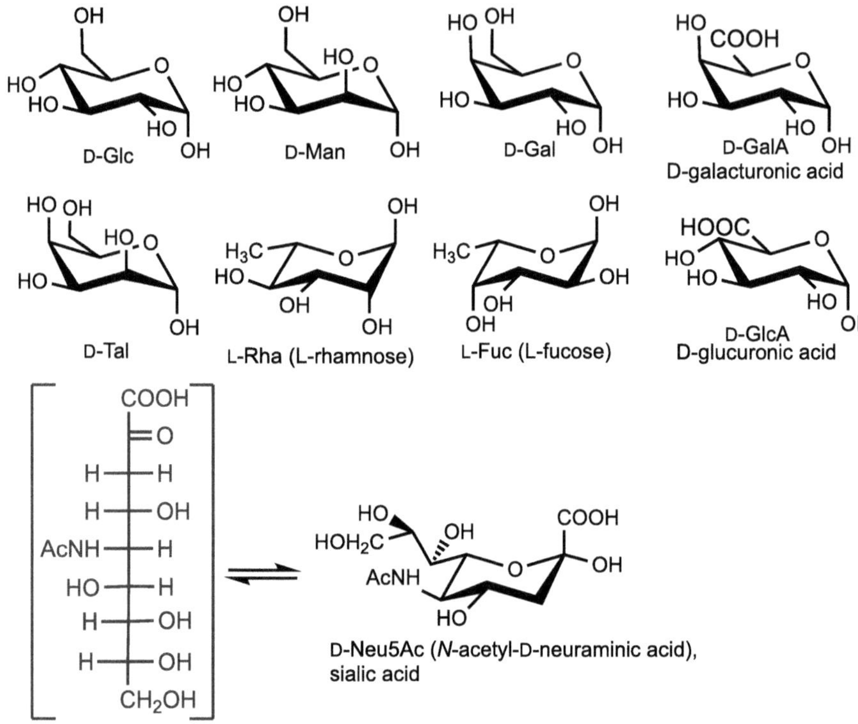

Fig. 4.6 Some important monosaccharide building blocks in complex oligosaccharide chains, shown as α-anomers

sources by chemical or enzymatic methods or by biosynthesis in a microorganism, possibly with subsequent enzymatic steps. However, as of today, none of these production methods is sufficiently developed [42]. As mentioned at the beginning of Sect. 4.2, the production of defined carbohydrates in microorganisms or even cell culture is not possible in the same way as for proteins and nucleic acids.

Lignocellulose, a By-product of Agriculture

Lignocellulose, a renewable raw material, essentially consists of three components: 10–30% lignin, 20–40% hemicellulose, and 30–45% cellulose (depending on plant species). Lignin is a rigid framework of polyphenols, cellulose is the polymer of β-(1 → 4)-D-glucose, and hemicellulose is a branched polysaccharide made up of pentoses (xylose and arabinose), hexoses (mainly mannose, glucose, and galactose), and glucuronic and galacturonic acid. This hemicellulose can also be used as a source of oligosaccharides for food applications, and this development has already begun [43]. In addition, lignocellulose is of course also economically important as a source for biofuels, chemicals, and materials—especially with regard to reducing the use of fossil fuels.

4.5 Questions

Let's Test Your Understanding of This Chapter's Content

4.1 Why are the synthesis and analysis of carbohydrates more complex than those of peptides and nucleic acids?

4.2 Can you specify the number of isomeric disaccharides that can be formed from galactose and mannose (both in the pyranose form), and write them in abbreviated form?

4.3 As a review from Chap. 2: What is the general type of bond represented by the glycosidic bond?

4.4 Is the glycosidic linkage a directional bond?

4.5 What types of protecting groups are used for hydroxyl groups in oligosaccharide synthesis?

4.6 Why are multiple temporary protecting groups sometimes needed for monosaccharide building blocks in oligosaccharide synthesis (as opposed to peptide synthesis)?

4.7 Name three activation methods for the glycosyl donor that are predominantly used today!

4.8 In oligosaccharide solid-phase synthesis (SPOs), is the glycosyl donor or the glycosyl acceptor (free HO group) anchored to the solid phase?

4.9 What is meant by capping and why is it usually applied in the synthesis cycle?

4.10 Name methods for the analysis of oligosaccharides!

4.11 Is there an automated sequencing method for oligo- and polysaccharides?

4.12 What are alduronic acids?

True or **F**alse?

4.13 A linear trisaccharide always has a reducing end, whereas a disaccharide does not necessarily! ☐

4.14 There are eight diastereomeric ketohexoses. ☐

4.15 Every monosaccharide exists in solution as an equimolar mixture of α- and β-anomers. ☐

References

1. Schmidt, R.R.: Neue Methoden zur Glycosid- und Oligosaccharidsynthese – gibt es Alternativen zur Koenigs-Knorr-Methode? Angew. Chem. **98**(3), 213–236 (1986). https://doi.org/10.1002/ange.19860980305; Schmidt, R.R.: New methods for the synthesis of glycosides and oligosaccharides – are there alternatives to the Koenigs-Knorr method? [New synthetic methods (56)]. Angew. Chem. Int. Ed. **25**(3), 212–235 (1986). https://doi.org/10.1002/anie.198602121

2. Laine, R.A.: Invited commentary: a calculation of all possible oligosaccharide isomers both branched and linear yields 1.05×10^{12} structures for a reducing hexasaccharide: the isomer barrier to development of single-method saccharide sequencing or synthesis systems. Glycobiology. **4**(6), 759–767 (1994). https://doi.org/10.1093/glycob/4.6.759
3. Seeberger, P.H., Ge, Y., Szymanski, C.M., …, Strasser, R.: What comes next in glycobiology. Cell. **187**(11), 2628–2632 (2024). https://doi.org/10.1016/j.cell.2024.04.028
4. Zhu, X., Schmidt, R.R.: Neue Prinzipien für die Bildung von glycosidischen Bindungen. Angew. Chem. **121**(11), 1932–1967 (2009). https://doi.org/10.1002/ange.200802036; Zhu, X., Schmidt, R.R.: New principles for glycoside-bond formation. Angew. Chem. Int. Ed. **48**(11), 1900–1934 (2009). https://doi.org/10.1002/anie.200802036
5. Nielsen, M.M., Pedersen, C.M.: Catalytic glycosylations in oligosaccharide synthesis. Chem. Rev. **118**(17), 8285–8358 (2018). https://doi.org/10.1021/acs.chemrev.8b00144; Krasnova, L., Wong, C.-H.: Oligosaccharide synthesis and translational innovation. J. Am. Chem. Soc. **141**(9), 3735–3754 (2019). https://doi.org/10.1021/jacs.8b11005
6. Panza, M., Pistorio, S.G., Stine, K.J., Demchenko A.V.: Automated chemical oligosaccharide synthesis: novel approach to traditional challenges. Chem. Rev. **118**(17), 8105–8150 (2018). https://doi.org/10.1021/acs.chemrev.8b00051
7. Lin, M.-H., Wolf, J.B., Sletten, E.T., …, Seeberger, P.H.: Enabling technologies in carbohydrate chemistry: automated glycan assembly. ChemBioChem. **24**(4), e202200607 (2023). https://doi.org/10.1002/cbic.202200607
8. Niggemeyer, G.B., Danglad-Flores, J.A., Seeberger, P.H.: Automated synthesis of C1-functionalized oligosaccharides. J. Am. Chem. Soc. **147**(2), 1649–1655 (2025). https://doi.org/10.1021/jacs.4c11798
9. Wuts, P.G.M., Greene, T.W.: Greene's Protective Groups in Organic Synthesis, 4. Aufl. Wiley, New York (2007). https://doi.org/10.1002/0470053488
10. Koeners, H.J., Verhoeven, J., Van Boom, J.H.: Synthesis of oligosaccharides by using levulinic ester as an hydroxyl protecting group. Tetrahedron Lett. **21**(4), 381–382 (1980). https://doi.org/10.1016/S0040-4039(01)85479-4; van Boom, J.H., Burgers, P.M.J.: Use of levulinic acid in the protection of oligonucleotides via the modified phosphotriester method: synthesis of decaribonucleotide u-a-u-a-u-a-u-a-u-a. Tetrahedron Lett. 17(52), 4875–4878 (1976). https://doi.org/10.1016/S0040-4039(00)78935-0
11. Agoston, K., Watt, G.M., Fugedi, P.: A new set of orthogonal protecting groups on a monosaccharide scaffold. Tetrahedron Lett. **56**(35), 5010–5012 (2015). https://doi.org/10.1016/j.tetlet.2015.07.015
12. Danglad Flores, J., Leichnitz, S., Sletten, E.T., …, Seeberger, P.H.: Microwave-assisted automated glycan assembly. ChemRxiv, 1–16 (2021). https://doi.org/10.26434/chemrxiv.14292476.v2
13. Love, K.R., Andrade, R.B., Seeberger, P.H.: Linear synthesis of a protected H-type II pentasaccharide using glycosyl phosphate building blocks. J. Org. Chem. **66**(24), 8165–8176 (2001). https://doi.org/10.1021/jo015987h
14. Opatz, T., Kunz, H.: A facile cleavage of allyl ethers on solid phase. Tetrahedron Lett. **41**(52), 10185–10188 (2000). https://doi.org/10.1016/S0040-4039(00)01835-9
15. Xia, J., Abbas, S.A., Locke, R.D., …, Matta, K.L.: Use of 1,2-dichloro 4,5-dicyanoquinone (DDQ) for cleavage of the 2-naphthylmethyl (NAP) group. Tetrahedron Lett. **41**(2), 169–173 (2000). https://doi.org/10.1016/S0040-4039(99)02046-8; Lloyd, D., Bylsma, M., Bright, D.K., …, Bennett, C.S.: Mild method for 2-naphthylmethyl ether protecting group removal using a combination of 2,3-Dichloro-5,6-dicyano-1,4-benzoquinone (DDQ) and β-Pinene. J. Org. Chem. **82**(7), 3926–3934 (2017). https://doi.org/10.1021/acs.joc.7b00065; Volbeda, A.G., Kistemaker, H.A.V., Overkleeft, H.S., …, Codée, J.D.C.: Chemoselective cleavage of p-meth-

oxybenzyl and 2-naphthylmethyl ethers using a catalytic amount of HCl in hexafluoro-2-propanol. J. Org. Chem. **80**(17), 8796–8806 (2015). https://doi.org/10.1021/acs.joc.5b01030

16. Boltje, T.J., Li, C., Boons, G.-J.: Versatile set of orthogonal protecting groups for the preparation of highly branched oligosaccharides. Org. Lett. **12**(20), 4636–4639 (2010). https://doi.org/10.1021/ol101951u
17. Protecting Groups: Strategies and Applications in Carbohydrate Chemistry. In: Vidal, S. (Hrsg.), Wiley-VCh (2019). ISBN: 9783527697014, https://doi.org/10.1002/9783527697014
18. Plante, O.J., Andrade, R.B., Seeberger, P.H.: Synthesis and use of glycosyl phosphates as glycosyl donors. Org. Lett. **1**(2), 211–214 (1999). https://doi.org/10.1021/ol9905452
19. Danishefsky, S.J., Bilodeau, M.T.: Glycale in der organischen Synthese: die Entwicklung umfassender Strategien zum Aufbau von Oligosacchariden und Glycokonjugaten mit biologischer Bedeutung. Angew. Chem. **108** 1482–1522 (1996). https://doi.org/10.1002/ange.19961081304; Danishefsky, S.J., Bilodeau, M.T.: Glycals in organic synthesis: the evolution of comprehensive strategies for the assembly of oligosaccharides and glycoconjugates of biological consequence. Angew. Chem. Int. Ed. **35**(13–14), 1380–1419 (1996). https://doi.org/10.1002/anie.199613801
20. Seeberger, P.H., Danishevsky, S.J.: Solid-phase synthesis of oligosaccharides and glycoconjugates by the glycal assembly method: a five year retrospective. Acc. Chem. Res. **31**(10), 685–695 (1998). https://doi.org/10.1021/ar9600648
21. Gervay-Hague, J.: Taming the reactivity of glycosyl iodides to achieve stereoselective glycosidation. Acc. Chem. Res. **49**(1), 35–47 (2016). https://doi.org/10.1021/acs.accounts.5b00357
22. Weishaupt, M., Eller, S., Seeberger, P.H.: Solid phase synthesis of oligosaccharides. Methods Enzymol. **478**(Glycomics), 463–484 (2010). https://doi.org/10.1016/S0076-6879(10)78022-8
23. Routenberg Love, K., Seeberger, P.H.: Automated solid-phase synthesis of protected tumor-associated antigen and blood group determinant oligosaccharides. Angew. Chem. **116**(5), 612–615 (2004). https://doi.org/10.1002/ange.200352539; Routenberg Love, K., Seeberger, P.H.: Automated solid-phase synthesis of protected tumor-associated antigen and blood group determinant oligosaccharides. Angew. Chem. Int. Ed. **43**(5), 602–605 (2004). https://doi.org/10.1002/anie.200352539
24. Tuck, O.T., Sletten, E.T., Danglad-Flores, J., Seeberger, P.H.: Towards a systematic understanding of the influence of temperature on glycosylation reactions. Angew. Chem. Int. Ed. **61**(15), e202115433 (2022). https://doi.org/10.1002/anie.202115433
25. Wen, L., Edmunds, G., Gibbons, C., …, Wang, P.G.: Toward automated enzymatic synthesis of oligosaccharides. Chem. Rev. **118**(17), 8151–8187 (2018). https://doi.org/10.1021/acs.chemrev.8b00066
26. Sletten, E.T., Fittolani, G., Hribernik, N., …, Delbianco, M.: Phosphates as assisting groups in glycan synthesis. ACS Central Sci. **10**(1), 138–142 (2024). https://doi.org/10.1021/acscentsci.3c00896
27. Goto, K., Mizuno, M.: Application of fluorous chemistry for oligosaccharide synthesis. Trends Glycosci. Glycotechnol. **25**(146), 203–213 (2013). https://doi.org/10.4052/tigg.25.203; **Englisch**/japanische Fassung: https://www.jstage.jst.go.jp/browse/tigg/25/146/_contents. Zugegriffen am 20.12.2025; Huang, W., Gao, Q., Boons, G.-J.: Assembly of a complex branched oligosaccharide by combining fluorous-supported synthesis and stereoselective glycosylation using anomeric sulfonium ions. Chem. Eur. J. **21**(37), 12920–12926 (2015). https://doi.org/10.1002/chem.201501844
28. Mahmud, I.H., Goekjian, P.G.: Applications of fluorous tag methodology in carbohydrate synthesis. In: Pilar Rauter, A., Lindhorst, T.K., Queneau, Y. (Hrsg.) Carbohydrate Chemistry: Chemical and Biological Approaches, Bd. 45., S. 1–56. The Royal Society of Chemistry (2021). https://doi.org/10.1039/9781839164538-00001

29. Harvey, D.J.: Derivatization of carbohydrates for analysis by chromatography; electrophoresis and mass spectrometry. J. Chromatogr. B. **879**(17–18), 1196–1225 (2011). https://doi.org/10.1016/j.jchromb.2010.11.010; Ruiz-Matute, A.I., Hernandez-Hernandez, O., Rodriguez-Sanchez, S., …, Martínez-Castro, I.: Derivatization of carbohydrates for GC and GC-MS analyses. J. Chromatogr. B. **879**(17–18), 1226–1240 (2011). https://doi.org/10.1016/j.jchromb.2010.11.013; Becker, M., Zweckmair, T., Forneck, A., …, Liebner F.: Evaluation of different derivatisation approaches for gas chromatographic-mass spectrometric analysis of carbohydrates in complex matrices of biological and synthetic origin. J. Chromatogr. A. **1281**, 115–126 (2013). https://doi.org/10.1016/j.chroma.2013.01.053
30. Björndal, H., Hellerqvist, C.G., Lindberg, B., Svensson, S.: Gas-liquid chromatography and mass spectrometry in methylation analysis of polysaccharides. Angew. Chem. Int. Ed. **9**(8), 610–619 (1970). https://doi.org/10.1002/anie.197006101; Björndal, H., Hellerqvist, C.G., Lindberg, B., Svensson, S.: Gas-Flüssigkeits-Chromatographie und Massenspektrometrie bei der Methylierungsanalyse von Polysacchariden. Angew. Chem. **82**(16), 643–652 (1970). https://doi.org/10.1002/ange.19700821602; Prehm, P.: Methylation of carbohydrates by methyl trifluoromethanesulfonate in trimethyl phosphate. Carbohydr. Res. **78**(2), 372–374 (1980). https://doi.org/10.1016/0008-6215(80)90018-X; Ciucanu, I., Kerek, F.: A simple and rapid method for the permethylation of carbohydrates. Carbohydr. Res. **131**(2), 209–217 (1984). https://doi.org/10.1016/0008-6215(84)85242-8
31. Michalski, R., Kończyk, J.: Ion chromatography and related techniques in carbohydrate analysis: a review. Molecules. **29**(14), 3413 (2024). https://doi.org/10.3390/molecules29143413
32. Kurzyna-Szklarek, M., Cybulska, J., Zdunek, A.: Analysis of the chemical composition of natural carbohydrates – an overview of methods. Food Chem. **394**, 133466 (2022). https://doi.org/10.1016/j.foodchem.2022.133466
33. Grabarics, M., Lettow, M., Kirschbaum, C., …, Pagel, K.: Mass spectrometry-based techniques to elucidate the sugar code. Chem. Rev. **122**(8), 7840–7908 (2022). https://doi.org/10.1021/acs.chemrev.1c00380
34. Hatvany, J.B., Gallagher, E.S.: Hydrogen/deuterium exchange for the analysis of carbohydrates. Carbohydr. Res. **530**, 108859 (2023). https://doi.org/10.1016/j.carres.2023.108859
35. Fontana, C., Widmalm, G.: Primary structure of glycans by NMR spectroscopy. Chem. Rev. **123**(3), 1040–1102 (2023). https://doi.org/10.1021/acs.chemrev.2c00580
36. Speciale, I., Notaro, A., Garcia-Vello, P., …, De Castro, C.: Liquid-state NMR spectroscopy for complex carbohydrate structural analysis: a hitchhiker's guide. Carbohydr. Polym. **277**, 118885 (2022). https://doi.org/10.1016/j.carbpol.2021.118885
37. Zhao, Y., Su, Z., Zhang, X., …, Li, G.: Recent advances in nanopore-based analysis for carbohydrates and glycoconjugates. Anal. Meth. **16**(10), 1454–1467 (2024). https://doi.org/10.1039/D3AY02040A
38. Li, Q., Lei, T., Sun, D.-W.: Analysis and detection using novel terahertz spectroscopy technique in dietary carbohydrate-related research: principles and application advances. Crit. Rev. Food Sci. Nutr. **63**(13), 1793–1805 (2023). https://doi.org/10.1080/10408398.2023.2165032
39. Yu, Y., Delbianco, M.: Conformational studies of oligosaccharides. Chem. Eur. J. **26**(44), 9814–9825 (2020). https://doi.org/10.1002/chem.202001370
40. Zhu, L., Li, H., Luo, T., …, Zhang, B.: Human milk oligosaccharides: a critical review on structure, preparation, their potential as a food bioactive component, and future perspectives. J. Agricult. Food Chem. **71**(43), 15908–15925 (2023). https://doi.org/10.1021/acs.jafc.3c04412
41. Yu, Y., Song, Y., Zhao, Y., …, Wang, H.: Quality control, safety assessment and preparation approaches of low molecular weight heparin. Carbohydr. Polym. **339**, 122216 (2024). https://doi.org/10.1016/j.carbpol.2024.122216

42. Zhao, S., Zhang, T., Kan, Y., …, Li, J.-P.: Overview of the current procedures in synthesis of heparin saccharides. Carbohydr. Polym. **339**, 122220 (2024). https://doi.org/10.1016/j.carbpol.2024.122220
43. Bhatia, L., Sharma, A., Bachheti, R.K., Chandel, A.K.: Lignocellulose derived functional oligosaccharides: production, properties, and health benefits. Prep. Biochem. Biotechnol. **49**(8), 744–758 (2019). https://doi.org/10.1080/10826068.2019.1608446

5 Glycoconjugates, Glycoproteins

Abstract

Analysis and synthesis of oligosaccharides are currently of particular interest, especially in connection with glycoconjugates. It is increasingly recognized that, for many essential biological functions and also for diseases, it is precisely the carbohydrate portion of proteins or lipids that is of great importance. However, materials isolated from natural sources usually exhibit microheterogeneities, since the biosynthesis of carbohydrates is not as template-driven as protein biosynthesis. Therefore, homogeneous chemically synthesized glycoconjugates would be especially valuable. This section illustrates this issue using examples. On the one hand, the glycopeptidolipids of mycobacteria are discussed; on the other hand, the types of glycoproteins and their synthesis, as well as their biosynthesis, are addressed.

▶ For a new comprehensive overview of the significance of glycoconjugates in general, see: Shivatare, S. S., Shivatare, V. S. and Wong, C.-H.: Glycoconjugates: Synthesis, Functional Studies, and Therapeutic Developments. *Chem. Rev.*, **122** (20), 15603–15671, *doi*: 10.1021/acs.chemrev.1c01032 (2022) [1].

5.1 Glycolipids

Mycobacteria are a special group of microorganisms that are distinguished primarily by their uniquely structured cell wall [2, 3]. Outside the actual cell membrane, there are layers of peptidoglycan (PG), arabinogalactan (AG), and mycolic acids (long-chain

D. Krois, *Organic Chemical Methods*, https://doi.org/10.1007/978-3-662-73573-2_5

branched fatty acids), which form a repellent envelope, making the bacteria resistant to many disinfectants as well as antibiotics. Mycobacteria are responsible for a number of serious diseases, especially tuberculosis (*Mycobacterium tuberculosis*) and, more rarely, leprosy (*M. leprae*), but representatives of the MAC (*Mycobacterium avium complex*) are increasingly becoming a health problem, particularly in immunocompromised individuals [4]. The development of specific antibiotics first requires precise knowledge of the structure of this special cell wall and the possible targets for therapy. This was recognized early on, and methods for the analysis of the carbohydrate component in particular were developed and applied [5, 6].

In the following, we will discuss, as an example of glycoconjugates, the variable glycopeptidolipids (GPL) on the outermost surface of the cell wall of mycobacteria, specifically of *Mycobacterium avium*. This selection is purely arbitrary, but the aim here is not to provide an exhaustive treatment of such structures, but rather to give an impression of their complexity and variability. What is the significance of these structural elements? The oligosaccharide chains on the surface of *Mycobacterium avium* are, as we will see, highly variable and thus probably serve primarily to evade the immune system, but are also crucial for motility and colony formation. In the referenced publication [The variable surface glycolipids of mycobacteria: structures, synthesis of epitopes, and biological properties, Aspinall, G. O., Chatterjee, D. and Brennan, P. J., *Adv. Carbohydr. Chem. Biochem.*, 51, 169–242 (1995)] [5], these glycolipids are first analyzed, and based on the oligosaccharide chain a classification into different serovariants of the bacterium is made, in order to then develop syntheses of structural elements that may be useful for immunization purposes.

The structure of the glycopeptidolipids (GPL) was elucidated after their isolation using the methods of carbohydrate analysis described in Sect. 4.2 (especially FAB-MS and NMR) [5, 6]. Figure 5.1 shows the basic structure of the lipopeptide part with the linkage site to the oligosaccharide chain. Table 5.1 lists the individual variants of the oligosaccharide part that have been found and groups them into three main serovariants.

What is striking here is the great variability of the carbohydrate part and also the extensive derivatization at the non-reducing end of the oligosaccharide. Identifying all these structural variations, even after isolation and derivatization, certainly places high demands on the procedures used in carbohydrate analysis.

Apart from the GPLs just discussed, lipooligosaccharides (LOS) also occur as surface antigens, especially in *Mycobacterium tuberculosis*, although the glycan part is often reduced to a trehalose unit [Glc*p*(α1↔α1)Glc*p*] (Fig. 5.2).

As already mentioned, mycobacteria are resistant to many antibiotics, and there is therefore great interest in developing specific vaccines. Accordingly, oligosaccharide sequences were selected to be produced homogeneously and purely by chemical synthesis. In some cases, synthesis was also the only method to unambiguously determine the absolute configuration [7–9].

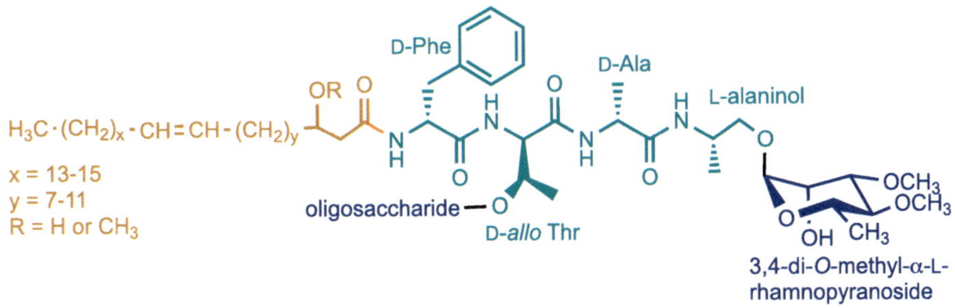

Fig. 5.1 General structure of the glycopeptidolipids (GPL) of *Mycobacterium avium*. (According to [2, 5])

Table 5.1 Variable oligosaccharide chains of the glycopeptidolipids of *Mycobacterium avium*. (According to [5])

Serovariant	Oligosaccharide sequence
Group 1	
2	4-*O*-Ac-2,3-di-*O*-Me-α-L-Fuc*p*-(1→3)-α-L-Rha*p*-(1→2)-6-deoxy-L-Tal
4	4-*O*-Me-α-L-Rha*p*-(1→4)-2-*O*-Me-α-L-Fuc*p*-(1→3)-α-L-Rha*p*-(1→2)-6-deoxy-L-Tal
9	4-*O*-Ac-2,3-di-*O*-Me-α-L-Fuc*p*-(1→4)-β-D-Glc*p*A-(1→4)-2,3-di-*O*-Me-α-L-Fuc*p*-(1→3)-α-L-Rha*p*-(1→2)-6-deoxy-L-Tal
14	4-formamido-4,6-didesoxy--2-O-Me-3-C-Me-α-D-Man*p*-(1→3)-2-*O*-Me-α-D-Rha*p*-(1→3)-2-*O*-Me-α-L-Fuc*p*-(1→3)-α-L-Rha*p*-(1→2)-6-deoxy-L-Tal
20	2-*O*-Me-α-D-Rha*p*-(1→3)-2-*O*-Me-α-L-Fuc*p*-(1→3)-α-L-Rha*p*-(1→2)-6-deoxy-L-Tal
25	2-*O*-Me-α-D-Fuc*p*4*N*Ac-(1→4)-β-D-Glc*p*A-(1→4)-2-*O*-Me-α-L-Fuc*p*-(1→3)-α-L-Rha*p*-(1→2)-6-deoxy-L-Tal
26	2,4-di-*O*-Me-α-L-Fuc*p*-(1→4)-β-D-Glc*p*A-(1→4)-2-*O*-Me-α-L-Fuc*p*-(1→3)-α-L-Rha*p*-(1→2)-6-deoxy-L-Tal
Group 2	
12	4-(2-hydroxy)propanamido--4,6-didesoxy-3-*O*-Me-β-D-Glc*p*-(1→3)-4-*O*-Me-α-L-Rha*p*-(1→3)-α-L-Rha*p*-(1→3)-α- L-Rha*p*-(1→2)-6-deoxy-L-Tal
17	3-(3-hydroxy-2-methyl)--butanamido-3,6-didesoxy-β-D-Glc*p*-(1→3)-4-*O*-Me-α-L-Rha*p*-(1→3)-α- L-Rha*p*-(1→3)-α-L-Rha*p*-(1→2)-6-deoxy-L-Tal
19	3,4-di-*O*-Me-β-D-Glc*p*A-(1→3)-3-*C*-Me-2,4-di-*O*-Me-α-L-Rha*p*-(1→3)-α-L-Rha*p*-(1→3)-α-L-Rha*p*-(1→2)-6-deoxy-L-Tal
Group 3	
8	4,6-*O*-(1-carboxyethylidene)-3-*O*-Me-β-D-Glc*p*-(1→3)-α-L-Rha*p*-(1→2)-6-deoxy-L-Tal
21	4,6-*O*-(1-carboxyethylidene)-β-D-Glc*p*-(1→3)-α-L-Rha*p*-(1→2)-6-deoxy-L-Tal

Tal …talose, Rha*p* …rhamnopyranose, Fuc*p* …fucopyranose, Glc*p*…glucopyranose, Glc*p*A…glucuronic acid (as pyranose), Man*p*…mannopyranose; 3-*C*-methyl…additional to the hydroxy group also amethyl group at C3 ; 4,6-*O*-(1-carboxyethylidene)-…CH_3-(COOH)C<…ketal; e.g. serovariant 17: the terminal Glc*p* has the modification 3-hydoxy-2-methyl-butanamido … CH_3-CH(OH)-CH(CH_3)-CO-NH- at C3 (configuration at carbons 2 and 3 was not established).

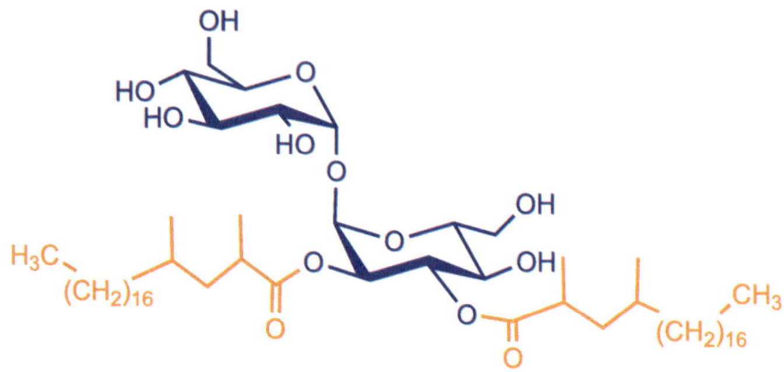

Fig. 5.2 LOS from *Mycobacterium tuberculosis* with a trehalose unit and two lipid chains

5.2 Glycopeptides and Glycoproteins

5.2.1 Analysis of Glycoproteins

More than 30 years ago, glycoproteins were primarily studied with regard to their structure, function, and biosynthesis [10]. Today, the number of proteins present in glycosylated form is estimated to be much higher than previously thought. Thus, the importance of the carbohydrate part of glycoproteins is undisputed. The identification and characterization of glycoproteins, despite greatly improved instrumental methods, remains a challenge, as the microheterogeneities also need to be elucidated—especially with regard to diagnosis of diseases. In order to obtain analyzable material at all, enrichment methods are often used prior to mass spectrometric analysis. Sometimes so-called *hydrophilic interaction liquid chromatography* (HILIC) [11] is used to separate glycoproteins, but often chemical methods must also be used to concentrate glycoproteins [12, 13]. One method utilizes the reaction of hydrazides with aldehydes. For this, an aldehyde group must first be generated within the oligosaccharide chain—either by an enzymatic reaction or by periodate oxidation (suitable for glycoproteins on the cell surface). In the example in Fig. 5.3a, the hydrazide group is bound to a polymer. A polymer that is soluble or precipitates depending on the conditions has proven particularly advantageous, as the binding reaction proceeds better in a homogeneous liquid phase. After separation, the glycoprotein can be released again for MS analysis. In part **b** of the figure, the example of a special boric acid derivative is shown [13], which reacts with 4,6-dihydroxy groups of a monosaccharide in the chain. Other enrichment methods used are based on the general methodology for conjugate preparation, see Chap. 7.

The methods presented in [12] can elucidate glycoproteins from biological material and analyze their heterogeneities. However, to study their biological significance, certain amounts of homogeneous products are needed, which can only be obtained by targeted

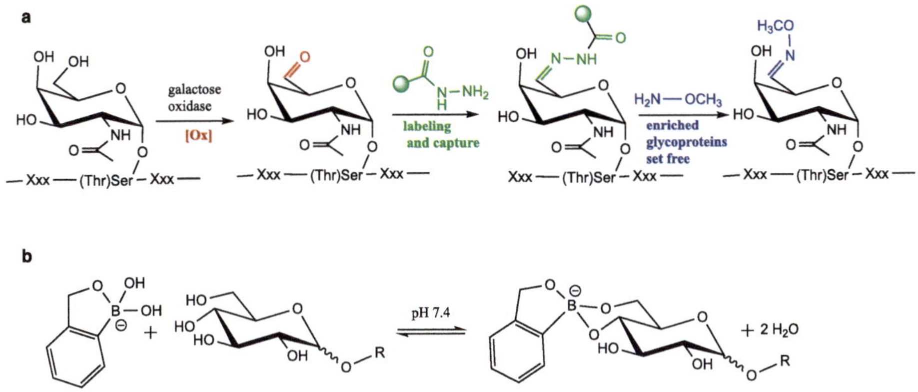

Fig. 5.3 Chemical enrichment methods for glycoproteins for MS analysis: **a** via hydrazone formation; **b** by reversible reaction with boric acid derivatives

synthesis. Therefore, there is great interest in the chemical synthesis of glycoproteins, which is even more complex than the synthesis of proteins themselves.

5.2.2 Types of Glycoproteins

First, a brief note on the different types of glycoproteins: If the oligosaccharide chain is *O*-glycosidically linked to the side chain of the amino acid serine or threonine, it is called an *O*-glycoprotein; if it is *N*-glycosidically linked to asparagine, it is an *N*-glycoprotein. Depending on this, the structure of the carbohydrate part also differs. Figure 5.4 shows the core structures of **a** *O*- and **b** *N*-glycoproteins (*N*-glycan, see also Sect. 8.2.2) [14].

An Example of the Synthesis of the Core Oligosaccharide Part of N-Glycoproteins

Continuing from the chapter on oligosaccharides (Sect. 4.1), I would like to discuss here the final steps of the oligosaccharide synthesis of the core region of *N*-glycoproteins from a 1984 publication [14]. In Fig. 5.5, the two identical protected β-D-Gal(1→4)-2-deoxy-2-amino-D-Glc units are first attached to the branched, appropriately protected tetrasaccharide by the Koenigs-Knorr method. Then, the protecting groups are removed in a total of seven (!) steps. Sodium methoxide removes the acetyl protecting groups, and hydrazinolysis releases the amino groups from the phthalimido protecting groups. Then, all free hydroxyl and amino functions are acetylated again. The cyclic acetal on the first protected amino glucose is opened with TFA. The subsequent hydrogenolysis removes the benzyl ethers (OBn) and reduces the azido group to an amino group. The now free functional groups are acetylated again, and finally, with potassium carbonate/methanol, the acetic acid esters are cleaved again, but the acetamido groups remain intact under these conditions. This sophisticated sequence of deprotection was certainly selected as the best option after many preliminary experiments. This example clearly shows how complex carbohydrate synthesis can be.

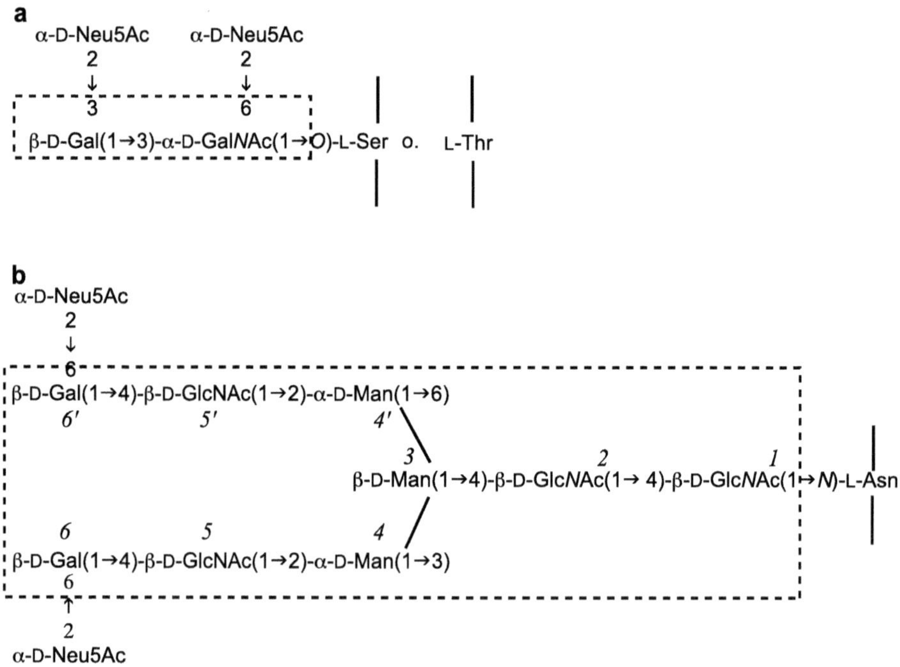

Fig. 5.4 Core structures of the oligosaccharide chain of (**a**) *O*-glycoproteins and (**b**) *N*-glycoproteins (Neu5Ac … 5-*N*-acetylneuraminic acid). (According to [14])

5.2.3 Synthesis of Glycopeptides and Glycoproteins

The synthesis of glycopeptides and glycoproteins remains a challenge even today. This is mainly because, on the one hand, oligosaccharides are sensitive to acidic conditions—the glycosidic bond, as an acetal bond, is easily cleaved under acid—and on the other hand, peptides and proteins undergo rearrangements (e.g., Sect. 3.2.4) and other modifications under strongly basic conditions. In addition, the *O*-glycosidic bond to serine or threonine is particularly base-labile, leading to β-elimination and thus also to "racemization" of the amino acid. This combination makes the choice of protecting groups considerably more difficult [15]. Particularly suitable for this purpose are, among others, those PGs based on allyl (Aloc and –OAll, see Tables 3.1 and 3.2), which can be cleaved under neutral conditions (Sect. 9.1). The synthesis of glycopeptides was subsequently further improved and is now also carried out as a solid-phase synthesis [16]. The synthesis of glycoproteins is achieved by ligation (see Chap. 8) of appropriate glycopeptide fragments with the remainder of the protein produced by molecular biology methods [17]. The synthesis of glycoproteins without chemical-synthetic methods is even less developed [18].

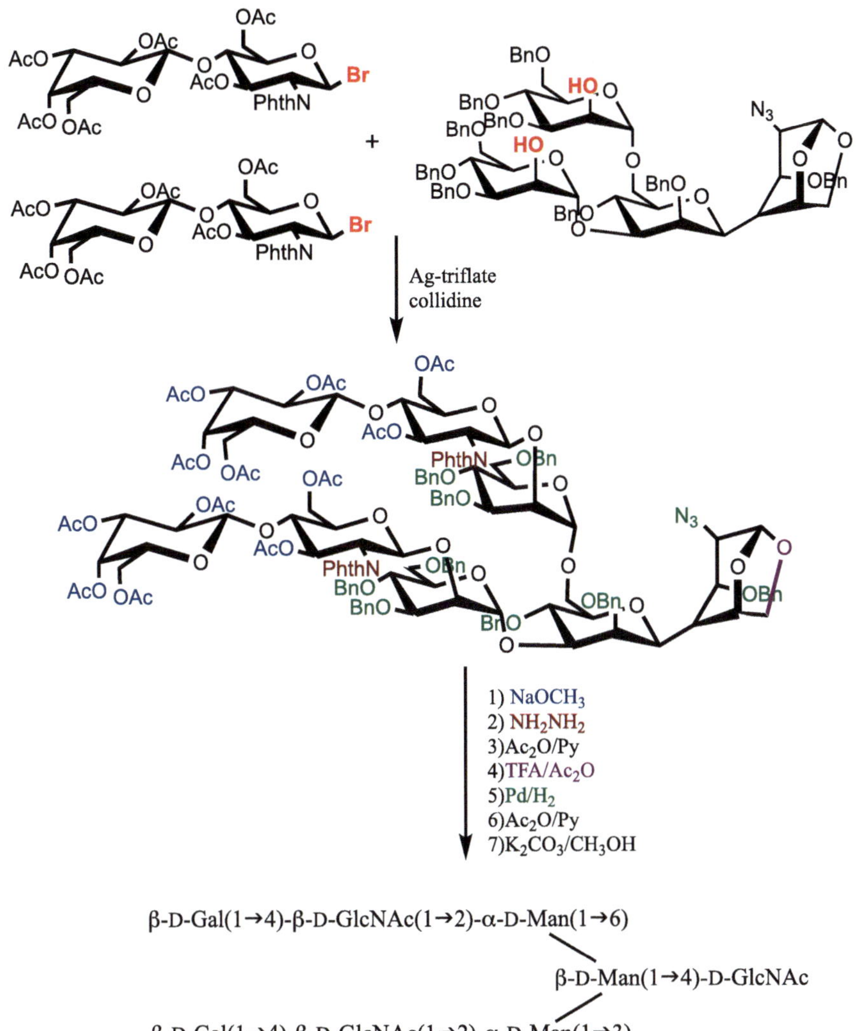

Fig. 5.5 Final phase of the synthesis of the oligosaccharide core region of *N*-glycoproteins. (According to [14])

Examples of Glycoprotein Syntheses

In [19], the semisynthesis of human interleukin IL-6 is described, in which a relatively small, chemically synthesized glycopeptide was assembled with two recombinantly produced protein fragments to form the complete glycoprotein. This made it possible to produce several variants in the carbohydrate part for biological studies.

Erythropoietin (EPO) is a glycoprotein that has been approved by the US Food and Drug Administration (FDA) for the treatment of anemia since 1989 and is produced recombinantly in mammalian cells. However, it is a complex mixture with respect to the oligosaccharide chains. Now, however, there are some partially synthetic approaches to homogeneous variants (Sect. 8.2.2), which are primarily used to study the different biological activities [20].

5.2.4 Biosynthesis of *N*-Glycoproteins

In this textbook, we generally do not deal with biosynthetic pathways. However, the posttranslational attachment of an oligosaccharide chain to asparagine seems interesting to me even from the perspective of mechanistic organic chemistry and will therefore be briefly discussed here. It has long been known that only Asn units that occur in a sequence of the following type are glycosylated: –Asn–Xxx–Thr/Ser–. The hydroxyl group of the amino acid two residues downstream must therefore play a very special role in the enzymatic transfer. In [21], the probable mechanism of this reaction is discussed under the dramatic title *The Clash of the Titans*. The oligosaccharide chain is activated by conjugation with the lipid dolichol via a diphosphate bridge [10], see Fig. 5.6a. But why can the amide group of the Asn side chain act as a nucleophile to attack the oligosaccharide? The amide nitrogen atom is hardly nucleophilic (Sect. 2.1, properties of the peptide bond = amide bond). The mechanism shown in Fig. 5.6b explains this: The above-mentioned amino acid sequence can form a so-called *Asx-turn*, with a hydrogen bond extending from the side chain of Asn (or Asp) to the peptide bond **and** to the hydroxyl group of Thr or Ser (Fig. 5.6). This greatly facilitates the removal of a proton from the amide NH_2 group by the enzyme, as can also be illustrated by the resonance structures shown, and thus this N atom becomes a strong nucleophile (Fig. 5.6).

5.3 Questions

Let us test understanding of the content of this chapter

5.1 What are glycoconjugates and what is their significance?
5.2 What is alaninol?
5.3 Which monosaccharide does the abbreviation "Rha" stand for and what is it derived from?
5.4 What does the abbreviation "Glc*p*A" mean?
5.5 What is trehalose?
5.6 On which amino acids can oligosaccharide chains be found in glycoproteins?
5.7 Can glycans be *N*-glycosidically bound to all asparagines in a protein sequence?
5.8 Name two enrichment methods for glycoproteins that are used for MS analysis!
5.9 Why is the chemical synthesis of glycopeptides so difficult?

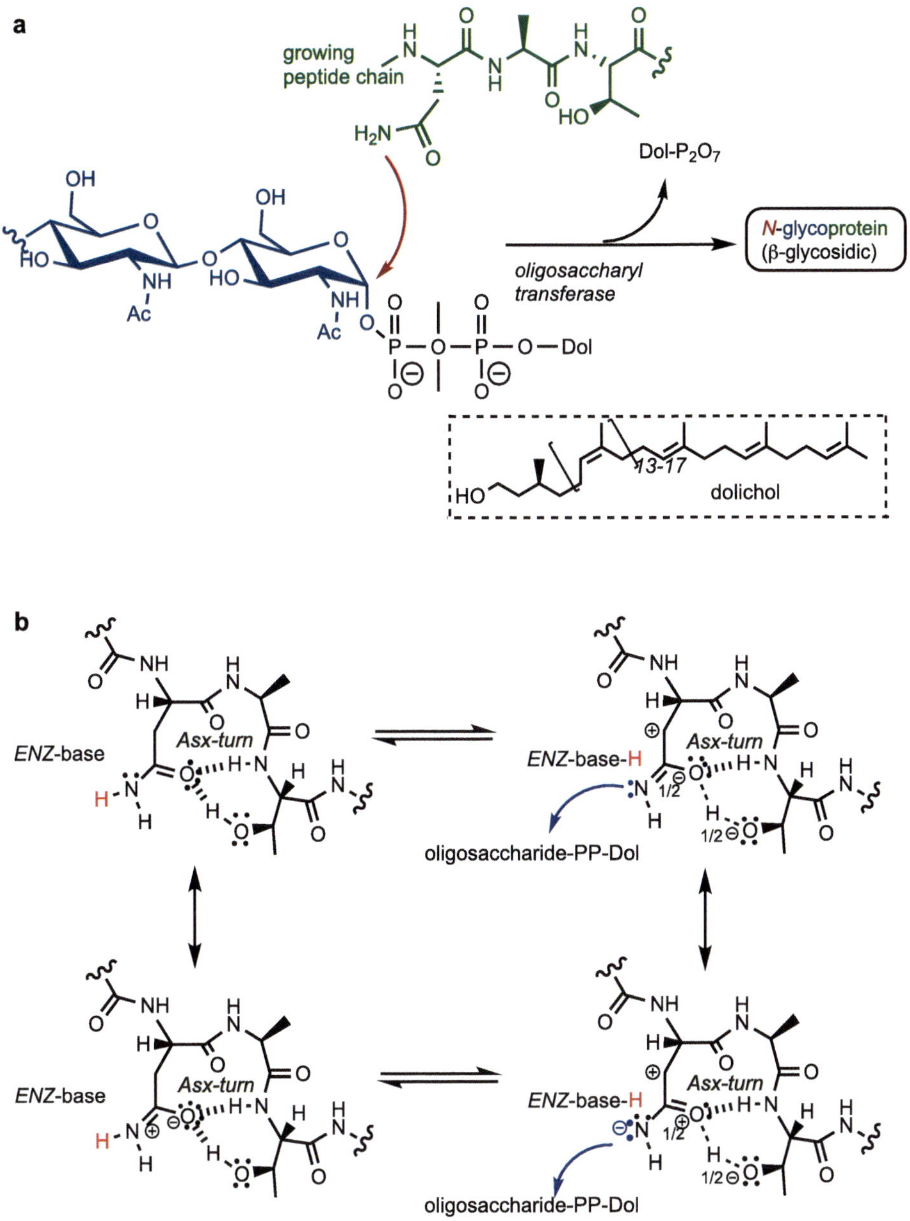

Fig. 5.6 Mechanism of *N*-glycosylation of proteins: **a** general scheme; **b** proposed mechanism of nucleophilic activation of the Asn amide group. (According to [21])

References

1. Shivatare, S.S., Shivatare, V.S., Wong, C.-H.: Glycoconjugates: synthesis, functional studies, and therapeutic developments. Chem. Rev. **122**(20), 15603–15671 (2022). https://doi.org/10.1021/acs.chemrev.1c01032
2. Chatterjee, D., Khoo, K.H.: The surface glycopeptidolipids of mycobacteria: structures and biological properties. Cell. Molecular. Life. Sci. CMLS. **58**(14), 2018–2042 (2001). https://doi.org/10.1007/PL00000834 Schorey, J.S., Sweet, L.: The mycobacterial glycopeptidolipids: structure, function, and their role in pathogenesis. Glycobiology. **18**(11), 832–841 (2008). https://doi.org/10.1093/glycob/cwn076; Chatterjee, D., Brennan, P.J.: Chapter 9 – Glycosylated components of the mycobacterial cell wall: structure and function. In: Moran, O.H.P.J.B., Mark von Itzstein, A.P. (Hrsg) Microbial Glycobiology, S. 147–167. Academic Press, Elsevier (2010). https://doi.org/10.1016/B978-0-12-374546-0.00009-2
3. Batt, S.M., Minnikin, D.E., Besra, G.S.: The thick waxy coat of mycobacteria, a protective layer against antibiotics and the host's immune system. Biochem. J. **477**(10), 1983–2006 (2020). https://doi.org/10.1042/bcj20200194; Jones, A., Lee, O.Y.C., Minnikin, D.E., Baird, M.S., Al-Dulayymi, J.R.: A re-investigation of the mycolic acids of Mycobacterium avium. Chem. Phys. Lipids. **230**,104928 (2020). https://doi.org/10.1016/j.chemphyslip.2020.104928; Parmar, S., Tocheva, E. I.: The cell envelope of Mycobacterium abscessus and its role in pathogenesis. PLOS Pathogens. **19**(5), e1011318 (2023). https://doi.org/10.1371/journal.ppat.1011318; Chauhan, M., Barot, R., Yadav, R., …, Murumkar, P.R..: The mycobacterium tuberculosis cell wall: an alluring drug target for developing newer anti-TB drugs – a perspective. Chem. Biol. Drug Design. **104**(3), e14612 (2024). https://doi.org/10.1111/cbdd.1461
4. Busatto, C., Vianna, J.S., da Silva, L.V., …, da Silva, P.E.A.: Mycobacterium avium: an overview. Tuberculosis. **114**, 127–134 (2019). https://doi.org/10.1016/j.tube.2018.12.004; Mattoo, R.: Targeting emerging mycobacterium avium infections: perspectives into pathways and antimicrobials for future interventions. Fut. Microbiol. **16**(10), 753–764 (2021). https://doi.org/10.2217/fmb-2021-0016; Kaczmarkowska, A., Didkowska, A., Kwiecień, E., …, Anusz, K.: The mycobacterium avium complex – an underestimated threat to humans and animals. Ann. Agricul. Env. Med. **29**(1), 22–27 (2022). https://doi.org/10.26444/aaem/13639
5. Aspinall, G.O., Chatterjee, D., Brennan, P.J.: The variable surface glykolipids of mycobacteria: structures, synthesis of epitopes, and biological properties. Adv. Carbohydr. Chem. Biochem. **51**, 169–242 (1995). https://doi.org/10.1016/S0065-2318(08)60194-8
6. Grice, I.D., Wilson, J.C.: Analytical approaches towards the structural characterization of microbial wall glycopolymers. In: Moran, O.H.P.J.B., Mark von Itzstein, A.P. (Hrsg.) Microbial Glycobiology., S. 233–252. Academic Press, Elsevier (2010). https://doi.org/10.1016/B978-0-12-374546-0.00013-4
7. Liu, K., Wang, L. and Guo, Z.: An extensive review of studies on mycobacterium cell wall polysaccharide-related oligosaccharides – part III: synthetic studies and biological applications of arabinofuranosyl oligosaccharides and their analogs, derivatives and conjugates. J. Carbohydr. Chem., **38** (7-8), 414-469 (2019). https://doi.org/10.1080/07328303.2019.1630841
8. Heidelberg, T., Martin, O.R.: Synthesis of the glycopeptidolipid of Mycobacterium avium Serovar 4: first example of a fully synthetic C-Mycoside GPL. J. Org. Chem. **69**(7), 2290–2301 (2004). https://doi.org/10.1021/jo030304e; Fekete, A., Gyergyói, K., Kövér, K.E., …, Liptak, A.: Preparation of the pentasaccharide hapten of the GPL of Mycobacterium avium serovar 19 by achieving the glycosylation of a tertiary hydroxyl group. Carbohydrate Res. **341**(10), 1312–1321,(2006). https://doi.org/10.1016/j.carres.2006.04.038
9. Murthy Sabbavarapu, N., Seeberger, P.: H.: Automated Glycan Assembly of Mycobacterial Hexaarabinofuranoside and Docosasaccharide Arabinan (Araf23) Motifs found on Myco-

bacterium tuberculosis. Chem. Eur. J. **29**(23), e202300032 (2023). https://doi.org/10.1002/chem.202300032

10. Montreuil, J., Glycoproteins. In: Neuberger, A., von Deenen, L.L.M. (Hrsg.) Comprehensive Biochemistry, Bd.: 19B, part II, S. 1–188, Elsevier, Amsterdam (1982); Schwarz, R.T., Datema, R.: The lipid pathway of protein glycosylation and its inhibitors: the biological significance of protein-bound carbohydrates. Adv. Carbohydr. Chem. Biochem. **40**, 287–379 (1982). https://doi.org/10.1016/S0065-2318(08)60111-0
11. Tengattini, S., Massolini, G., Rinaldi, F., …, Temporini, C.: Hydrophilic interaction liquid chromatography (HILIC) for the analysis of intact proteins and glycoproteins. TrAC Trends Anal. Chem. **174**, 117702 (2024). https://doi.org/10.1016/j.trac.2024.117702
12. Yin, H. and Zhu, J.: Methods for quantification of glycopeptides by liquid separation and mass spectrometry. Mass. Spectromet. Rev. **42**(2), 887–917 (2023). https://doi.org/10.1002/mas.21771; Suttapitugsakul, S., Sun, F., Wu, R.: Recent advances in glycoproteomic analysis by mass spectrometry. Analyt. Chem. **92**(1), 267–291 (2020). https://doi.org/10.1021/acs.analchem.9b04651; Sun, F., Suttapitugsakul, S. and Wu, R.: Systematic characterization of extracellular glycoproteins using mass spectrometry. Mass. Spectromet. Rev. **42**(2), 519–545 (2023). https://doi.org/10.1002/mas.21708
13. Dowlut, M., Hall, D.G.: An improved class of sugar-binding boronic acids, soluble and capable of complexing glycosides in neutral water. J. Am. Chem. Soc. **128**(13), 4226–4227 (2006). https://doi.org/10.1021/ja057798c
14. Paulsen, H.: Haworth memorial lecture: synthesis of complex oligosaccharide chains of glycoproteins. Chem. Soc. Rev., **13** (1), 15-45 (1984). https://doi.org/10.1039/CS9841300015
15. Kunz, H.: Synthese von Glycopeptiden, Partialstrukturen biologischer Erkennungskomponenten. Angew. Chem. **99**(4), 291–311 (1987). https://doi.org/10.1002/ange.19870990405; Kunz, H.: Synthesis of glycopeptides, partial structures of biological recognition components [New Synthetic Methods (67)]. Angew. Chem. Int. Ed. **26**(4), 294–308 (1987). https://doi.org/10.1002/anie.198702941; Garg, H.G., von dem Bruch, K., Kunz, H.: Developments in the synthesis of glycopeptides containing glycosyl-l-asparagine, l-serine, and l-threonine. Adv. Carbohydr. Chem. Biochem. **50**, 277–310 (1994). https://doi.org/10.1016/S0065-2318(08)60153-5
16. Ricardo, M.G., Seeberger, P.H.: Merging solid-phase peptide synthesis and automated glycan assembly to prepare lipid-peptide-glycan chimeras. Chem. Eur. J. **29**(54), e202301678 (2023). https://doi.org/10.1002/chem.202301678
17. Unverzagt, C., Kajihara, Y.: Chemical assembly of N-glycoproteins: a refined toolbox to address a ubiquitous posttranslational modification. Chem. Soc. Rev. **42**(10), 4408–4420 (2013). https://doi.org/10.1039/C3CS35485G; Kajihara, Y., Murakami, M. and Unverzagt, C.: Chemical Glycoprotein Synthesis. In: Hung, S.-C., Zulueta, M.M.L. (Hrsg.) Glycochemical synthesis, S. 263–292, Wiley (2016). https://doi.org/10.1002/9781119006435.ch11; Liu, Y., Nomura, K., Abe, J., Kajihara, Y.: Recent advances on the synthesis of N-linked glycoprotein for the elucidation of glycan functions. Curr. Opin. Chem. Biol. **73**, 102263 (2023). https://doi.org/10.1016/j.cbpa.2023.102263; Zhao, J., Ye, F., Huang, P., Wang, P.: Recent advances in chemical synthesis of O-linked glycopeptides and glycoproteins: an advanced synthetic tool for exploring the biological realm. Curr. Opin. Chem. Biol. **77**, 102405 (2023). https://doi.org/10.1016/j.cbpa.2023.102405
18. Hug, I., Zheng, B., Reiz, B., …, Feldman, M.F.: Exploiting bacterial glycosylation machineries for the synthesis of a Lewis Antigen-containing glycoprotein. J. Biol. Chem. **286**(43), 37887–37894, S37887/37881-S37887/37885 (2011). https://doi.org/10.1074/jbc.M111.287755; Palma, J. A., Bunyatov, M. I., Hulbert, S. W., …, DeLisa, M.P.: Bacterial glycoengineering: cell-based and cell-free routes for producing biopharmaceuticals with cus-

tomized glycosylation. Curr. Opin. Chem. Biol. **81**, 102500 (2024). https://doi.org/10.1016/j.cbpa.2024.102500

19. Reif, A., Siebenhaar, S., Tröster, A., …, Unverzagt, C.: Semisynthesis of biologically active glycoforms of the human cytokine interleukin 6. Angew. Chem. **126** (45), 12321–12327 (2014). https://doi.org/10.1002/ange.201407160; Reif, A., Siebenhaar, S., Tröster, A., …, Unverzagt, C.: Semisynthesis of biologically active glycoforms of the human cytokine interleukin 6. Angew. Chem. Int. Ed. **53**(45), 12125–12131 (2014). https://doi.org/10.1002/anie.201407160
20. Wang, Y., Yang, S.H., Brimble, M.A., Harris, P.W.R.: Recent progress in the synthesis of homogeneous Erythropoietin (EPO) glycoforms. ChemBioChem. **21**(23), 3301–3312 (2020). https://doi.org/10.1002/cbic.202000347; Hessefort, H., Gross, A., Seeleithner, S., …, Unverzagt, C.: Chemical and enzymatic synthesis of sialylated glycoforms of human erythropoietin. Angew. Chem. Int. Ed. **60**(49), 25922–25932 (2021). https://doi.org/10.1002/anie.202110013
21. Imperiali, B.: Protein glycosylation: the clash of the titans. Acc. Chem. Res. **30**(11), 452–459 (1997). https://doi.org/10.1021/ar950226k

Oligonucleotides and Nucleic Acids

6

Abstract

This chapter deals with methods for sequencing nucleic acids, beginning with the groundbreaking work of Maxam and Gilbert on the one hand, and Sanger on the other. However, new methods developed only after the completion of the sequencing of the human genome are also discussed. The second part addresses oligonucleotide synthesis, covering both the phosphoramidite and the hydrogen phosphonate methods. Finally, various approaches to the development of modified nucleic acids are discussed.

6.1 DNA Sequencing

The sequencing of deoxyribonucleic acids (DNA) was, of course, originally an enormously important goal of biochemistry and genetics. In 1980, two independent groups were awarded the Nobel Prize for their DNA sequencing methods [1, 2]. While Frederick Sanger developed an enzymatic method that, in principle, is still used today, Walter Gilbert's method is primarily a chemical approach. Therefore, we will mainly discuss his procedure here. In both cases, analytical evaluation is performed by gel electrophoresis.

6.1.1 Maxam-Gilbert DNA Sequencing

This chemical method is based on four different reaction sequences, in which the first step selectively and with only low yield (about 1 in 100) labels specific nucleobases, while the second step, then carried out under drastic conditions, quantitatively cleaves the DNA strand at the labeled site. Gel electrophoresis (as in the Sanger method) finally

D. Krois, *Organic Chemical Methods*, https://doi.org/10.1007/978-3-662-73573-2_6

identifies the length of the resulting fragments with single-nucleotide resolution. The prerequisite for all of this is the isolation of a double-stranded DNA fragment that is labeled at only one 5′ end with a radioactive phosphorus isotope (Fig. 6.1). In the first reaction sequence, the purine bases **A** or **G** are initially cleaved off (with low yield), and the strand is then cleaved at these positions. In the second sequence, guanines are methylated at N-7 with dimethyl sulfate, and the chain is then cleaved at these sites. The action of hydrazine in the third reaction variant leads to derivatization of the pyrimidine bases **C** and **T**, which then, under the influence of piperidine, results in cleavage at these positions. If this selective reaction (with hydrazine) is carried out in the presence of 1 M sodium chloride solution, labeling and thus cleavage occurs only at the cytosine base (4th sequence) (see example in Table 6.1). The latter reaction condition was based on a more or less accidental observation [1]. In this way, the nucleobase sequence for the labeled oligonucleotide can then be deduced after parallel gel electrophoresis.

Further details on the method originally used by Maxam and Gilbert

The radioactive labeling of the DNA fragment was carried out by enzymatic methods, as shown in Fig. 6.1, utilizing the presence of the *lac* operon in the strand.

Methylation of the purines occurs at guanine at N-7 (accessible from the *major groove* of the double strand) and at adenine at N-3 (in the less accessible *minor groove*), and therefore more slowly. The reason for the regioselectivity (N-3 or N-7) of methylation can only be due to the influence of the adjacent substituents C=O or $–NH_2$. In Fig. 6.2, the mechanism of cleavage of the methylated purines by the action of bases is shown. Once the nucleobase has been cleaved off, the deoxyribose, no longer glycosidically bound, very easily loses the phosphate residues, resulting in strand cleavage.

The hydrazinolysis of nucleosides with pyrimidine bases is a more complex reaction and is shown in Fig. 6.3, with the detected products shown in bold. The cleavage of urea from the remain-

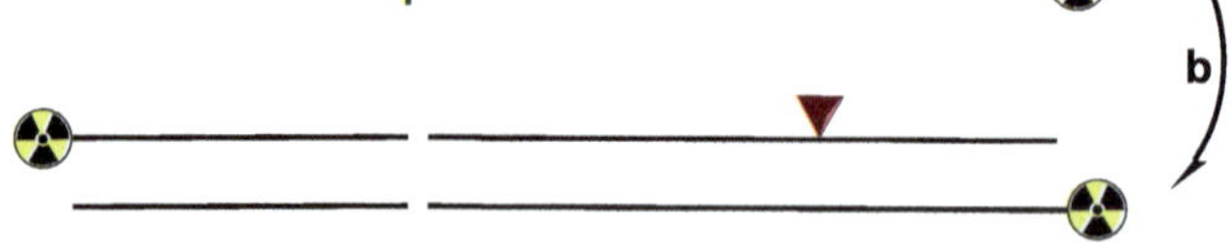

Fig. 6.1 Preparation of the homogeneous labeled DNA fragment according to Gilbert [1]: **a** radioactive labeling by polynucleotide kinase (→ radioactive phosphate at the 5′ end), **b** action of a restriction enzyme, so that only one radioactively labeled strand carries the binding site (▼) for the repressor

Table 6.1 Degradation according to Maxam-Gilbert using the example of a labeled DNA strand. The sequence can be reconstructed from the fragments obtained under the four different reaction conditions

*P----A-C-A-G-G-T-G-G-C-T-A-C-A-T-G-T-C-C-C-T-A-	*P----A-C-A-G-G-T-G-G-C-T-A-C-A-T-G-T-C-C-C-T-A-
1. Depurination + cleavage (A, G)	3. Hydrazine treatment + cleavage with piperidine (C, T)
*P----A-C	*P----A
*P----A-C-A	*P----A-C-A-G-G
*P----A-C-A-G	*P----A-C-A-G-G-T-G-G
*P----A-C-A-G-G-T-	*P----A-C-A-G-G-T-G-G-C
*P----A-C-A-G-G-T-G	*P----A-C-A-G-G-T-G-G-C-T-A
*P----A-C-A-G-G-T-G-G-C-T	*P----A-C-A-G-G-T-G-G-C-T-A-C-A
*P----A-C-A-G-G-T-G-G-C-T-A-C	*P----A-C-A-G-G-T-G-G-C-T-A-C-A-T-G
*P----A-C-A-G-G-T-G-G-C-T-A-C-A-T	*P----A-C-A-G-G-T-G-G-C-T-A-C-A-T-G-T
*P----A-C-A-G-G-T-G-G-C-T-A-C-A-T-G-T-C-C-C-T	*P----A-C-A-G-G-T-G-G-C-T-A-C-A-T-G-T-C
2. Methylation of G + cleavage (G)	*P----A-C-A-G-G-T-G-G-C-T-A-C-A-T-G-T-C-C
*P----A-C-A	*P----A-C-A-G-G-T-G-G-C-T-A-C-A-T-G-T-C-C-C
*P----A-C-A-G	4. Hydrazine treatment (1 M NaCl) + cleavage (piperidine) (C)
*P----A-C-A-G-G-T-	*P----A
*P----A-C-A-G-G-T-G	*P----A-C-A-G-G-T-G-G
*P----A-C-A-G-G-T-G-G-C-T-A-C-A-T-	*P----A-C-A-G-G-T-G-G-C-T-A
	*P----A-C-A-G-G-T-G-G-C-T-A-C-A-T-G-T
	*P----A-C-A-G-G-T-G-G-C-T-A-C-A-T-G-T-C
	*P----A-C-A-G-G-T-G-G-C-T-A-C-A-T-G-T-C-C

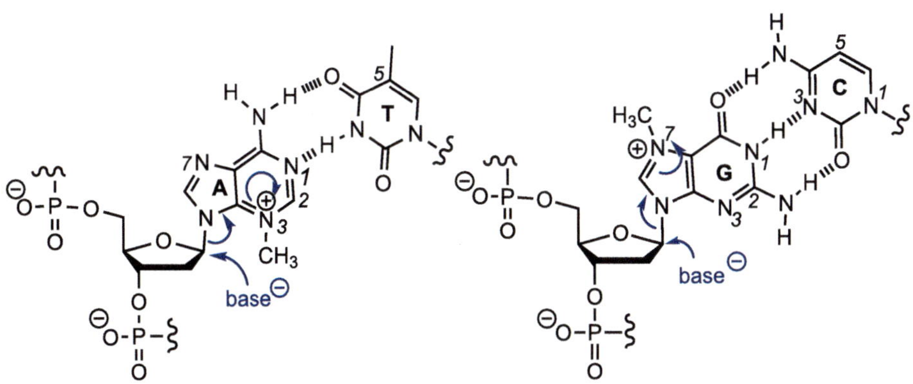

Fig. 6.2 Cleavage of the methylated purine bases adenine and guanine by bases

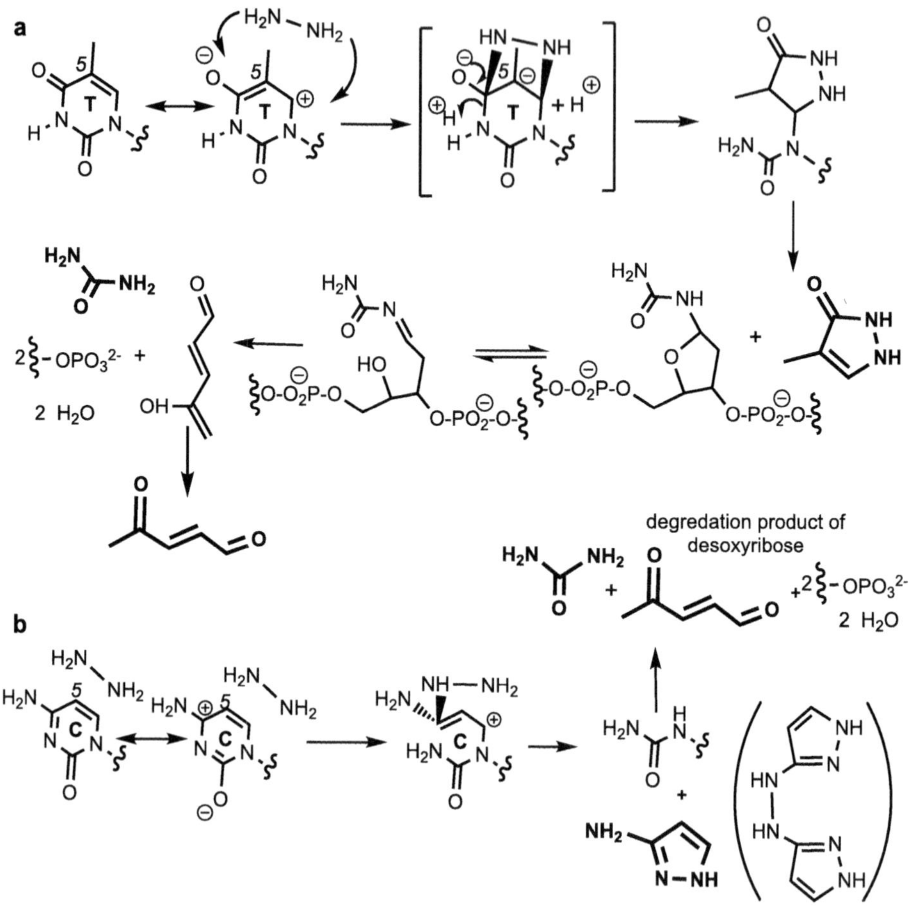

Fig. 6.3 Mechanisms of hydrazinolysis of (**a**) thymidine (T) and (**b**) cytidine (C) in DNA. The detected cleavage products are highlighted (bold)

ing deoxyribose derivative occurs in each case under the action of piperidine, primarily forming piperidine derivatives, but these have been omitted from Fig. 6.3 for clarity.

For example, 5-methylcytidine (5mC) can be easily identified by the hydrazine reaction, as it does not react. This is important because this rare modification can lead to dangerous mutations. In general, the base cytosine (**C**) is relatively easily converted to uracil (**U**) by deamination. However, this does not play such a significant role at the DNA level, because such a change is recognized by enzymes, since only thymine (**T**) and no uracil occurs in DNA. The same reaction (deamination), however, leads in the case of **5-methyl-**cytosine to **thymine** (which escapes repair) and thus to a completely altered genetic information. As early as 1978, A. M. Maxam conducted investigations in this regard [3]. In the review [4] from 1998, various methods for identifying methylated nucleobases are compared, including several chemical methods such as hydrazinolysis or the bisulfite reaction. Methylation of RNA is, of course, much more common and has a regulatory function. For the elucidation of methylation sites in RNA, a recent publication [5] uses an LC-MS method.

But does this method still is of significance today? Certainly not for routine genome sequencing. However, as recently as 2002, an improved labeling with the radioactive ^{35}S isotope was published for this purpose [6]. And even now, Maxam-Gilbert sequencing (or chemical sequencing) is used for the characterization of modified oligonucleotides, for example after UV irradiation or by anticancer drugs based on Pt complexes [7]. In 2018, a review appeared dealing with the sequencing of L-DNA (consisting of L-2-deoxyribose), the carrier of genetic information in a mirror-image world, a topic in synthetic biology. Sequencing methods such as those of Sanger (Section 6.1.2) or *sequencing by synthesis* (SBS, Section 6.1.3.1) require enzymes that are specific for D-DNA. Therefore, the chemical method according to Maxam-Gilbert, which works with achiral reagents, was adapted, with some of the labeling and degradation reactions modified, but the principle remained as described above (Liu, X., Zhu, T.F.: Sequencing Mirror-Image DNA Chemically. Cell Chem. Biol. 25(9), 1151–1156.E3 (2018). https://doi.org/10.1016/j.chembiol.2018.06.005)

6.1.2 Sanger Method

The essence of this method is surely well known [2]. Sanger first produced a single-stranded template DNA (*single stranded,* ssDNA) using a bacteriophage vector. In this case, the same primer can always be used. Then, using DNA polymerase, the complementary strand is synthesized, with four parallel reactions, each containing a small amount of **one** dideoxynucleoside triphosphate (ddNTP) in addition to a mixture of the four deoxynucleoside triphosphates (dNTP). This results in chain termination at the positions of this nucleotide, each with low yield. Sequencing is then again performed by parallel electrophoresis of all four reactions and detection by radioactive labeling with ^{32}P or ^{35}S at a dNTP. Subsequently, this method was improved [8, 9]. By using fluorescence detection, a different fluorophore can be attached to the primer in each of the four reactions. Thus, all four reactions can be recombined before electrophoresis (including capillary electrophoresis), and the color sequence then correlates with the sequence. This enables automatic evaluation.

Labeling of Sanger sequencing reactions with fluorophores

In Fig. 6.4, the structures of some fluorophores (see also Section 14.1) used for this purpose are shown.

6.1.3 Sequencing of Oligonucleotides Newer Sequencing Methods (next-generation sequencing NGS)

The new methods now used by commercially available platforms are mainly based on the incorporation of dNTPs by polymerases, but the evaluation is no longer based on electrophoretic separation; instead, specific signals are generated for each nucleotide

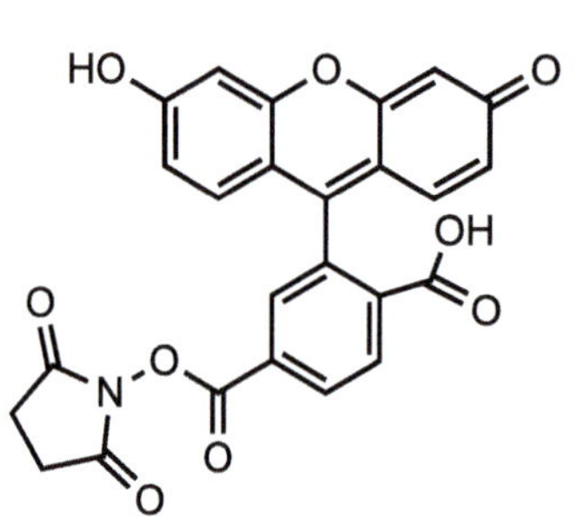

6-carboxyfluorescein succinimidyl ester (6-FAM, SE)
excitation-λ: 494 nm
emission-λ: 518 nm

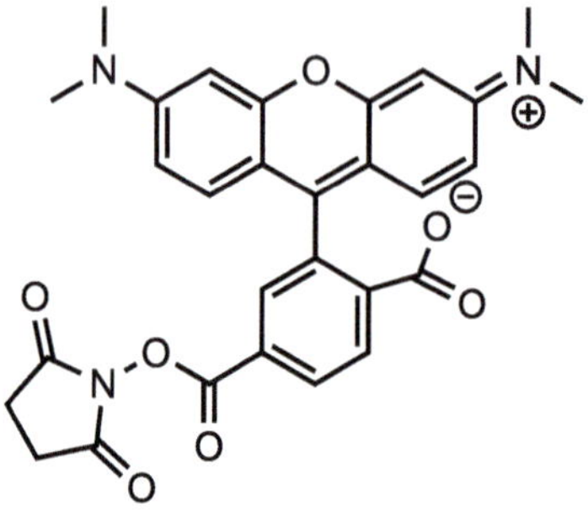

6-carboxytertramethylrhodamine succinimidyl ester (6-TAMRA, SE)
excitation-λ: 555 nm
emission-λ: 580 nm

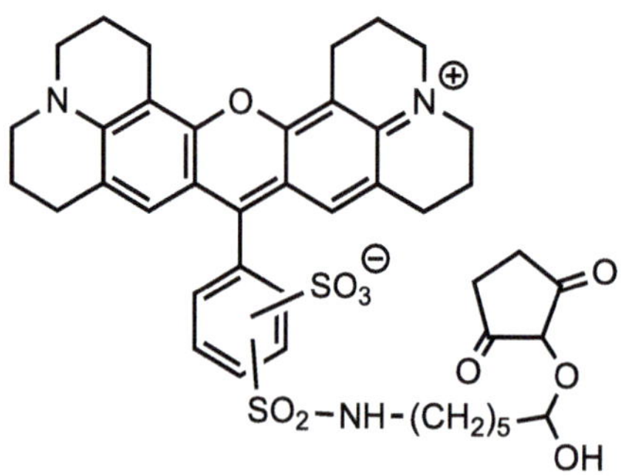

Texas Red,-X, succinimidylester
excitation-λ: 595 nm
emission-λ: 615 nm

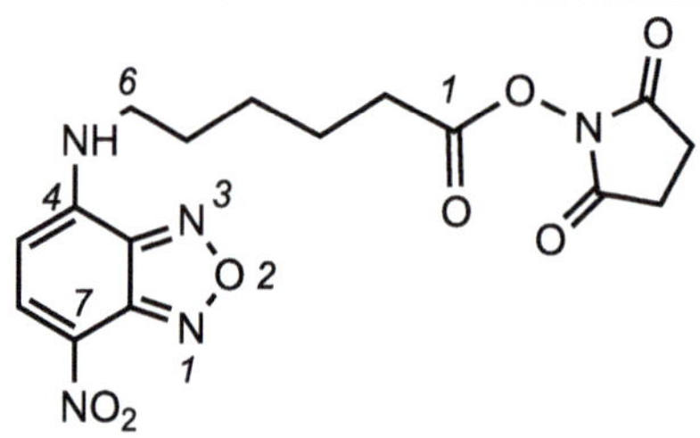

succinimidyl 6-(N-(7-nitrobenz-2-oxa-1,3-diazol-4-yl)amino)hexanoate (NBD-X, SE)
excitation-λ: 465 nm
emission-λ: 535 nm

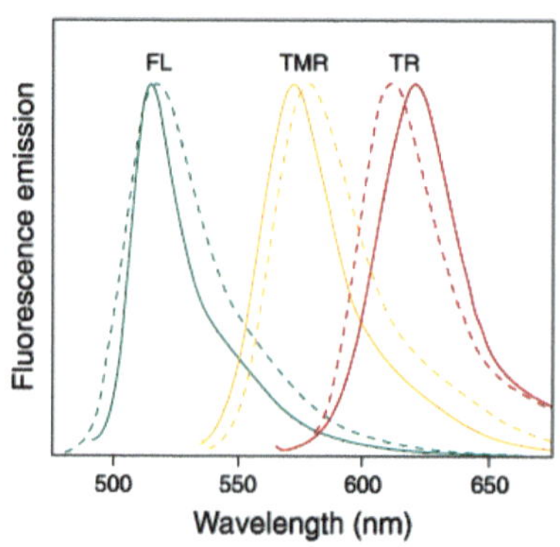

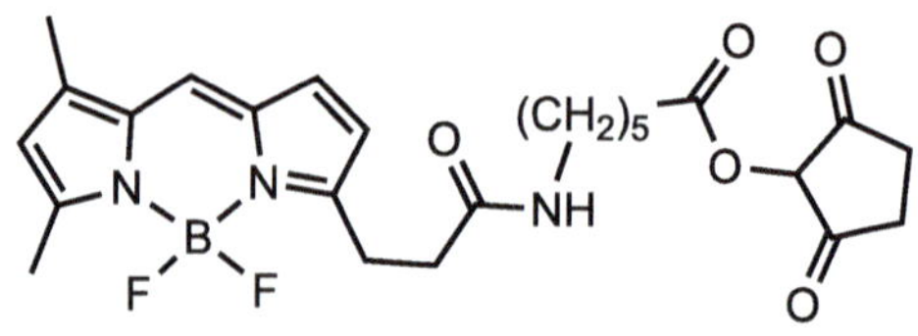

6-((4,4-difluoro-5,7-dimethyl- 4-bora-3a,4a-diaza-s-indacene- 3-propionyl)amino)hexanoic acid, succinimidyl ester (BODIPY® FL-X, SE

Spectra from the handbook of: Molecular Probes (Invitrogen, Thermo Fisher Scientific)

Normalized emission spectra of the goat-anti-mouse-IgG conjugates of fluorescein (FL), tetramethylrhodamine (TMR) und *Texas Red* (TR) (- - - -) compared to the goat-anti-mouse-IgG conjugates of BODIPY FL, BODIPY TMR and BODIPY TR (——)

Fig. 6.4 Examples of fluorophores that are also used for DNA sequencing. According to the manufacturers, BODIPY fluorophores are characterized by higher emission yield and, in some cases, greater stability. All are shown here as *N*-hydroxysuccinimide esters (Section 3.1.2.3)

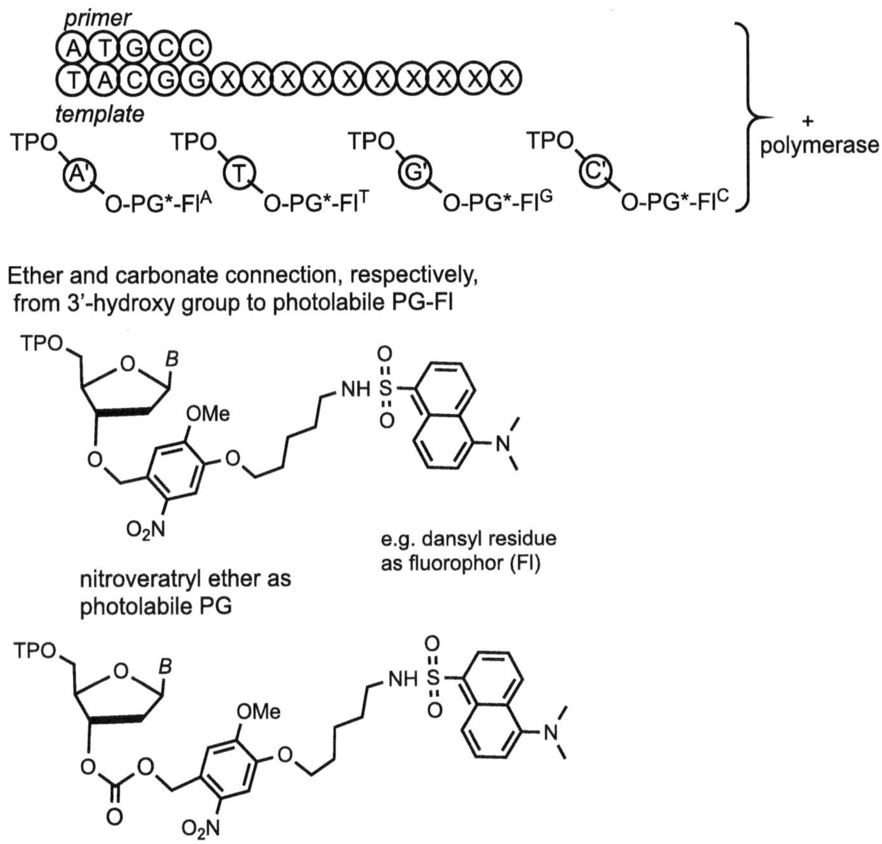

Fig. 6.5 Scheme of "sequencing by synthesis" (SBS) [*B* ≙ nucleobase, PG ≙ protecting group, Fl ≙ fluorophore]. (After Ref. [12])

incorporation [10, 11]. In the following, I will briefly discuss only four principal methods.

6.1.3.1 Sequencing by cyclic reversible termination (CRT), also sequencing by synthesis (SBS)

An early example of the principle of this method is described in [12] [Syntheses of Nucleosides Designed for Combinatorial DNA Sequencing, Welch, M.B., Martinez, C.I., Zhang, A.J., Jin, S., Gibbs R., Burgess, K. *Chem. Eur. J.*, **5**(3): 951–960 (1999)]: The template strand to be sequenced and a short primer are present. Also present are the 5′-triphosphates (TP) of all four nucleobases, each with a photolabile protecting group (which also contains a fluorophore) at the 3′-OH group (Fig. 6.5). A polymerase should now incorporate the next complementary nucleotide—but then the reaction stops, and the fluorescence, characteristic for the respective nucleotide, is recorded. Then the conjugate of protecting group (PG) and fluorophore (Fl) is cleaved off by irradiation, and the next cycle can begin.

Photochemical cleavage of nitroveratryl protecting groups
In Fig. 6.6, the mechanism for the cleavage of this photolabile protecting group is shown. We will encounter it again in Part 2 of the book.

So much for the beautiful theory. But the nucleoside triphosphates with protecting group and fluorophore (PG-Fl) were not accepted as substrates by any of the common polymerases, which is not so surprising given the greatly increased size. As a piquant detail, the authors had to admit in a *Corrigendum* [12] six years later that some derivatives did not even have the correct structure, but rather PG-Fl was bound to the nucleobase itself. A somewhat different variant was presented by the same group somewhat later [13], but corresponding results with perhaps modified polymerases that accepted these substrates were still lacking. As motivation to pursue this method further, it was stated that electrophoresis, as the limiting step in conventional sequencing, would thus be avoided.

Meanwhile, this principle is actually applied in commercial devices, and there are two variants:

I. Reversibly blocked dNTPs are used at the 3′ position, which are usually fluorescence-labeled at the base (thus similar to the example above). However, mutated polymerases had to be developed for this, which may not work as error-free and selectively as the natural ones. The cleavage of the blocking group at 3′ then occurs simultaneously with the cleavage of the fluorophore [14]. Recently, dNTPs with blocker and fluorophore at 3′ have also been developed that are accepted by a modified polymerase [15] (for the method, see also Wikipedia and Illumina [16]). A special form of this method, *single-molecule real-time sequencing* (SMRT), can also be used to identify methylated nucleobases [17].

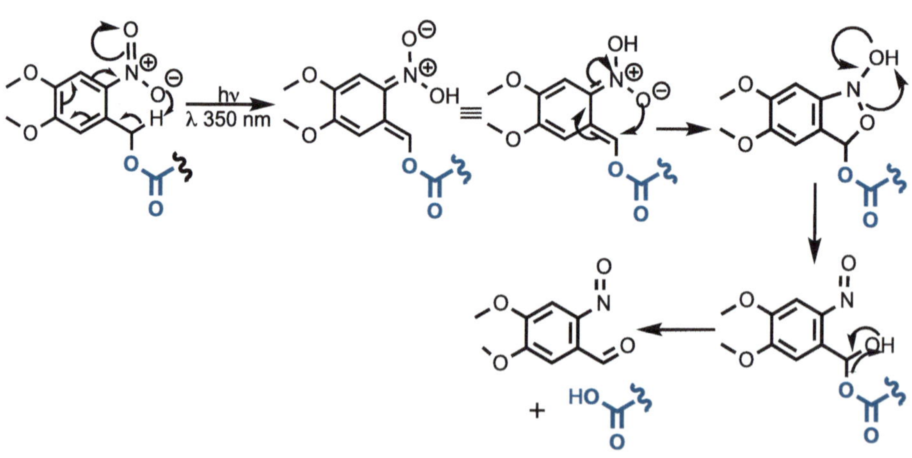

Fig. 6.6 Mechanism of the photochemical cleavage of a nitroveratryloxycarbonyl (Nvoc) protecting group, but also valid analogously for other nitroveratryl derivatives

II. Non-blocked dNTPs are used at the 3′ position, but they act as *inhibitors* of the polymerase due to a group at the base (PG-Fl). When this group is cleaved off, synthesis continues. In this case, natural polymerases can be used.

6.1.3.2 Nanopore - Sequencing of Oligonucleotides

In this technology, developed by "Oxford Nanopore Technology" (ONT) (for the history, see Ref. [18]), ssDNA (*single stranded* DNA) molecules are threaded through a protein nanopore embedded in a lipid membrane, and the measured electric current, influenced by about six nucleobases in the pore, allows for precise sequence information. The speed of the method and the savings in expensive reagents make this method very popular [19]. With these devices, RNA can also be sequenced and methylated nucleobases can also be detected [17]. Thus, unlike the other examples described here, no DNA synthesis using polymerases takes place.

6.1.3.3 Pyrosequencing

In this method, in each cycle, only **one** specific dNTP is added to the template and polymerase, and a reaction occurs with the release of diphosphate or not. In the presence of other immobilized enzymes and reagents, a light signal is generated by chemiluminescence for each released diphosphate ion, which is displayed as a peak. If several identical nucleotides are present in succession in the template strand, several dNTPs are, of course, incorporated very quickly one after the other. The light signal is linearly proportional to the amount of diphosphate ions produced, up to about six nucleotides. For longer homopolymeric stretches, this method then no longer works accurately.

6.1.3.4 Sequencing by pH detection

This method works in principle similarly to pyrosequencing: In each cycle, only **one** specific dNTP is offered to the polymerase. The signal, however, is based on the release of a proton with each nucleotide incorporation. Using a special semiconductor chip, the voltage change is registered, as with a pH electrode. This signal again indicates whether one or more nucleotides of this type have been incorporated. The device is very compact, but only allows sequencing of relatively short DNA fragments (100–200 bp).

6.2 Oligonucleotide Syntheses

The synthesis of oligodeoxynucleotides is carried out almost exclusively on the solid phase, following the same principle as SPPS, which we have already encountered. Of course, the chemical reactions and the solid phase itself are different. Primarily, two different reaction sequences are used, which we will discuss below. But first, a brief overview of the most important protecting groups (PGs) used in this context.

6.2.1 Temporary and Permanent Protecting Groups in Oligonucleotide Synthesis

- **Temporary** protecting groups at the **5′-OH** group: The temporary protecting group (PG) most commonly used is the dimethoxytrityl group (DMT). This requires less strongly acidic conditions for cleavage than the trityl or monomethoxytrityl (MMT) group and can therefore be removed with highly diluted dichloroacetic acid or the Lewis acid zinc bromide. The trimethoxytrityl group (TMT), which is also sometimes used, is almost too acid-labile, as each *p*-methoxy substituent further stabilizes the carbocation formed during acid cleavage, thus facilitating removal. As a photolabile temporary PG in oligonucleotide synthesis for DNA *microarrays*, the 2-(2-nitrophenyl)-propoxycarbonyl group (NPPOC) (Fig. 6.7) is most commonly used [20].
- Protecting group **Phosphoric acid ester**: To protect the hydroxyl group at P(III) or P(V), either the methyl ester (–OMe) is used (cleavage by thiophenol/triethylamine) or the β-cyanoethyl ester (–O–CH_2–CH_2–CN, CNE–), which is cleaved by pyridine/triethylamine.
- Protection of the **NH_2** functions of the **nucleobases** of **A**, **G**, and **C**: **A** and **C** are usually protected by a benzoyl group (Ph–CO–), **G** by an isobutyryl group [$(CH_3)_2$–CH–CO–]; cleavage with concentrated ammonia or methylamine. However, there are also attempts to dispense with these protecting groups.

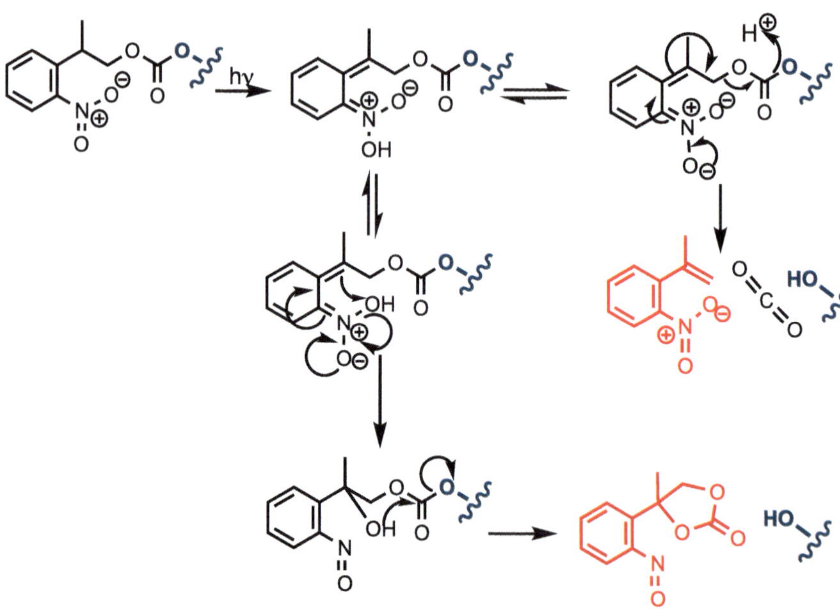

Fig. 6.7 Cleavage of the NPPOC protecting group by UV light

- "Protection" of the **3′ terminus**: in solid-phase synthesis, this is linked to the solid phase via the anchor—silica gel or *controlled pored glass* (CPG); in solution, often as a levulinic acid ester (cleavage with hydrazine).
- In the synthesis of oligo**ribo**nucleotides, protection of the **2′-OH** is, of course, also necessary: For this, the *tert*-butyldimethylsilyl (TBDMS) group is most commonly used—cleavage by tetrabutylammonium fluoride (TBAF) [21]. However, steric hindrance then slows reactions at the 3′-O atom; there is also always a risk of isomerization 2′ ⇆ 3′ (Section 2.3). The **T**riisopropylsilyl**oxym**ethyl group (tom) is used as a less sterically hindering alternative, which is also cleaved by TBAF/THF [22].

Photochemical Cleavage of the NPPOC Protecting Group

Figure 6.7 shows the mechanism of cleavage of the photolabile 2-(2-nitrophenyl)-propoxycarbonyl protecting group (NPPOC), with one of the two pathways predominating depending on the conditions [20].

6.2.2 The Phosphoramidite Method

This method was developed as early as 1980 [23–25] and is still widely used today [26, 27]. Figure 6.8 shows the general reaction scheme. In solid-phase synthesis, the first nucleoside is attached via its 3′-OH group to the solid support—usually *controlled pored glass* (CPG)—using an anchor, which serves as the permanent protecting group. The subsequent nucleosides are added as phosphoramidites with trivalent phosphorus, each coupled to the 5′-hydroxyl group of the growing oligonucleotide, and the resulting triesters of phosphorous acid are then oxidized to phosphates. The DMT protecting group is predominantly used as the temporary protecting group at the 5′-OH group. In each cycle, a *capping* step is performed to irreversibly block any unreacted 5′-OH groups (usually by acetylation). This prevents the formation of "truncated" oligonucleotides with properties (such as length) very similar to the desired end product, as their synthesis is terminated earlier. Oxidation to the phosphoric acid triester is rapid and straightforward, using iodine or *tert*-butyl hydroperoxide (tBuOOH).

Mechanism of the Coupling Reaction of Phosphoramidites under Tetrazole "Catalysis"

Although this method has been used for so long, it was not until 2004 that a paper appeared examining the course and mechanism in detail [28]. Based on the results obtained so far, protonation of the nitrogen atom is considered a catalytically important step, as this weakens the P–N bond. Protonation at the phosphorus atom, on the other hand, would strengthen the P–N bond (Fig. 6.9). The key factor is the appropriate acid strength of the activator, even in the aprotic solvent acetonitrile. At the same time, nucleophilic catalysis is also effective. These mechanistic insights are used to identify even better catalysts than tetrazole (the standard). Tetrazole is a catalyst in the sense that it is ultimately regenerated in the reaction, but it is required in equimolar amounts, as the tetrazolide intermediate is formed to a large extent. Today, 4,5-dicyanoimidazole (DCI) is often used as the activator.

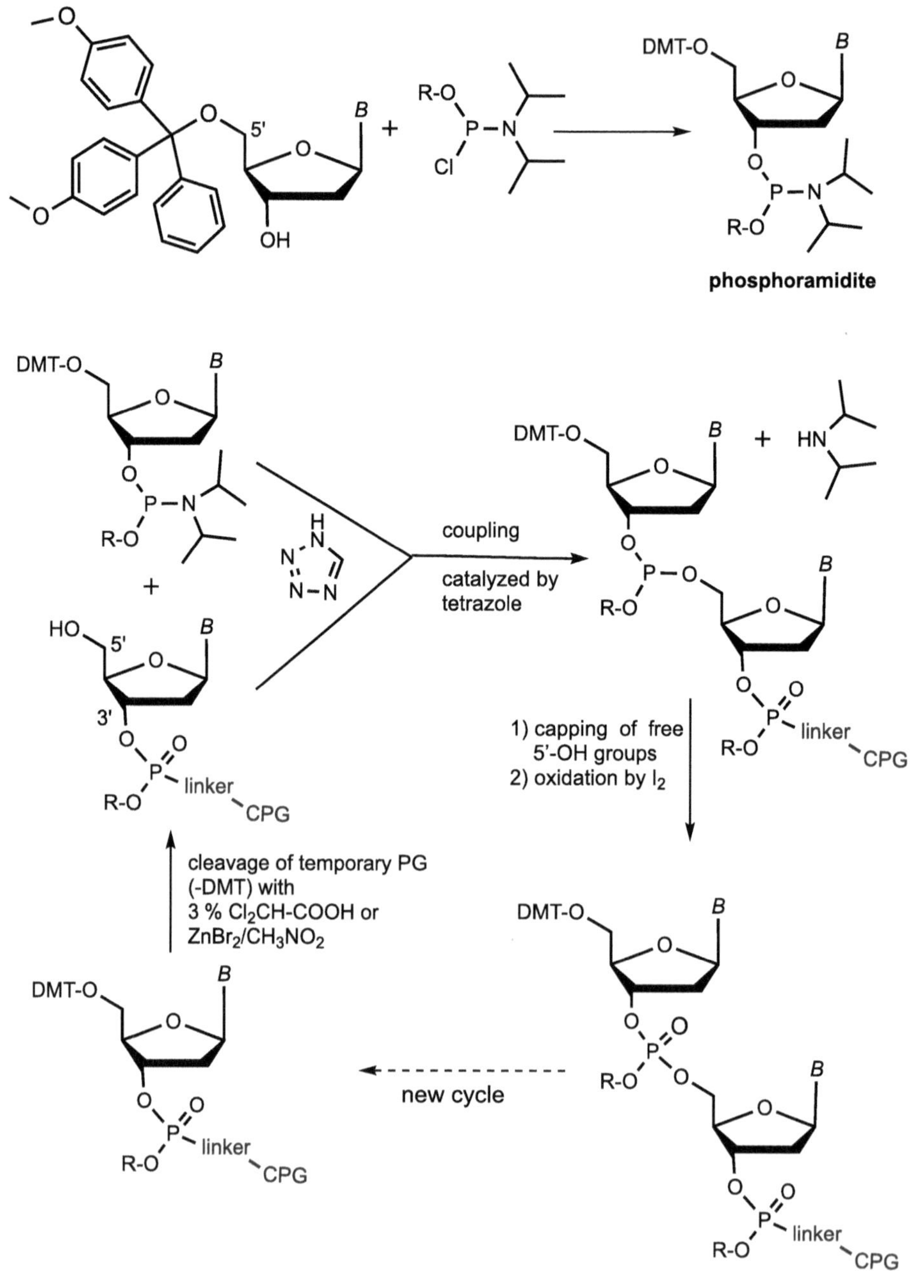

Fig. 6.8 Scheme of deoxyoligonucleotide synthesis via phosphoramidites [*B* ... nucleobase (possibly with NH_2 protecting groups, see Section 6.2.1), CPG ... *controlled pored glass*]

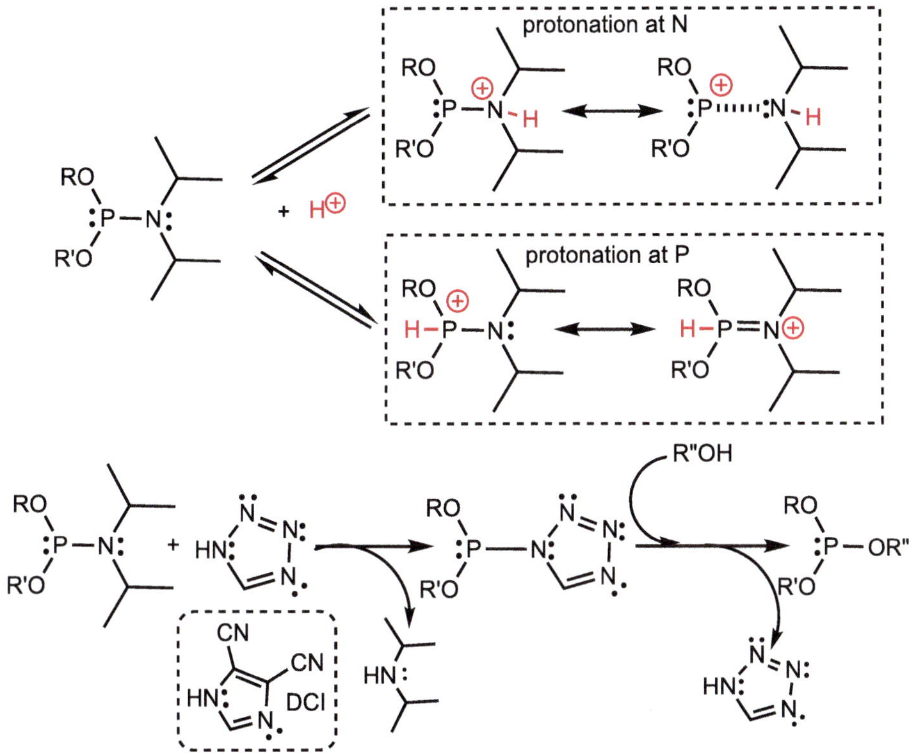

Fig. 6.9 On the mechanism of substitution of phosphoramidites by alcohols under tetrazole "catalysis". (After Ref. [28])

Since cleavage from the solid phase can sometimes cause problems, various special anchors (linkers) have also been introduced [27].

The synthesis of oligonucleotides is essential for many molecular biology techniques (e.g., as primers), where only small amounts are needed, which can be quickly synthesized and purified using the methods described above. However, RNA oligonucleotides are increasingly of interest as therapeutics. For this purpose, larger quantities must be produced in high, reproducible purity. A very detailed study from 2022 [29] describes the synthesis of an oligonucleotide on a kilogram scale in solution using fragment couplings.

From a synthetic chemistry perspective, an interesting method for obtaining pure ribooligonucleotides without chromatographic or electrophoretic separation is based on the fact that only the finished oligonucleotide with a free 5′-OH group, after cleavage from the polymer, is briefly bound to another polymer. This allows separation from all truncated side products, which, due to the *capping* reaction, do not possess a free OH group [30].

Background Information on the Isolation of Pure Ribooligonucleotides without Chromatography
Figure 6.10 shows the exact process: After completion of the solid-phase synthesis, a tetrazine derivative is attached to the free 5′-OH end. The oligonucleotide is then photolytically cleaved from the solid phase (mechanism see Fig. 6.6). This results in a solution containing a mixture of

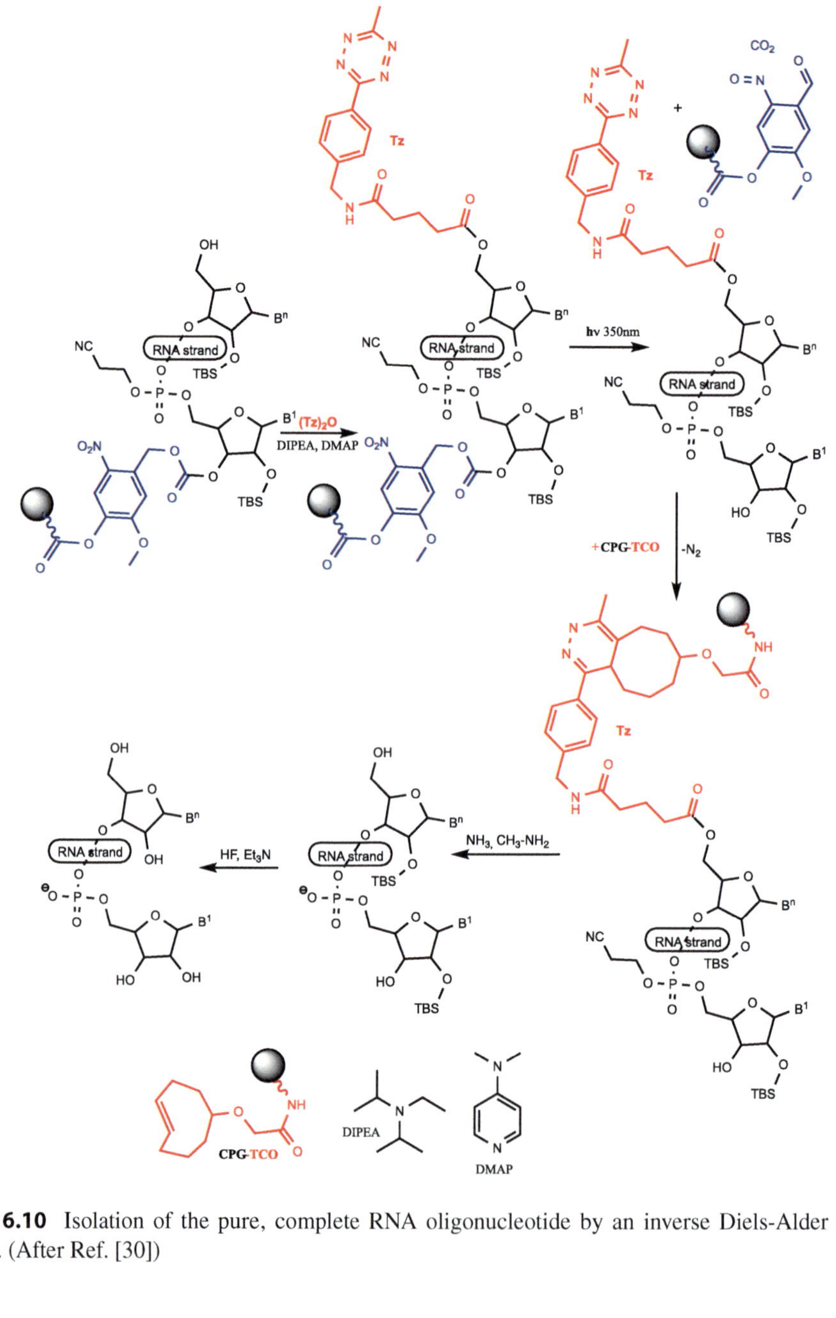

Fig. 6.10 Isolation of the pure, complete RNA oligonucleotide by an inverse Diels-Alder reaction. (After Ref. [30])

truncated oligonucleotides with an acetyl group at the 5′-OH (from the *capping* with acetic anhydride) and the desired product with the tetrazine derivative. Only the latter then reacts as an electron-deficient diene in an inverse Diels-Alder reaction with the strained *trans*-cyclooctene attached to a polymer. This allows the product to be easily separated from the dissolved side products. Finally, all auxiliary and protecting groups are removed.

6.2.3 The H-Phosphonate Method

In addition to the use of phosphoramidites, H-phosphonates [also derivatives of P(III)] are also used for the solid-phase synthesis of oligonucleotides [31, 32]. A 5′-DMT-protected 3′-H-phosphonate is activated with an acid chloride, usually pivaloyl chloride, in the presence of a free 5′-OH group of the growing oligonucleotide, forming an H-phosphonate diester. The cycle then begins anew. Only at the end of the synthesis, before removal of the permanent protecting groups, all H-phosphonates are oxidized to phosphates with iodine (I_2), see Fig. 6.11.

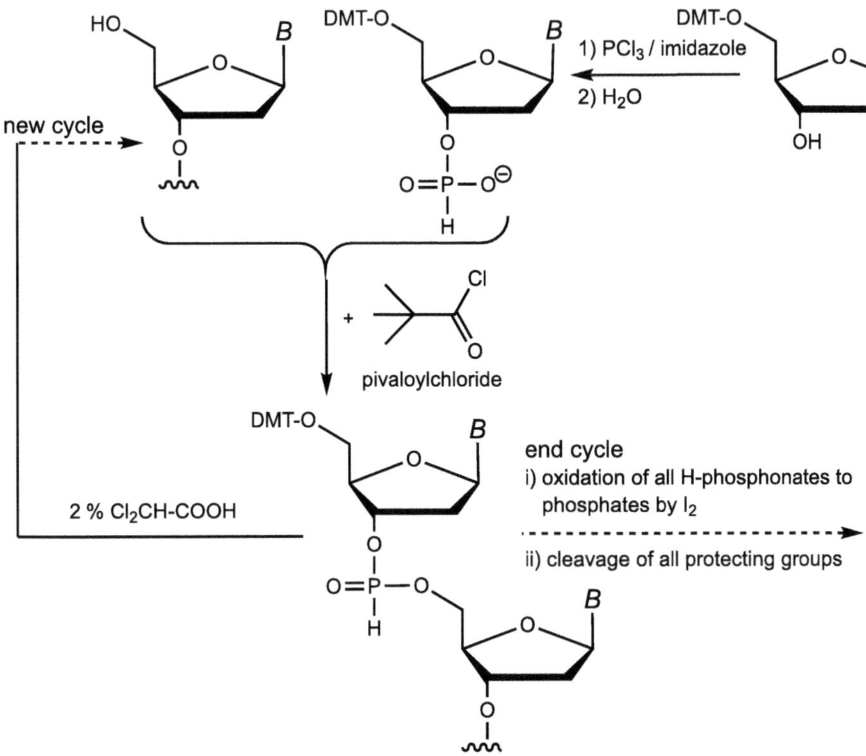

Fig. 6.11 Oligonucleotide synthesis by the H-phosphonate method [*B* … nucleobase]

Table 6.2 Scheme of the steps of a solid-phase synthesis by the H-phosphonate method (continuous-flow, in column, see Section 3.1.3.2). (After Ref. [32])

Step		Reagents and Solvents (Flow rate 3 mL/min)	Time (min)
Standard cycle for chain elongation			
1	Cleavage of DMT PG	2% dichloroacetic acid in CH_2Cl_2	2
2	Washing	Acetonitrile	1
3	Washing	Acetonitrile/pyridine (1:1)	1
4	Coupling	Monomer (20–30 equivalents, 20 mM solution) and coupling reagent (e.g., pivaloyl chloride) (100–150 equivalents, 100 mM) in acetonitrile/pyridine (1:1) are alternately pumped through the column in 6–9 segments (each for 3 s)	≤ 1
5	Washing	Acetonitrile/pyridine (1:1)	0.5
6	Washing	Acetonitrile	1.5
Final cycle			
1	Oxidation	2% I_2 in pyridine/water (98:2)	10
2	Washing	Acetonitrile	2
3	Cleavage of DMT PG	2% dichloroacetic acid in CH_2Cl_2	2
4	Washing	Acetonitrile	3

This method yields somewhat lower product yields and has therefore generally only been used for shorter oligonucleotides. Subsequently, for synthesis via H-phosphonates, especially in solution, an activation reagent bound to a solid phase may be used. If the final oxidation is carried out with suitable sulfur-containing reagents, phosphorothioate oligonucleotides can be obtained directly [33].

Table 6.2 shows an example time schedule for the steps of a solid-phase synthesis via H-phosphonates. The short reaction times are striking.

6.2.4 Synthesis of Cyclic Oligonucleotides

For certain biological applications, cyclic oligonucleotides are of interest. Accordingly, solid-phase syntheses have been adapted for this purpose [21, 34]. Without going into experimental details, here is a brief general explanation of the scheme in Fig. 6.12: A special anchor is used for the solid phase at the 3′ end. The first nucleotide has a phosphoric acid ester protecting group orthogonal to the others (e.g., CNE- instead of Me-, or vice versa), and after completion of the normal linear synthesis, this phosphoric acid ester protecting group is selectively removed following cleavage of the 5′-DMT

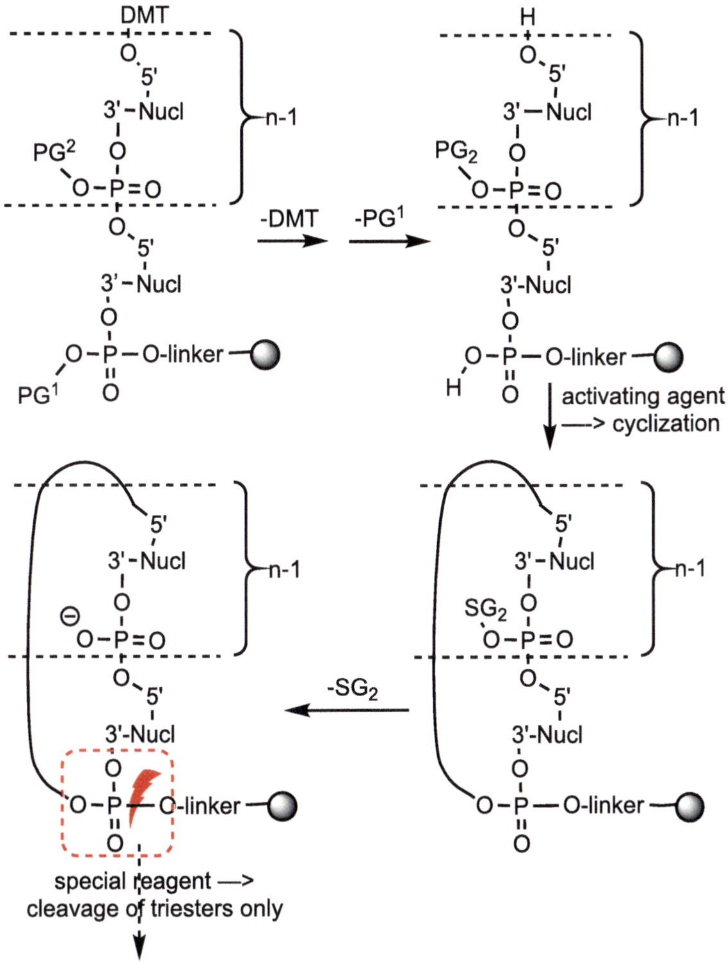

Fig. 6.12 Schematic sequence of a cyclic oligonucleotide synthesis on the solid phase [Nucl … nucleoside, PG … protecting group]. (After Ref. [34])

protecting group. After addition of an activating reagent that induces ring closure, the cyclic oligonucleotide is then cleaved from the anchor, and specifically ***only*** the cyclized product, which is present at the 3′ end as a phosphoric acid **tri**ester, because phosphoric acid diesters are not cleaved in this way. Thus, even with lower cyclization yields, the desired product can be obtained relatively uniformly.

6.3 Modified Nucleic Acids

Why have a number of modified, non-natural nucleic acids been designed and synthesized? There are primarily two motives: On the one hand, the aim was to investigate to what extent natural DNA and RNA are particularly well suited for their functions compared to alternative repetitive structures. On the other hand, there has been and continues to be a search for metabolically more stable and better membrane-permeable analogs, with the goal of "gene therapy." Finally, synthetic biology has set out to find "orthogonal" genetic carriers for "orthogonal" organisms.

Modifications have been and are being made at the level of the ribose-phosphate backbone and at the bases themselves. The following are some examples.

6.3.1 Modifications to the Carbohydrate Moiety

Pyranosyl-RNA and Homo-DNA

In the research group of Albert Eschenmoser at ETH Zurich, alternative nucleic acids with β-D-ribopyranose (6-ring instead of 5-ring) and phosphate linkages at the 2′- and 4′-OH groups (pyranosyl-RNA), as well as with β-D-2′,3′-dideoxyglucopyranose with phosphate bridges at the 4′- and 6′-OH groups (homo-DNA), were synthesized and their properties investigated (Fig. 6.13a). The base pairing properties, the so-called backbone inclination, and the resulting lower stability of a double helix were seen as confirmation of nature's choice for a furanose carbohydrate moiety in nucleic acids [35].

An Example of Arabinonucleosides

In Ref. [36], the conformation of the furanose rings of fluorine-substituted arabinonucleosides is compared. When transitioning from a 2′-monofluoro to a 2′,4′-difluoro derivative, the conformation changes from *south pucker* to *north pucker* (Fig. 6.13b, c). This influences duplex formation with DNA or RNA. As already mentioned above, the primary aim of oligonucleotide modification is to improve stability and affinity, with the goal of developing diagnostics and therapeutics. In the following section, we will see a wealth of bridged ribonucleosides, so-called *locked nucleic acids* (LNA) or *bridged nucleic acids* (BNA), which can strongly alter the properties of DNA or RNA even when only a few units are incorporated.

LNA (Locked Nucleic Acids)

By connecting the 2′ and 4′ positions of ribose with a 2- or possibly 3-atom bridge, LNAs are obtained in which the conformation of the furanose ring is now fixed (Fig. 6.13d). If the bridge is a –CH_2–O–, it is also referred to as "oxy-LNA," but there are also corresponding analogs with S or NR instead of O. The significant stiffening of the backbone allows for special pairing properties with natural DNA and RNA strands. Ref. [37] describes the often elaborate syntheses of these LNAs. Review [38] focuses

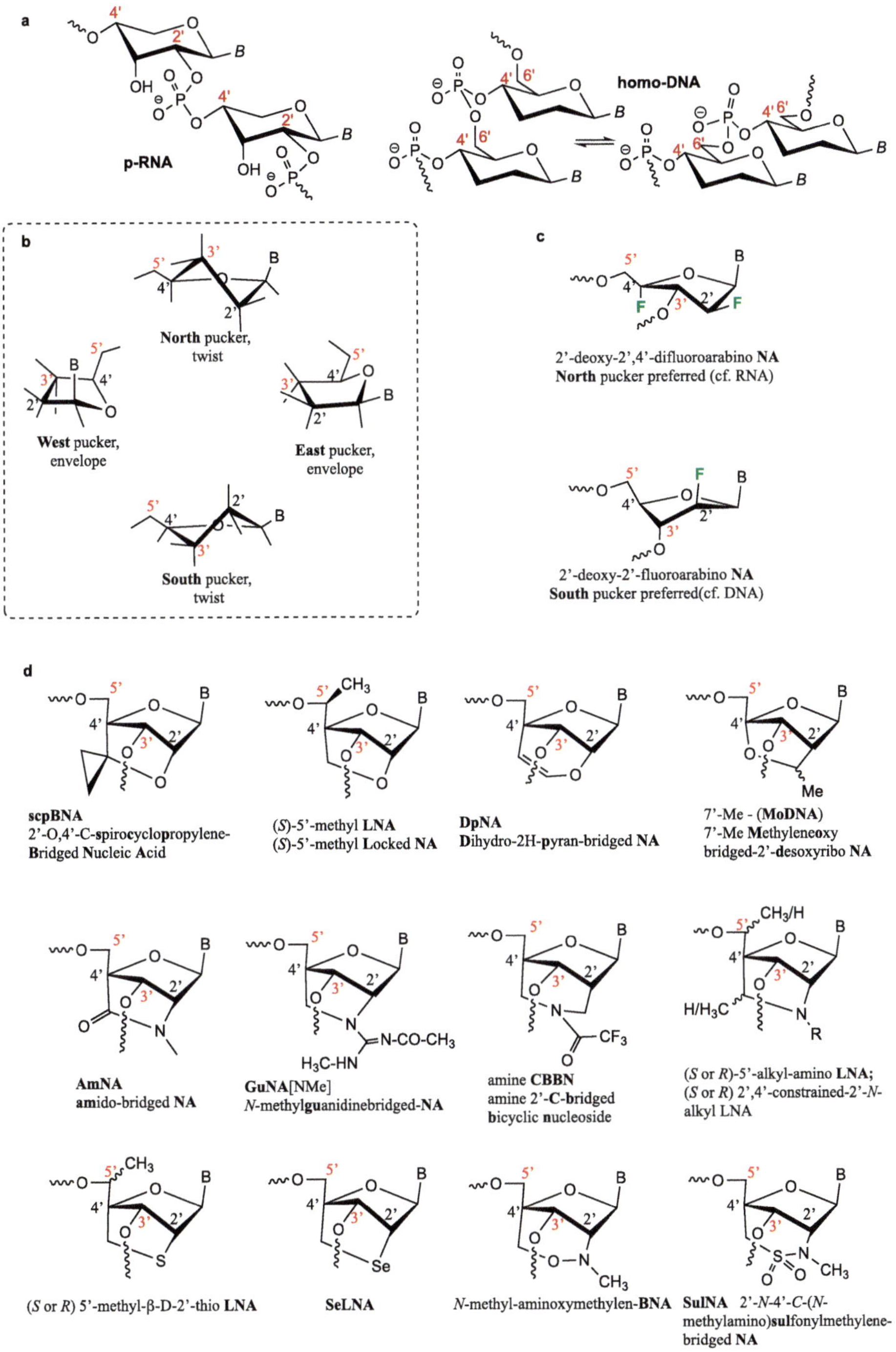

Fig. 6.13 Nucleic acids with modified carbohydrate moieties [*B* … nucleobase]. **a** Pyranosyl-RNA and homo-DNA from the group of A. Eschenmoser [35]; **b** designations for the various conformations of the furanose ring [36]; **c** two examples of fluorine-substituted arabinonucleosides [36]; **d** overview of synthesized locked nucleic acids (LNA) to date [37]

primarily on the application of these modifications for the diagnosis and treatment of cancer.

As a counterpart to the rigidly fixed ribofuranose ring in LNAs, the other extreme has also been developed: the open-chain variant, the so-called *unlocked nucleic acids* (UNA), and the properties of both modifications have been compared [39].

6.3.2 Peptide Nucleic Acids (PNA)

Nielsen PNA

It was actually surprising to find that a DNA mimic with a peptide backbone, but with natural nucleobases, exhibits remarkably good pairing properties with natural DNA or RNA strands. The simple variant introduced by Peter Nielsen uses a glycine-ethylenediamine unit instead of the phosphate-ribose unit, with nucleobases each attached via a C2 bridge through an amide bond to the glycine nitrogen ("**A**mino**e**thylen**g**lycine" PNA, aeg-PNA, "Nielsen" PNA) (Fig. 6.14a) [40]. The advantages of these PNAs are obvious: the synthesis is relatively simple (SPPS), the corresponding oligonucleotides are uncharged and therefore, in principle, more membrane-permeable, and they have a lifespan greater by orders of magnitude, since no corresponding degrading enzymes exist. How could they be used [41]?

There are a variety of pairing possibilities with DNA and RNA (Fig. 6.14b): PNA can form triplex helices with DNA via Hoogsteen hydrogen bonds (indicated by red lines in the figure) and thus recognize and mark specific sites (A); or, through Hoogsteen and Watson-Crick pairing, undergo strand invasion (B and C, where both PNA parts can be connected). Strand invasion can also be forced solely by Watson-Crick pairing (D). Using specially designed bases, the Janus-wedge triple helix is also possible (E).

PNA is already widely used *in vitro* for assays, but the hoped-for development for use as diagnostics and therapeutics *in vivo* has not yet been achieved. The main problem is the low solubility and poor uptake into cells [42, 43].

Accordingly, a variety of variants have been developed. On the one hand, attempts have been made to further strengthen binding to RNA or DNA single strands by rigidifying the backbone, e.g., by a cyclopentane ring (cpPNA) or even by two five-membered rings in pyrrolidinyl-PNA (acpcPNA) (Fig. 6.14c). On the other hand, the aim was to improve solubility and cell permeability—for this, a guanidino group (γ-GPNA) or a short diethylene glycol unit (miniPEG-PNA) appear particularly promising. Modifications have also been made to the bases, especially to improve Hoogsteen recognition of DNA sequences or even to achieve double Watson-Crick recognition for Janus-wedge strand invasion. Reviews [42, 44] provide a detailed discussion of all modifications to PNAs and their possible applications. Synthetic aspects in particular are covered in Ref. [44].

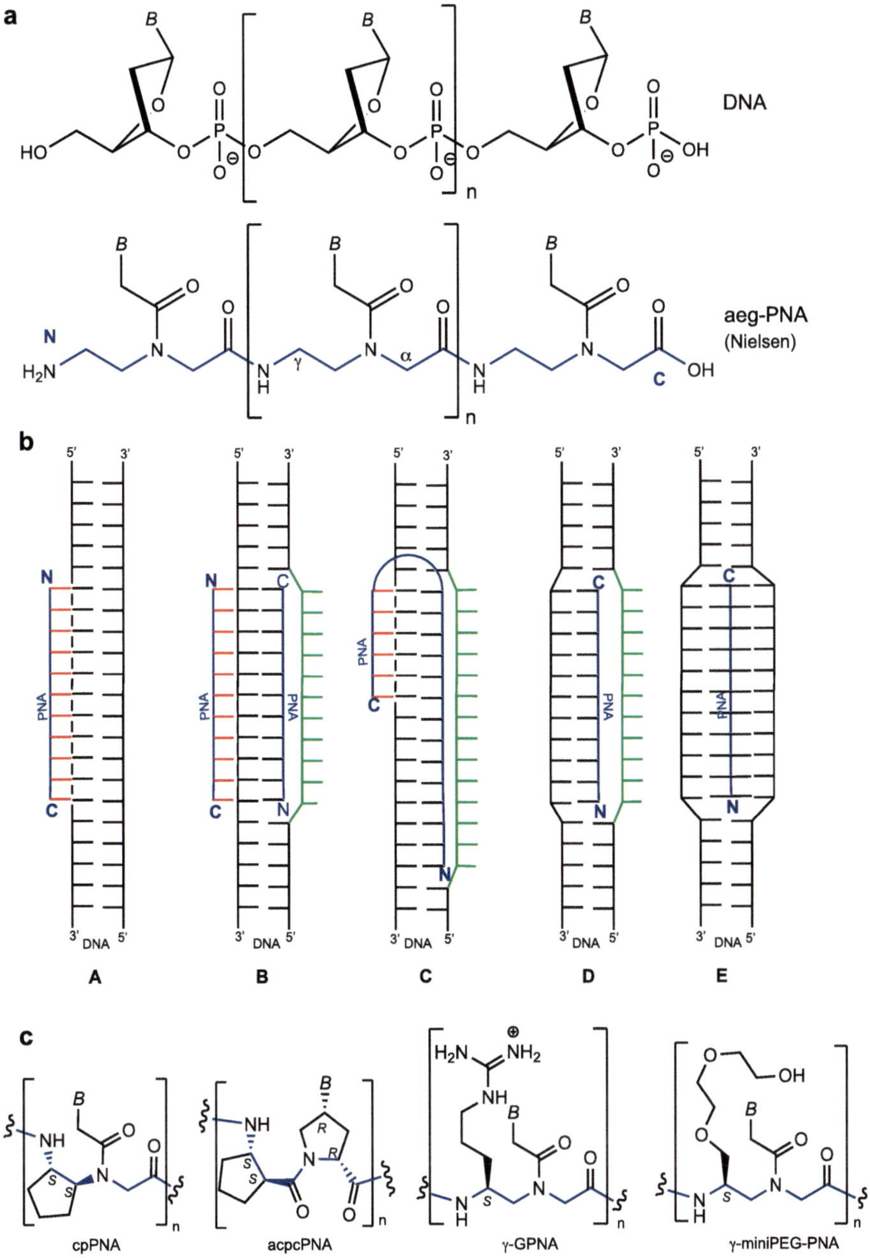

Fig. 6.14 **a** Comparison of DNA with "Nielsen" peptide nucleic acid (aeg-PNA); **b** various pairing possibilities of PNA and DNA, red … Hoogsteen H-bonds, green … part of the DNA strand whose Watson-Crick H-bonds have been broken by strand invasion; **c** some examples of modifications of aeg-PNA to improve pairing properties or solubility and uptake into cells [*B* … nucleobase] [42]

6.3.3 Modifications to the Nucleobases

Modifications to the nucleobases using the normal nucleic acid backbone have been made to investigate the driving force for base pairing according to Watson and Crick, both from a thermodynamic perspective and with regard to the polymerase reaction, i.e., the kinetic process of selectively building the complementary strand. More on this can be found in Sect. 11.1. Modifications to individual nucleobases can also be used for detection reactions, e.g., by fluorescence or fluorescence quenching (more on this in Sect. 11.2) or for the targeted introduction of destructive influences, such as radicals (Sect. 11.3) (further literature on the topic: [45–47])

6.3.4 Unnatural Nucleic Acids (XNA)

There are several research groups working on the design of unnatural base pairings. What is the goal?—Primarily, the addition of another base pair would significantly expand the information storage capacity of nucleic acids and also enable the selective incorporation of multiple unnatural amino acids (see also Chap. 12). However, a prerequisite for this is faithful replication by naturally occurring polymerases. Review [45] provides a very detailed description of the approach to developing such an unnatural base pair. Countless nucleotides had to be synthesized in order to test their incorporation into DNA. The possible applications of such unnatural nucleic acids (XNA) range from gene therapy to materials science. A final goal of "synthetic biology," however, is a DNA that is unnatural at all base pairs for new, "orthogonal," living organisms. This would even be safer than making changes to existing life forms, as claimed in an essay published in 2010 [48].

Summary and Outlook

In this chapter, we have seen that the contribution of organic synthesis to this field can be very significant. Starting from reagents (modified nucleoside triphosphates) for sequencing, to the provision of any desired oligonucleotides (as primers, templates, etc.), up to entirely novel nucleobases and nucleic acids. Above all, the modification of nucleic acids offers plenty of motivation for organic chemists on the one hand and for biologists and physicians on the other.

6.4 Questions

Let's test your understanding of this chapter's content

6.1 Explain the principle of the Maxam-Gilbert DNA sequencing method!
6.2 What is the significance of dideoxynucleoside triphosphates in the Sanger sequencing method?
6.3 What do the methods "pyrosequencing" and "sequencing by pH detection" have in common?
6.4 What is the temporary protecting group in oligonucleotide synthesis, and which position is protected?
6.5 What is a phosphoramidite?
6.6 When does the oxidation of P(III) to P(V) occur in the phosphoramidite method, and how?
6.7 Why do you think solid-phase oligonucleotide synthesis is generally faster and easier than SPPS?
6.8 What is the significance of LNA, and what is the structural difference?
6.9 The simplest peptide nucleic acids (PNA) have which backbone?
6.10 What significance could the development of a third base pair have?

References

1. Gilbert, W.: DNA sequencing and gene structure. Science. **214**(4527), 1305–1312 (1981). https://doi.org/10.1126/science.7313687
2. Sanger, F.: Determination of nucleotide sequences in DNA. Science. **214**(4526), 1205–1210 (1981). https://doi.org/10.1126/science.7302589
3. Ohmori, H., Tomizawa, J.I., Maxam, A.M.: Detection of 5-methylcytosine in DNA sequences. Nucl. Acids. Res. **5**(5), 1479–1485 (1978). https://doi.org/10.1093/nar/5.5.1479
4. Rein, T., DePamphilis, M.L., Zorbas, H.: Identifying 5-methylcytosine and related modifications in DNA genomes. Nucl. Acids. Res. **26**(10), 2255–2264 (1998). https://doi.org/10.1093/nar/26.10.2255
5. Fu, L., Amato, N.J., Wang, P., …, Wang, Y.: Simultaneous quantification of methylated cytidine and adenosine in cellular and tissue RNA by nano-flow liquid chromatography-tandem mass spectrometry coupled with the stable isotope-dilution method. Anal. Chem. **87**(15), 7653–7659 (2015). https://doi.org/10.1021/acs.analchem.5b00951
6. Yang, X., Grant, K.B.: Chemical sequencing of restriction fragments 3'-end-labeled with [35S] dATPαS. J. Biochem. Biophys. Methods. **50**(2–3), 123–128 (2002). https://doi.org/10.1016/S0165-022X(01)00179-8
7. Zhu, J., Zhao, Y., Zhu, Y., …, Guo, Z.: DNA cross-linking patterns induced by an antitumor-active trinuclear platinum complex and comparison with its dinuclear analogue. Chem. Eur. J. **15**(21), 5245–5253, S5245/5241-S5245/5211 (2009). https://doi.org/10.1002/chem.200900217; Ito, K., Kawanishi, S.: Sequence specificity of ultraviolet A-induced DNA damage in the presence of photosensitizer. Methods Enzymol. **319**, 417–427 (2000). https://doi.org/10.1016/S0076-6879(00)19039-1

8. Smith, L.M., Sanders, J.Z., Kaiser, …, Hood, L.E.: Fluorescence detection in automated DNA sequence analysis. Nature. **321**, 674–679 (1986). https://doi.org/10.1038/321674a0
9. Smith, L.M., Brumley Jr., R.L., Buxton,…, Tong, X.: High-speed automated DNA sequencing in ultrathin slab gels. Methods Enzymol. **271**, 219–237 (1996). https://doi.org/10.1016/S0076-6879(96)71012-1
10. Technology feature: DNA sequencing: generation next-next. Nature Meth. **5**(3), 267–274 (2008), https://doi.org/10.1038/nmeth0308-267
11. Kaur, R., Malik, C.P.: Next generation sequencing: a revolution in gene sequencing. CIB Tech J. Biotechnol. **2**(4), 1–20 (2013). http://www.cibtech.org/J-Biotechnology/PUBLICATIONS/2013/Vol-2-No-4/CJB-01-01-MALIK-NEXT-SEQUENCING.pdf. Zugegriffen am 29.12.2025; van Dijk, E.L., Auger, H., Jaszczyszyn, Y., Thermes, C.: Ten years of next-generation sequencing technology. Trends Genet. **30**(9), 418–426 (2014). https://doi.org/10.1016/j.tig.2014.07.001; Hui, P., Next generation sequencing: Chemistry, technology and applications. In: Tang, N. N. S., Poon, T. (Hrsg.) Top. Curr. Chem., Bd.: 336, Chemical Diagnostics, 1–18, Springer, Berlin Heidelberg (2014). https://doi.org/10.1007/128_2012_329
12. Welch, M.B., Martinez, C.I., Zhang, A.J., …, Burgess, K.: Syntheses of nucleosides designed for combinatorial DNA sequencing. Chem. Eur. J. **5**(3), 951–960 (1999). https://doi.org/10.1002/(SICI)1521-3765(19990301)5:33.0.CO;2-G; Welch, M., Martinez, C., Zhang, A., …, Burgess, K.: Corrigendum: Syntheses of nucleosides designed for combinatorial DNA sequencing. Chem. Eur. J. **11**(24), 7145–7145 (2005). https://doi.org/10.1002/chem.200590080
13. Lu, G., Burgess, K.: A diversity oriented synthesis of 3′-O-modified nucleoside triphosphates for DNA 'sequencing by synthesis'. Bioorg. Med. Chem. Lett. **16**(15), 3902–3905 (2006). https://doi.org/10.1016/j.bmcl.2006.05.035
14. Knapp, D.C., Serva, S., D'Onofrio, J., …, Engels, J.W.: Fluoride-cleavable, fluorescently labelled reversible terminators: synthesis and use in primer extension. Chem. Eur. J. **17**(10), 2903–2915 (2011). https://doi.org/10.1002/chem.201001952
15. Kim, D.-R., Kim, T.-S., Kim, E., …, Ahn, D.-R.: Synthesis of 3′-O-fluorescently mono-modified reversible terminators and their uses in sequencing-by-synthesis. Bioorg. Med. Chem. Lett. **24**(1), 209–213 (2014). https://doi.org/10.1016/j.bmcl.2013.11.040
16. Zur Illumina-Methgode siehe auch: https://en.wikipedia.org/wiki/Illumina_dye_sequencing. Zugegriffen am 29.12.2025 oder https://emea.illumina.com/science/technology/next-generation-sequencing.html. Zugegriffen am 29.12.2025
17. Beaulaurier, J., Schadt, E.E., Fang, G.: Deciphering bacterial epigenomes using modern sequencing technologies. Nature Rev. Genetics. **20**(3), 157–172 (2019). https://doi.org/10.1038/s41576-018-0081-3
18. https://nanoporetech.com/about/history. Zugegriffen am 29.12.2025
19. Cao, C., Long, Y.-T.: Biological nanopores: confined spaces for electrochemical single-molecule analysis. Acc. Chem. Res. **51**(2), 331–341. https://doi.org/10.1021/acs.accounts.7b00143 (2018); https://nanoporetech.com/platform/technology. Zugegriffen am 29.12.2025; https://en.wikipedia.org/wiki/Nanopore_sequencing. Zugegriffen am 29.12.2025
20. Wöll, D., Smirnova, J., Galetskaya, M., …, Steiner, U.E.: Intramolecular sensitization of photocleavage of the photolabile 2-(2-nitrophenyl)propoxycarbonyl (NPPOC) protecting group: photoproducts and photokinetics of the release of nucleosides. Chem. Eur. J. **14**(21), 6490–6497 (2008). https://doi.org/10.1002/chem.200800613; Agbavwe, C., Kim, C., Hong, D.G., Somoza, M.M.: Efficiency, error and yield in light-directed maskless synthesis of DNA microarrays. J. Nanobiotechnol. **9**, 57 (2011). https://doi.org/10.1186/1477-3155-9-57

21. De Napoli, L., Galeone, A., Mayol, L., …, Santacroce, C.: Facile preparation of cyclic oligoribonucleotides. J. Chem. Soc. Perkin Trans 1, 1993 (7), 747–749 (1993). https://doi.org/10.1039/P19930000747
22. Pitsch, S.: Protecting groups for the synthesis of ribonucleic acids. Chimia. **55**(4), 320–324 (2001). https://doi.org/10.2533/chimia.2001.320
23. Beaucage, S.L., Caruthers, M.H.: Deoxynucleosidephosphoramidites – a new class of key intermediates for deoxypolynucleotide synthesis. Tetrahedron Lett. **22**(20), 1859–1862 (1981). https://doi.org/10.1016/S0040-4039(01)90461-7
24. Matteucci, M.D., Caruthers, M.H.: Synthesis of deoxynucleotides on a polymer support. J. Am. Chem. Soc. **103**(11), 3185–3191 (1981). https://doi.org/10.1021/ja00401a041
25. Beaucage, S.L., Iyer, R.P.: Advances in the synthesis of oligonucleotides by the phosphoramidite approach. Tetrahedron. **48**(12), 2223–2311 (1992). https://doi.org/10.1016/S0040-4020(01)88752-4
26. Brazier, J.: Chemical synthesis of oligonucelotide sequences: phosphoramidite chemistry. In: Scarlett, G. (Hrsg.) DNA Manipulation and Analysis, Bd. 2633, S. 185–193, Springer US (2023). https://doi.org/10.1007/978-1-0716-3004-4_14
27. Yamamoto, K., Fuchi, Y., Ito, Y., Hari, Y.: Expansion of phosphoramidite chemistry in solid-phase oligonucleotide synthesis: rapid 3′-Dephosphorylation and strand cleavage. J. Org. Chem. **88**(5), 2726–2734 (2023). https://doi.org/10.1021/acs.joc.2c02195
28. Nurminen, E., Lönnberg, H.: Mechanisms of the substitution of phosphoramidites and their congeners. J. Phys. Org. Chem. **17**(1), 1–17 (2004). https://doi.org/10.1002/poc.681
29. Zhou, X., Kiesman, W. F., Yan, W., …, Shi, X.: Development of Kilogram-Scale Convergent Liquid-Phase Synthesis of Oligonucleotides. J. Org. Chem., **87** (4), 2087-2110 (2022). https://doi.org/10.1021/acs.joc.1c01756
30. He, M., Wu, X., Mao, S., … Royzen, M.: Bio-orthogonal chemistry enables solid phase synthesis and HPLC and gel-free purification of long RNA oligonucleotides. ChemComm, **57** (35), 4263–4266 (2021). https://doi.org/10.1039/D1CC00096A
31. Froehler, B.C., Matteucci, M.D.: Nucleoside H-phosphonates: valuable intermediates in the synthesis of deoxyoligonucleotides. Tetrahedron Lett. **27**(4), 469–472 (1986). https://doi.org/10.1016/S0040-4039(00)85507-0; Garegg, P.J., Lindh, I., Regberg, T., Stawinski, J., Strömberg, R., Henrichson, C.: Nucleoside H-phosphonates. III. Chemical synthesis of oligodeoxyribonucleotides by the hydrogenphosphonate approach. Tetrahedron Lett. **27**(34), 4051–4054 (1986). https://doi.org/10.1016/S0040-4039(00)84908-4
32. Garegg, P.J., Lindh, I., Regberg, T., Stawinski, J., Strömberg, Henrichson, C.: Nucleoside H-phosphonates. IV.: Automated solid phase synthesis of oligoribonucleotides by the hydrogenphosphonate approach. Tetrahedron Lett. **27**(34), 4055–4058 (1986). https://doi.org/10.1016/S0040-4039(00)84909-6
33. Stawinski, J., Strömberg, R., Di- and Oligonucleotide Synthesis Using H-Phosphonate Chemistry. In: Herdewijn, P. (Hrsg.) Methods in Molecular Biology, Bd. **288**, Oligonucleotide Synthesis, 6, 81–100, Humana Press (2005). https://doi.org/10.1385/1-59259-823-4:081; Adamo, I., Dueymes, C., Schonberger, A., …, Vasseur, J.-J.: Solution-phase synthesis of phosphorothioate oligonucleotides using a solid-supported acyl chloride with H-phosphonate chemistry. Europ. J. Org. Chem. (2), 436–448 (2006). https://doi.org/10.1002/ejoc.200500547; Reese, C. B., Song, Q.: The H-phosphonate approach to the synthesis of oligonucleotides and their phosphorothioate analogs in solution. J. Chem. Soc., Perkin Trans. 1, 1999 (11), 1477–1486 (1999). https://doi.org/10.1039/A901301F; Vlaho, D., Damha, M.J.: Synthesis of Chimeric Oligonucleotides Having Modified Internucleotide Linkages via an Automated H-Phosphonate/Phosphoramidite Approach. Curr. Prot. Nucleic Acid Chem. **73**(1), e53. (2018). https://doi.org/10.1002/cpnc.53

34. Alazzouzi, E., Escaja, N., Grandas, A., Pedroso, E.: Eine kurze Festphasensynthese für cyclische Oligodesoxyribonucleotide. Angew. Chem. **109** (13–14), 1564–1567 (1997). https://doi.org/10.1002/ange.19971091333; Alazzouzi, E., Escaja, N., Grandas, A., Pedroso, E.: A Straightforward solid-phase synthesis of cyclic oligodeoxyribonucleotides. Angew. Chem. Int. Ed. **36** (13–14), 1506–1508 (1997). https://doi.org/10.1002/anie.199715061; Pedroso, E., Escaja, N., Frieden, M., Grandas, A., Solid-phase synthesis of circular oligonucleotides. In: Herdewijn, P. (Hrsg.) Methods in Molecular Biology, Bd. **288**, Oligonucleotide Synthesis, 7, 101–125, Humana Press, (2005). https://doi.org/10.1385/1-59259-823-4:101
35. Micura, R., Kudick, R., Pitsch, S., Eschenmoser, A.: Opposite orientation of backbone inclination in pyranosyl-RNA and Homo-DNA correlates with opposite directionality of duplex properties. Angew. Chem. Int. Ed. **38**(5), 680–683 (1999). https://doi.org/10.1002/(SICI)1521-3773(19990301)38:5<680::AID-ANIE680>3.0.CO;2-C; Micura, R., Kudick, R., Pitsch, S., Eschenmoser, A.: Die gegensätzliche Orientierung der Rückgratneigung in Pyranosyl-RNA und homo-DNA korreliert mit einer entsprechend gegensätzlichen Orientierung der Duplexeigenschaften. Angew. Chem. **111** (5), 715–718, (1999). https://doi.org/10.1002/(SICI)1521-3757(19990301)111:53.0.CO;2-J; Pitsch, S., Wendeborn, S., Krishnamurthy, R., …, Eschenmoser, A.: Pentopyranosyl Oligonucleotide Systems. 9th Communication. Helv. Chim. Acta. **86**(12), 4270–4363 (2003). https://doi.org/10.1002/hlca.200390349; Egli, M., Pallan, P.S., Pattanayek, R., …, Eschenmoser, A.: Crystal structure of homo-DNA and nature's choice of pentose over hexose in the genetic system. J. Am. Chem. Soc. **128**(33) (2006). https://doi.org/10.1021/ja062548x; Egli, M., Lubini, P., Pallan, P.S.: The long and winding road to the structure of homo-DNA. Chem. Soc. Rev. **36**(1), 31–45 (2007). https://doi.org/10.1039/B606807C
36. Martínez-Montero, S., Deleavey, G.F., Dierker-Viik, A., …, Damha, M.J.: Synthesis and properties of 2′-Deoxy-2′,4′-difluoroarabinose-modified nucleic acids. J. Org. Chem. **80**(6), 3083–3091 (2015). https://doi.org/10.1021/jo502948t
37. Goswami, A., Prasad, A.K., Jyotirmoy, M., Khaneja, M.: Synthesis and applications of bicyclic sugar modified locked nucleic acids: a review. Nucleosides Nucleotides Nucleic Acids. **41**(5-6), 503–529 (2022). https://doi.org/10.1080/15257770.2022.2052316
38. Kamali, M.J., Salehi, M., Fatemi, S., …, Ahmadifard, M.: Locked nucleic acid (LNA): a modern approach to cancer diagnosis and treatment. Experim Cell Res. **423**(1), 113442 (2023). https://doi.org/10.1016/j.yexcr.2022.113442
39. Campbell, M.A., Wengel, J.: Locked vs. unlocked nucleic acids (LNA vs. UNA): contrasting structures work towards common therapeutic goals. Chem. Soc. Rev. **40**(12), 5680–5689 (2011). https://doi.org/10.1039/C1CS15048K
40. Brodyagin, N., Katkevics, M., Kotikam, V., Ryan, C.A., Rozners, E.: Chemical approaches to discover the full potential of peptide nucleic acids in biomedical applications. Beilstein J. Org. Chem. **17**, 1641–1688 (2021). https://doi.org/10.3762/bjoc.17.116
41. Nielsen, P.E.: Peptide nucleic acid. A molecule with two identites. Acc. Chem. Res. **32**(7), 624–630 (1999). https://doi.org/10.1021/ar980010t; Nielsen, P.E.: The many faces of PNA. Lett. Pept. Sci. **10**(3–4), 135–147 (2003). https://doi.org/10.1007/s10989-004-4860-7; Nielsen, P. E.: Peptide nucleic acids and the origin of life. Chem. Biodiv. **4**(9), 1996–2002 (2007). https://doi.org/10.1002/cbdv.200790166
42. Nielsen, P.E.: Peptide Nucleic Acids (PNA) in chemical biology and drug discovery. Chem. Biodiv. **7**(4), 786–804 (2010). https://doi.org/10.1002/cbdv.201000005
43. Pandey, P.N., Saini, N., Sapre, N., …, Tiwari, A.K.: Prioritising breast cancer theranostics: a current medical longing in oncology: Cancer Treatment Res. Comm. **29**, 100465 (2021). https://doi.org/10.1016/j.ctarc.2021.100465; Suparpprom, C., Vilaivan, T., Perspectives on conformationally constrained peptide nucleic acid (PNA): insights into the structural design,

properties and applications: RSC Chem. Biol. **3**(6), 648–697 (2022). https://doi.org/10.1039/D2CB00017B; Katkevics, M., MacKay, J.A., Rozners, E., Triplex-forming peptide nucleic acids as emerging ligands to modulate structure and function of complex RNAs: Chem. Comm. **60** (15), 1999–2008 (2024). https://doi.org/10.1039/D3CC05409H; Sannigrahi, A., De, N., Bhunia, D., Bhadra, J., Peptide nucleic acids: Recent advancements and future opportunities in biomedical applications: Bioorg. Chem. **155**, 108146 (2025). https://doi.org/10.1016/j.bioorg.2025.108146

44. Nandhini, K.P., Shaer, D.A., Albericio, F., de la Torre, B.G.: The challenge of peptide nucleic acid synthesis: Chem. Soc. Rev. **52**(8), 2764–2789 (2023). https://doi.org/10.1039/D2CS00049K
45. Feldman, A.W., Romesberg, F.E.: Expansion of the genetic alphabet: a chemist's approach to synthetic biology. Acc Chem Res. **51**(2), 394–403 (2018). https://doi.org/10.1021/acs.accounts.7b00403
46. Kha, T.-K., Zhao, Y., Zhu, R.-Y.: Site-selective modification and labeling of native RNA. Chem Eur. J. **31**(12), e202404244 (2025). https://doi.org/10.1002/chem.202404244
47. Marx, A., Betz, K.: The structural basis for processing of unnatural base pairs by DNA polymerases. Chem Eur. J. **26**(16), 3446–3463 (2020). https://doi.org/10.1002/chem.201903525
48. Schmidt, M.: Xenobiology: a new form of life as the ultimate biosafety tool. BioEssays. **32**(4), 322–331 (2010). https://doi.org/10.1002/bies.200900147

Preparation of Conjugates 7

Abstract

An overview is provided of the application of various organic chemical reactions for the derivatization and labeling of biomolecules. Such methods have recently become increasingly important, both for conducting assays in research and diagnostics, and for tracking certain molecules in the cell *in vivo*. However, these reactions must possess specific properties. Several specific applications are also discussed in greater detail.

Introducing specific labels into unprotected biomolecules requires highly specialized methodology. The reactions used must be as selective as possible, must proceed in aqueous media and at room temperature, and not at extreme pH values. Selecting such reactions and developing them so that they can be reproducibly applied even by personnel without extensive chemical training has become the domain of chemists who assemble such kits at relevant companies. An example of a company publication featuring a wide variety of conjugation and labeling methods can be found in Ref. [1]. There are also dedicated journals such as *Bioconjugate Chemistry* of the American Chemical Society (ACS), which was founded in 1990. The increase in the number of pages per year may be taken as an indication of how rapidly this field of research is growing, now with an *impact factor* of over 4 in 2024 (Fig. 7.1).

In general, when labeling biologically active molecules, one must consider whether derivatization might significantly alter the biological activity. For example, if we consider a protein with many lysines and a single free cysteine, there are two possible strategies. From the perspective of a precise chemist, we can aim for a uniform derivative and therefore choose a reaction specific for cysteines, or we can deliberately prefer a mixture of proteins labeled at different lysines. In such a mixture, the probability that at least some derivatives retain their natural bioactivity is relatively high, but we have no

D. Krois, *Organic Chemical Methods*, https://doi.org/10.1007/978-3-662-73573-2_7

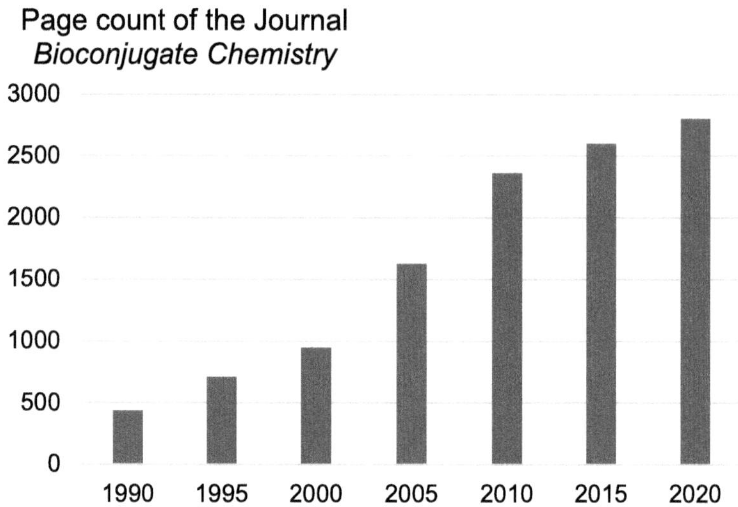

Fig. 7.1 Development of the number of pages per year for the journal *Bioconjugate Chemistry*, *impact factor* 4.3

information about which ones these are. If the single cysteine is derivatized, there is a clear yes/no answer.

It should be added here that nowadays, the possibility of introducing unnatural amino acids at selected positions in a protein offers a very interesting, albeit somewhat more elaborate, alternative to post-synthetic conjugation. More on this later in Chap. 12.

In the following, we will discuss the most important general methods as well as specific examples.

7.1 Derivatization of Amino Groups

For labeling proteins, the amino groups of the N-terminus and lysine side chains are suitable targets. Of course, amino functions on other biomolecules such as carbohydrates can also react in this way. Finally, amines can be generated from other functional groups in a first step.

7.1.1 Amide Formation via Activated Esters

We have already discussed the advantages of activated esters for forming amide bonds (Sect. 3.1.3). When it comes to biocompatibility, *N*-hydroxysuccinimide esters (–OSu or –NHS) have proven particularly effective. In fact, many labels, such as fluorophores, are routinely offered as *N*-hydroxysuccinimide esters. The advantages are high selectivity for amino groups, much slower reaction with hydroxyl groups, and the non-toxicity of the cleavage product, *N*-hydroxysuccinimide. The reaction can be carried out in buffer solution at pH 8–9. A slightly basic pH is necessary so that the amino group is not fully protonated (Fig. 7.2a).

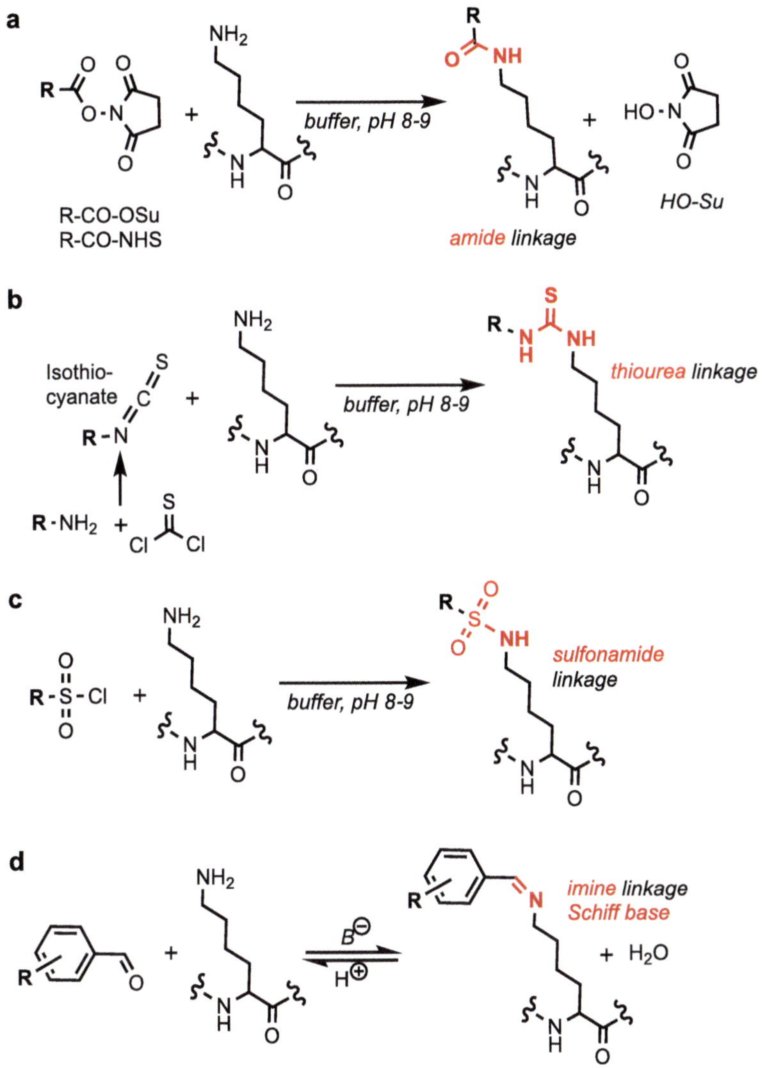

Fig. 7.2 Methods for derivatizing amino functions using lysine as an example

7.1.2 Thiourea Linkage with Isothiocyanates

The selective reaction of phenyl isothiocyanate with amines was already encountered in the Edman degradation (Sect. 3.2.4). Other isothiocyanates react similarly, and a pH between 8 and 9 is sufficient, as in the first example. This results in a stable thiourea linkage (Fig. 7.2b). It could rightly be argued that in acidic solution, degradation of the first amino acid could occur. However, such acidic conditions (such as pure trifluoroacetic acid) will not occur in this case, and in addition, a protein generally contains many more ε-amino groups of lysine, but only one N-terminus. But how are the corresponding isothiocyanates obtained?—Usually by addition of amines to thiophosgene ($CSCl_2$).

7.1.3 Other Methods

Less commonly used is the reaction with sulfonyl chlorides to form stable sulfonamides (Fig. 7.2c). This is somewhat less suitable due to the properties of acid chlorides. Moreover, this reaction is not as selective for amines, as alcohols are also derivatized. However, the resulting sulfonates are known—think of the use of tosylates for S_N2 reactions—to be relatively unstable and are rapidly cleaved again in aqueous media.

Occasionally, the reaction with aromatic aldehydes to form relatively stable imines, so-called Schiff bases, can also be of interest. This derivatization is reversible and can be cleaved again under acidic conditions (Fig. 7.2d).

7.2 Derivatization of Thiol Functions (e.g., Cysteine)

7.2.1 Preparation of Stable Thioethers

The chemistry of the thiol group is quite unique, and therefore thiols can also be derivatized very specifically. Thiols are characterized by high nucleophilicity but low basicity (Sect. 1.1). In contrast to amines, they thus react as nucleophiles even under non-basic conditions. This leads to an intermediate question: Do they therefore also react with *N*-hydroxysuccinimide esters (Sect. 7.1.1)?—Yes, in principle they do, but the resulting thioesters are themselves activated esters and then react with amino groups to ultimately form the amide.

This high nucleophilicity is the basis for the first two methods shown in Fig. 7.3. Nucleophilic addition (Michael addition) to the electron-deficient double bond in maleimide (Fig. 7.3a) under neutral pH conditions is truly highly specific for thiols. S_N2 reactions (b) with alkyl iodides or halides are also used for derivatization.

Recently, the formation of thioethers by thiol addition to specially activated C≡C triple bonds has become widespread, especially in relation to ADCs (*antibody-drug conjugates*), i.e., antibodies linked to a drug (usually a toxin) via a linker and thus used in

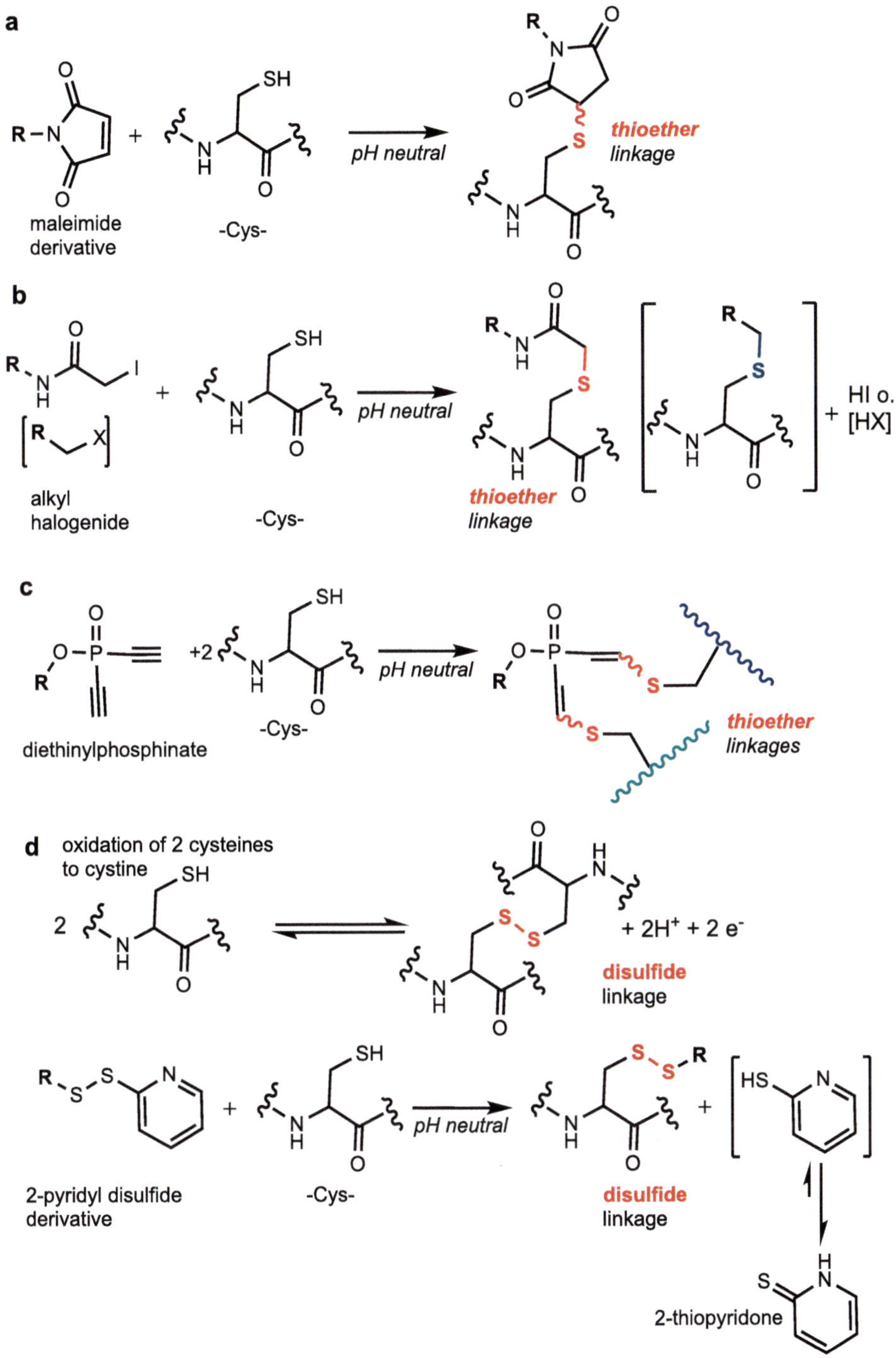

Fig. 7.3 Methods for derivatizing cysteine and other thiols

localized chemotherapy. In this context, the stability of the linkage is particularly important. Thioethers derived from maleimide precursors have proven not to be absolutely stable under physiological conditions. Therefore, alternative methods have been sought [2]. In example (c) in Fig. 7.3 a reagent with two terminal triple bonds is used. This allows either consecutive linkage with two different thiols (two different Cys-containing peptides), or a reduced disulfide bridge (Fig. 7.4) can be rebridged with this reagent (*rebridging*) and simultaneously labeled with the **R** group [4].

7.2.2 Reversible Linkage as Disulfides

The fourth reaction shown in Fig. 7.3d is based on another property. Thiols can be very easily oxidized to disulfides; the analogous reaction of alcohols to peroxides, on the other hand, requires very strong oxidizing agents. As is well known, many cysteines in proteins are present as disulfides. This is a key factor for the tertiary and quaternary structure of proteins (Sect. 2.1). For this oxidation, atmospheric oxygen is sufficient; a dilute cysteine solution oxidizes within a few hours if the solution is not degassed or saturated with argon. Disulfides form or are cleaved depending on whether an excess of another thiol or disulfide is added. Thus, cystines—the covalently linked cysteines—can be cleaved by an excess of β-mercaptoethanol. Disulfides with a 2-sulfhydryl-pyridine residue have a particularly interesting property. They are stronger oxidizing agents than other disulfides, and thus the reaction to the disulfide derivative of cysteine can be driven even with stoichiometric amounts. The reason is simple: The resulting "2-pyridinethiol" tautomerizes *in situ* to the much more stable 2-thiopyridone, and therefore this reaction is strongly favored. Incidentally, this property is also exploited in a method for determining free thiol groups in proteins (Fig. 7.4) [3].

Reduction of Disulfides and Determination of Free Thiols

As already mentioned, disulfide bridges can be opened by adding an excess of a thiol such as β-mercaptoethanol. Smaller excesses are required when using dithiothreitol (DTT), because the reaction is entropically favored. In addition, this reagent is less toxic and less odorous. However, a specific water-soluble phosphine reagent is also available, tris(2-carboxyethyl)phosphine hydrochloride (TCEP-HCl) (Fig. 7.4a, b).

One of the methods for determining whether and how many cysteines are present as thiols, as already mentioned, uses the formation of 2-thiopyridone, which also has an intense UV absorption in a wavelength range where few compounds absorb (λ approx. 340 nm), see Fig. 7.4c [3].

As we have seen, there are a number of highly specific conjugation methods for the amino acid side chains of lysine and cysteine. Nevertheless, attempts have been made to develop specific labeling methods for other amino acids as well. For the side chain of tyrosine, the high tendency for aromatic electrophilic substitution can be exploited (tryptophan may also react). Accordingly, there are some methods in use [5]. For the modification of other amino acids such as His, Arg, Asp, Glu, etc., there are already some approaches, but these have not yet been perfected [6].

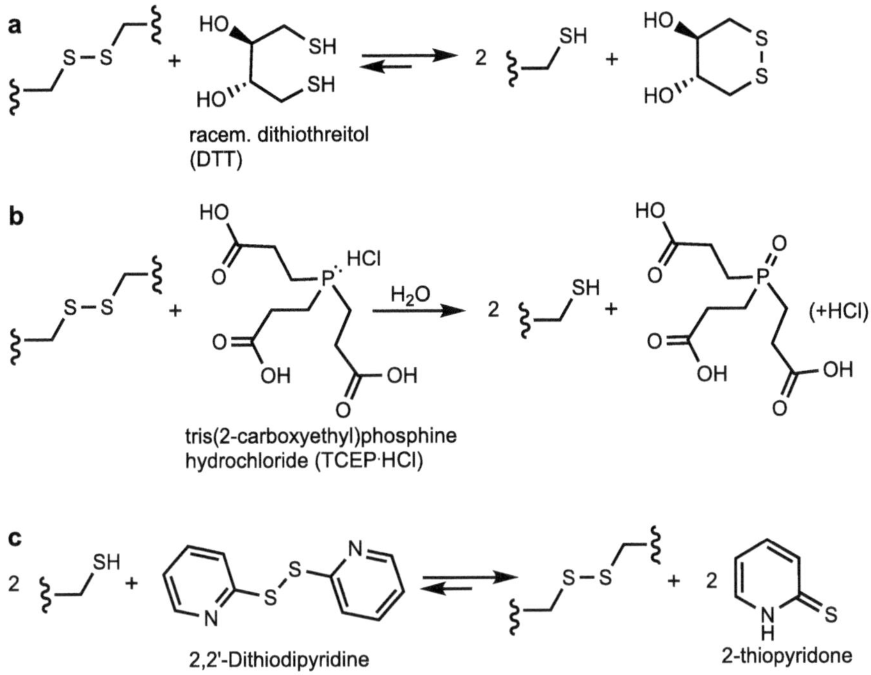

Fig. 7.4 **a**, **b** Methods for reduction of disulfide bridges, and **c** analytical determination of free thiols. (After Grassetti [3])

7.3 Derivatization of Alcohols and Carbonyl Compounds

7.3.1 Labeling of Alcohols—Directly or After Transformation

On the one hand, there are no reactions for alcohols as selective as for other functional groups; on the other hand, one may wish to label only a specific type of hydroxyl compound. Here are just two examples (Fig. 7.5).

Alcohols can be directly converted into stable urethanes using isocyanates carrying the labeling group. Since an isocyanate itself is not very stable, it is first thermally generated from the acyl azide in the kit offered by Molecular Probes® (reaction known as the Curtius rearrangement). The fluorescein molecule is initially blocked by acetylation and is only converted into the fluorophore at the end by hydroxylamine.

N-terminal serines, on the other hand, can be converted into aldehydes by periodate cleavage, which can then be more easily and specifically derivatized (Sect. 7.3.2).

Periodate cleavage of vicinal diols can also be used to introduce an aldehyde function at sialic acid residues in oligosaccharides (Sect. 7.9.1).

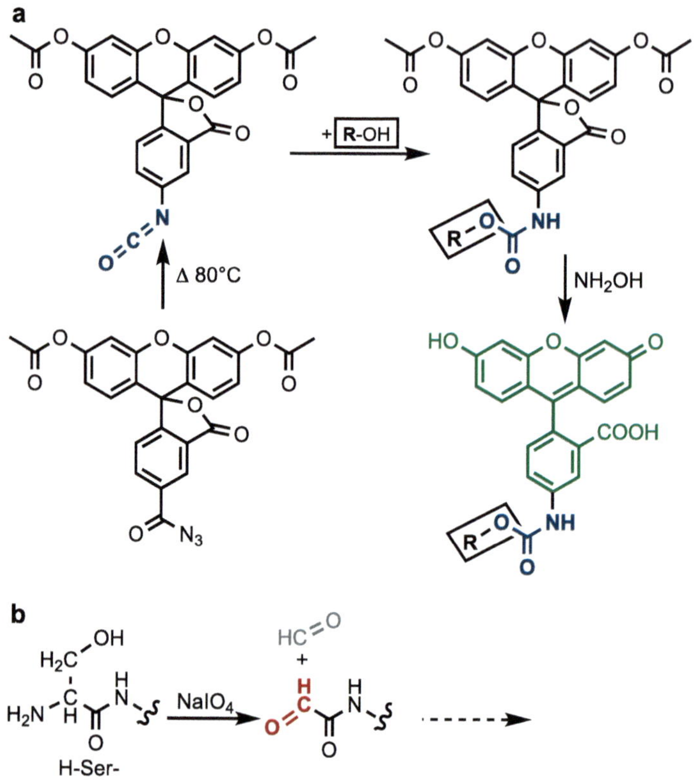

Fig. 7.5 Labeling of alcohols: **a** general method as urethane derivative; **b** conversion of an N-terminal serine into an aldehyde

7.3.2 Labeling of Aldehydes and Ketones

Ketones and aldehydes can be specifically converted to hydrazones with hydrazine groups. For this, hydrazide, semicarbazide, and carbohydrazide derivatives are mainly used. A reagent that is very well suited for sensitive detection of aldehydes and ketones is *N*-methyl-4-hydrazino-7-nitrobenzofurazan (NBD-methylhydrazine)—the corresponding hydrazones can be detected by fluorescence (excitation $\lambda = 470$ nm/emission $\lambda = 560$ nm) (Fig. 7.6a).

Another very clear-cut derivatization is the reaction of aldehydes or ketones with hydroxylamine derivatives (R–O–NH_2) to form oximes (Fig. 7.6b). For this reason, attempts are sometimes made to couple keto acids to the ε-amino group of lysines, in order to then have a specific site for selective conjugation with marker molecules containing hydroxylamine residues [7]. We will encounter examples of the application of this oxime conjugation later (Sect. 7.9.2 and 10.1).

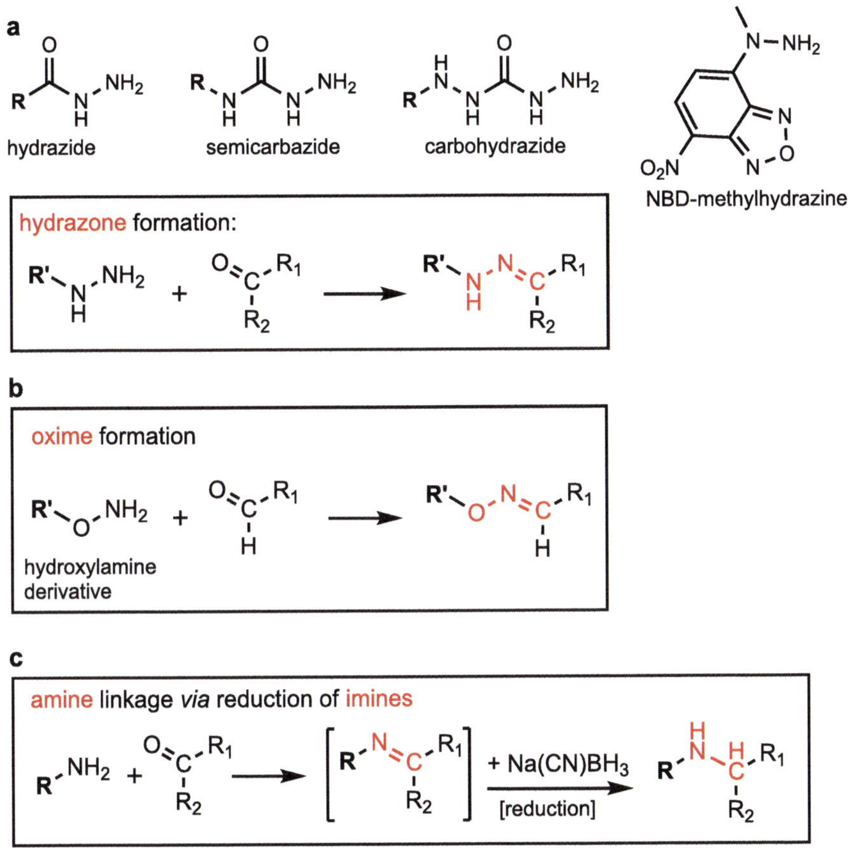

Fig. 7.6 Labeling methods for carbonyl compounds as used in commercial kits: **a** hydrazone formation; **b** oxime formation; **c** imine formation followed by reduction to the amine

Reaction of carbonyl compounds with primary amines usually yields unstable imines, which can be transformed into the more stable secondary amines by reduction with complex hydrides (Fig. 7.6c).

7.4 Selective Coupling of Azides with Terminal Alkynes ("Click" Reaction)

Sharpless et al. [8] defined "click" chemistry as organic reactions driven by a strong driving force (usually enthalpy or at least entropy gain) that lead to easily isolable products in high yields. They were intended primarily for the synthesis of many lead structures for pharmaceutical development. Among several other examples (mostly involving alkenes and strained three-membered rings), the 1,3-dipolar cycloaddition optimized by Sharpless best fulfills the criteria of click chemistry. It also proceeds well in aqueous media and is stereospecific—when using Cu^{+} catalysts, it is also practically regiospecific (Fig. 7.7a) and links an alkyne and an azide to form a stable 1,2,3-triazole derivative.

Since neither azides nor alkynes are common in nature, this reaction fulfills the criteria of a "bioorthogonal" reaction and has therefore quickly become established in the field of labeling. This allows specific conjugations to be carried out even in the cellular environment, provided that suitable *precursors* are introduced into the living cell or generated there in advance.

The examples in Fig. 7.7b–d show that this reaction has already been fully incorporated into the product lines of relevant companies and how versatile its applications are.

Fig. 7.8 shows an example from Ref. [1]: Whether cells are about to divide can be determined by their high DNA synthesis activity. If a thymine analog that can be detected later is introduced, it is partially incorporated into the growing DNA strands. For the detection of 5-ethynyl-2'-deoxyuridine (EdU), it is sufficient to fix the cells with formaldehyde and permeabilize them with a detergent so that the fluorophore with the azido group (e.g., Alexa Fluor 488, Fig. 7.7) can enter the cells and label the EdU-containing DNA. In the case of 5-bromo-2'-deoxyuridine (BrdU), which has been used for this assay for some time, the DNA must be completely denatured to make the BrdU units accessible to the corresponding antibody used to detect this thymine analog. The proportion of dividing cells can be determined with a flow cytometer, which measures the proportion of labeled DNA relative to total DNA.

The application of this and other click reactions for drug development is also discussed from the perspective of the pharmaceutical industry in the review [9].

7.5 Conjugate Formation via Staudinger Reaction

The so-called Staudinger reaction from 1919 has experienced a renaissance in recent years in the field of bioconjugation. Review [10] from 2020 provides a very comprehensive overview of this topic. Figure 7.9 illustrates the principle of the Staudinger reaction and the variants used for bioconjugation. The reaction offers the same advantages as the 1,3-dipolar cycloaddition discussed above: it is bioorthogonal and also fast. In addition, it can form a native amide linkage without leaving any trace (*traceless*). Such reactions are also of interest for the semisynthesis of proteins, as we will see in Chap. 8. More specialized variants are discussed in Ref. [11].

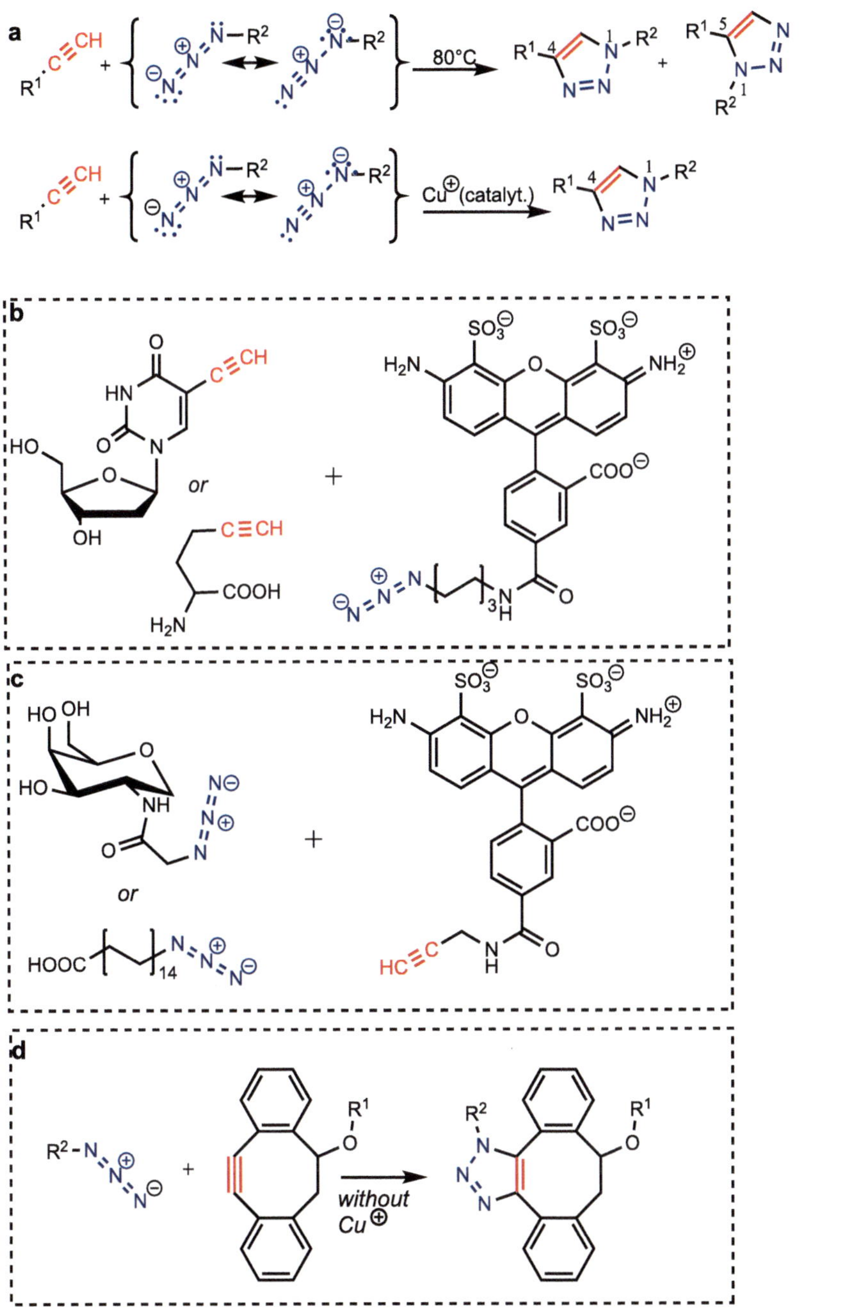

Fig. 7.7 The 1,3-dipolar cycloaddition reaction (so-called click reaction) for the preparation of conjugates: **a** general reaction without and with catalyst; **b** examples of conjugation of nucleobases and amino acids as the alkyne component with a fluorophore (Alexa Fluor 488) with an azido group under Cu^+ catalysis; **c** examples of conjugation of carbohydrates or lipids as the azide component with Alexa Fluor 488 with an alkyne group under Cu^+ catalysis; **d** example of the use of a special strained alkyne for application even in the absence of Cu^+ ions. (Examples from Ref. [1])

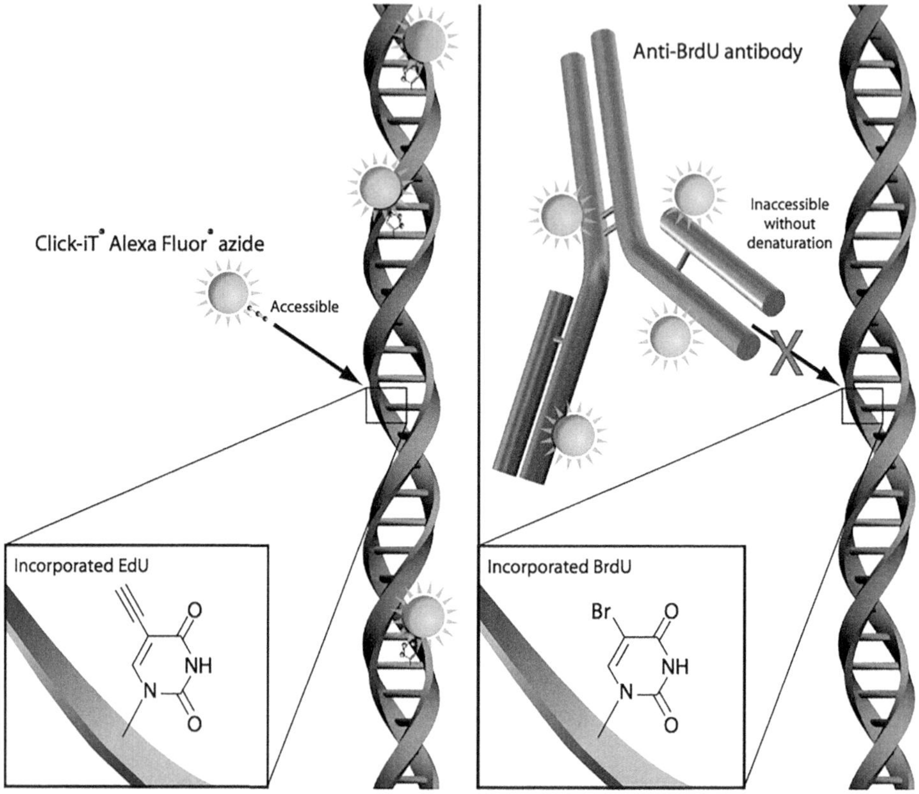

Fig. 7.8 Comparison between 5-ethynyl-2'-deoxyuridine (EdU) and 5-bromo-2'-deoxyuridine (BrdU) for determining cell proliferation activity (*cell proliferation assays*). (From Click-iT™ EdU Flow Cytometry Assay Kits—Molecular Probes®)

7.6 Conjugation via Activation by UV or Visible Light

7.6.1 Photoaffinity Labeling

The goal is to identify interaction partners of a specific molecule (protein, ligand, substrate) with a protein (carbohydrate, DNA) *in vivo*. For this purpose, the molecule is conjugated with a group that can be activated by irradiation and introduced into the relevant living system. Upon irradiation, a covalent bond should form instantaneously with the immediately adjacent molecules. The corresponding derivatives are then isolated and analyzed.

Several functionalities have proven suitable for this purpose, which are now presented (Fig. 7.10). Later, we will also discuss concrete applications. It is also essential that

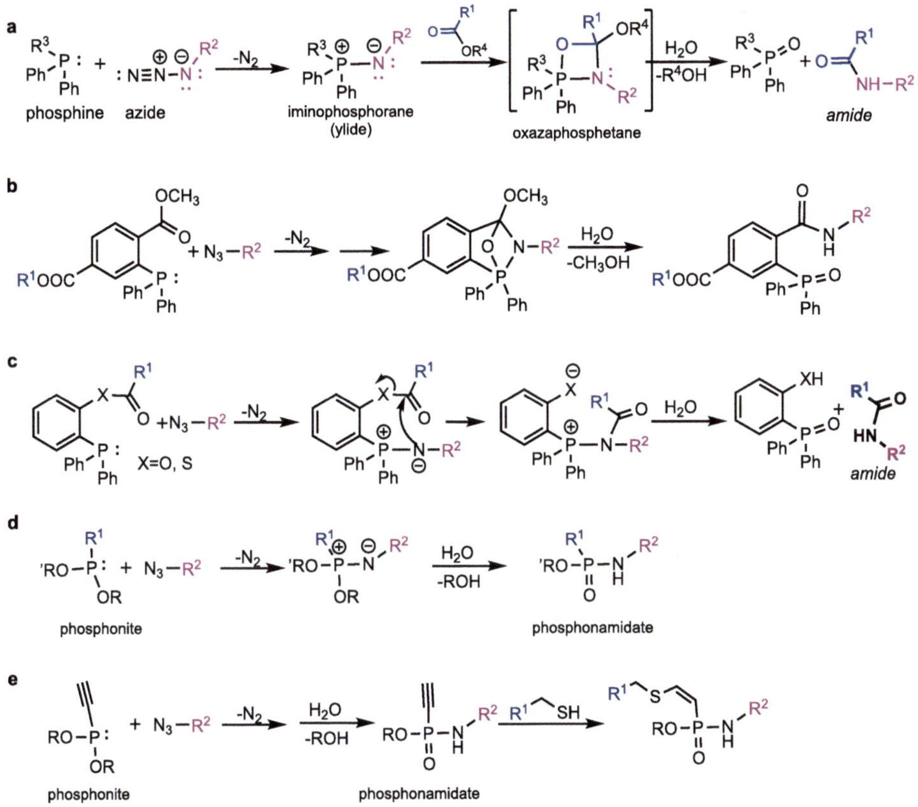

Fig. 7.9 The Staudinger reaction in its variants as a method of bioconjugation. **a** the principle; **b** the so-called "*non-traceless*" variant for conjugation, e.g., of a label R^1 with an azide-modified biomolecule R^2; **c** the traceless (*traceless*) Staudinger ligation, also for linking two peptides via a native peptide bond (after Ref. [10]); **d** the Staudinger-phosphonite reaction and **e** a variant thereof that also allows specific conjugation with thiols [11]

activation occurs by irradiation in the visible light range or at least near-visible UV light ($\lambda > 300$ nm), so that the living system is not destroyed.

As already became clear in the previous section, azides are stable in aqueous media at room temperature, despite their inherent tendency to release molecular nitrogen. When activated by irradiation, N_2 is immediately evolved and a nitrene is formed. Usually, the diradical triplet state is more stable (Fig. 7.10a), but with certain substituents, the nitrene can also react as a singlet (with a shorter half-life), e.g., via insertion into C–H bonds (b).

Coordination-unsaturated C atoms (carbenes) are also suitable for photoaffinity labeling. They can be photochemically generated from the remarkably stable diazirines (λ_{max} approx. 380 nm). Usually, especially with non-aromatic residues R, the singlet carbenes are more stable (Fig. 7.10c), which then react rapidly and selectively with C–H or O–H

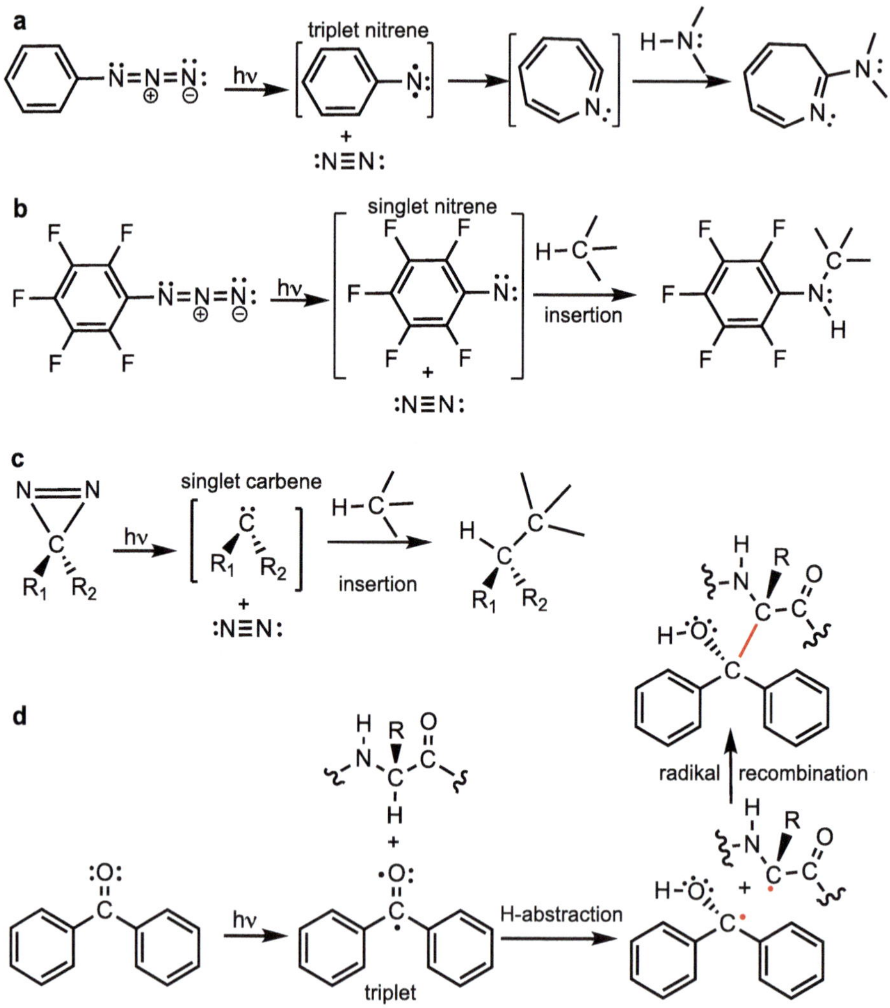

Fig. 7.10 Reactions of important functionalities used for photoaffinity labeling: **a** aromatic azides, which react as triplet nitrenes; **b** in the case of the fluorinated compound, the singlet nitrene is probably more stable; **c** diazirines usually react as singlet carbenes via C–H or O–H insertion; and **d** benzophenone exists in the excited state as a triplet diradical, which can react via H-abstraction followed by radical recombination

bonds via insertion. If C=C double bonds are present nearby, addition to form a cyclopropane may also occur.

The benzophenone chromophore (d) is excited by UV light at about $\lambda = 350–360$ nm. This leads to *intersystem crossing* (ISC) to the triplet state, which has diradical character.

This can then react, for example, by abstraction of an α-H atom from amino acid residues, and the two radicals usually stabilize by recombination. However, there are also other reaction pathways that do not result in a covalent bond between neighboring molecules and the benzophenone group [12].

Photoaffinity Crosslinks without Use of Conjugates
Another approach to elucidating protein-protein contacts by irradiation is described in Ref. [13]. A Ru(II) complex with three bipyridyl ligands is irradiated with visible light in the presence of ammonium persulfate. This generates radicals at spatially adjacent tyrosine residues, which can form covalent bonds with tyrosine or cysteine residues of a second protein in close contact. The added reagent here acts only as a kind of catalyst for linking two neighboring molecules (or for intramolecular crosslinking).

7.6.2 Photochemical Conjugation to Solid Surfaces

For the design of *microarrays*, biomolecules must be bound precisely to solid surfaces. Devices that work with laser light are particularly suitable for this purpose. Accordingly, functionalities similar to those used for photoaffinity labeling are also employed for this purpose. An article comparing the different types and various surfaces [14] concludes that for plastic surfaces such as cycloolefin copolymer, polystyrene, or poly(methyl methacrylate), diazirines or benzophenones are most suitable.

Because of its toxicity, anthraquinone is probably not suitable for use in living cells, but it can certainly be used for other conjugations, such as binding to solid surfaces, as a photoactivatable anchor. It reacts very similarly to benzophenone, also via abstraction of an H radical and recombination with the now-formed radical of the partner (Fig. 7.11). For an application, see Fig. 7.14.

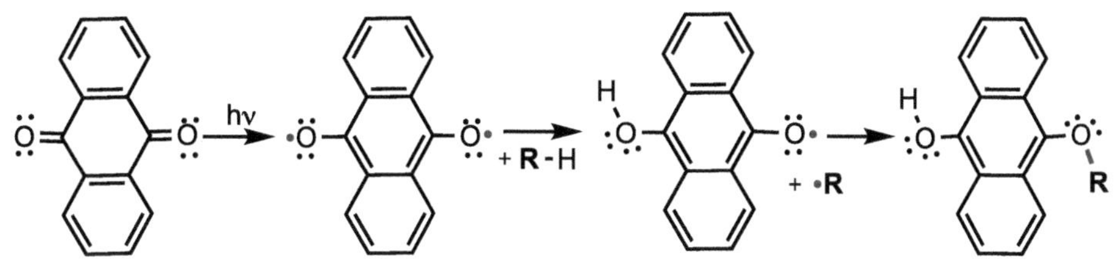

Fig. 7.11 Photochemically activatable conjugation of anthraquinone derivatives

7.7 Bifunctional Conjugation Reagents

7.7.1 Synthetic Applications

Based on the methods already discussed, reagents have also been designed that are suitable for sequentially linking two different molecules (heterobifunctional linkers).

Figure 7.12 shows some such reagents for linking two molecules with different functionalities. Maleoyl-β-alanine-*N*-hydroxysuccinimide ester (Mal>β-Ala-OSu) and the analogue with five instead of two CH_2 units as linker (Mal>ε-Capr-OSu) as well as the more conformationally restricted SMCC (Molecular Probes®) are first linked to an amine component and in a second step to a thiol component. The reagent SPDP (Molecular Probes®) first binds an amine as an amide (again via the *N*-hydroxysuccinimide ester) and can then form a disulfide bond with a thiol (which can again be cleaved reductively)

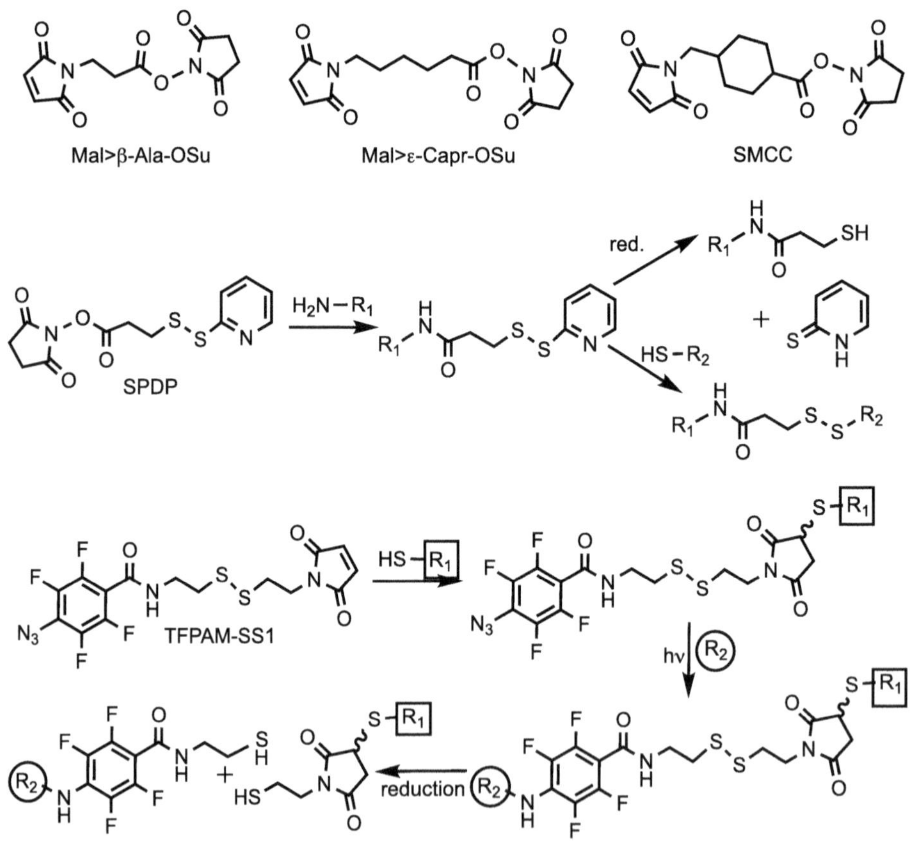

Fig. 7.12 Bifunctional conjugation reagents based on the reactions already discussed in this chapter

or, after reduction, itself become a thiol component. In both cases, the reaction can be monitored by measuring (UV-Vis absorption) the 2-thiopyridone formed. TFPAM-SS1 (Molecular Probes®) as an example is first bound to a thiol-containing component, whose interaction partner is then identified by photoaffinity labeling (via the azide). After isolation of the adduct, the two components could be reductively cleaved again, which should facilitate analysis.

Polyfunctional Linker Molecules

The development of new multifunctional linkers has become a popular research area. All sorts of combinations of the functional groups we have already discussed are conceivable here—as long as a feasible short synthesis can be developed. The examples shown in Fig. 7.13 are taken from

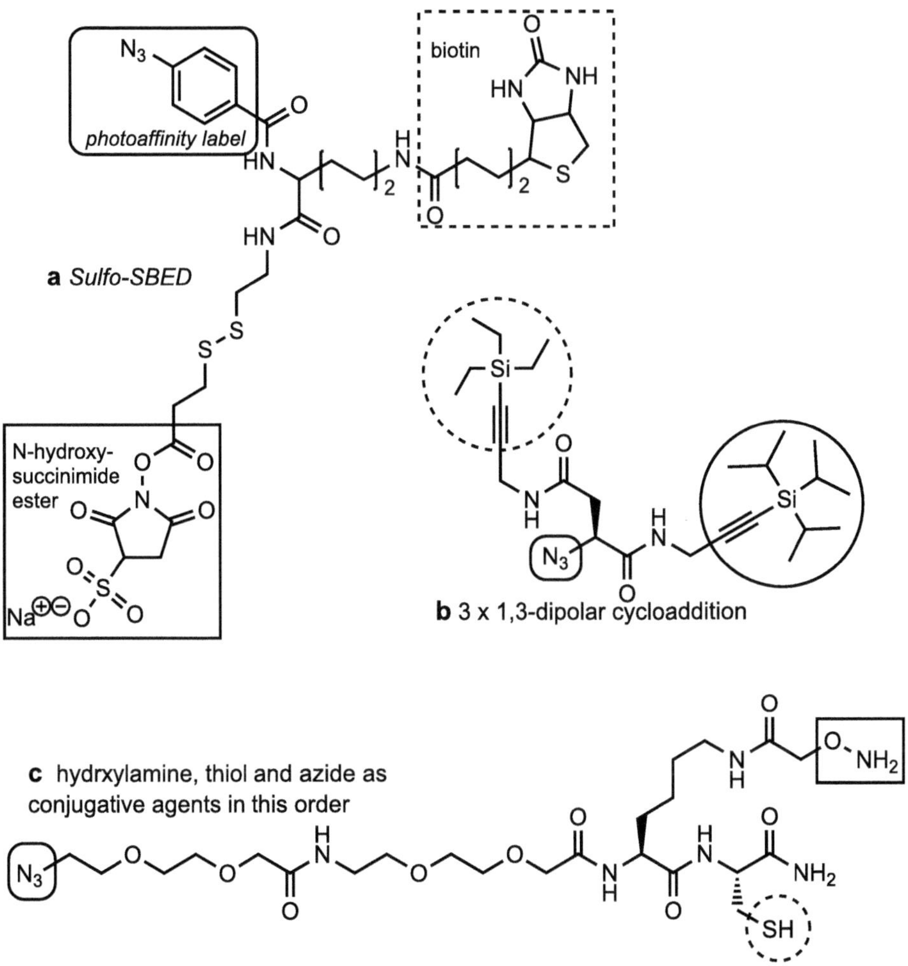

Fig. 7.13 Three examples of polyfunctional linkers: **a** a commercially available one with a biotin binding site for streptavidin conjugates; **b** after Ref. [16]; **c** after Ref. [17]

the literature [15–17]. Sulfo-SBED (Molecular Probes®) (**a**) is a commercial product that carries a biotin as an additional group and can therefore be detected with all kinds of streptavidin reporter conjugates (Sect. 7.8).

The molecule in Fig. 7.13b [16] can undergo *three* Cu-activated azide-alkyne cycloadditions in sequence: After the azide has reacted, the triethylsilyl protecting group can be removed with silver nitrate and this alkyne can react. Finally, the triisopropylsilyl protecting group is removed with tetrabutylammonium fluoride (TBAF), and the third click reaction occurs.

The multifunctional linker in Fig. 7.13c requires no protecting groups; the hydroxylamine-aldehyde, thiol-alkyl iodide, and azide-alkyne reactions are all orthogonal to each other.

Bifunctional conjugation molecules are especially important for binding test molecules to solid surfaces for *microarrays*. Ref. [15] describes the synthesis and use of such a compound for the placement of oligonucleotides on glass surfaces (Fig. 7.14). Immobilization on glass is achieved by the photochemical reaction of anthraquinone (Sect. 7.6.2), while binding of the oligonucleotide occurs via an S_N2 reaction of an S-nucleophile on the oligonucleotide with an iodoacetamide group (Sect. 7.2.1). The order of the two conjugation steps is freely selectable.

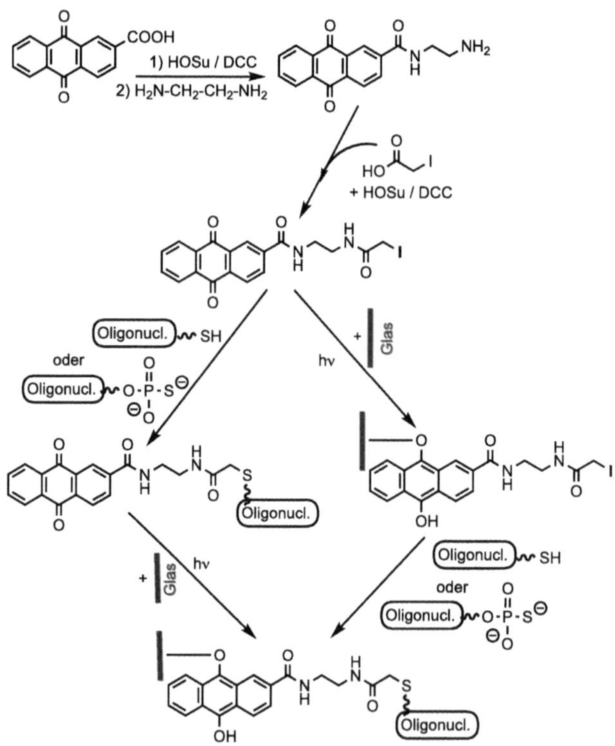

Fig. 7.14 Synthesis and application of a bifunctional reagent for binding oligonucleotides (oligonuc.) to glass slides [15]

7.7.2 Analytical Applications (in MS)

In the rapidly developing field of proteome analytics, efforts have been underway for some time to develop methods that allow documentation of protein-protein interactions, ideally with high throughput—that is, a high hit rate with minimal effort. MS methods are best suited for this. With common bifunctional linker molecules, this has not really proven effective, so new types of reagents have been designed that can be further analyzed by mass spectrometry.—What does this mean? Molecules were sought that, during MS/MS experiments, can be selectively cleaved by collision-induced dissociation (CID). Two articles on this topic have appeared in *Nature Methods* [19], focusing mainly on data analysis. For feasible evaluation, conjugates are essential that can be specifically cleaved in the MS instrument. This reduces the number of combinations from n^2 to $2n$. Examples of such specially developed reagents can be found in [20–24].

Figure 7.15 shows the method for detecting different linkages using such a reagent [20]. The two *N*-hydroxysuccinimide ester units can link lysines together. If this occurs between two proteins α and β, case Fig. 7.15a results. In the MS/MS experiment, CID cleaves almost exclusively at the two breakpoints at the sulfoxide. Four fragment ions are formed, which are now characteristic for the two proteins. In the case of no linkage (b) or an internal linkage within protein α (c), different patterns are found accordingly.

Ref. [21] describes another bifunctional *N*-hydroxysuccinimide ester (cyanurbiotindipropionylsuccinimide, CBDPS) (Fig. 7.16). It also contains a biotin unit to allow affinity chromatographic enrichment of the proteins with linker in advance. The reagent is also used as a 1:1 mixture with either eight ^{1}H or eight ^{2}H (=D) atoms. After CID in the MS instrument, the cleaved products each give two identical peaks with a mass difference of 4, while uncleaved products have a mass difference of 8. Another example uses the 1,3-dipolar cycloaddition (click chemistry) for enrichment of the proteins linked with the linker [25].

7.8 Biotin–Avidin (or –Streptavidin), an inseparable pair

We have already encountered biotin derivatives at least twice in connection with labeling strategies. The reason lies in its extremely strong binding to the glycoprotein avidin, which was first utilized around 1970. Avidin (mass approx. 66,000 daltons) has an isoelectric point (pI) of about 10.5, meaning it possesses many positively charged groups in its side chains and therefore, also due to its carbohydrate content, sometimes exhibits nonspecific interactions with negatively charged cell surfaces. The bacterial analog of the glycoprotein avidin was found in *Streptomyces avidinii*. Streptavidin contains no carbohydrate and has a pI of about 5. The widely used product from Thermo Scientific Pierce™ is a recombinant protein (mass = 53,000 daltons, approximately neutral pI). This streptavidin is a tetramer, and each subunit (Fig. 7.17) possesses a biotin binding site [26].

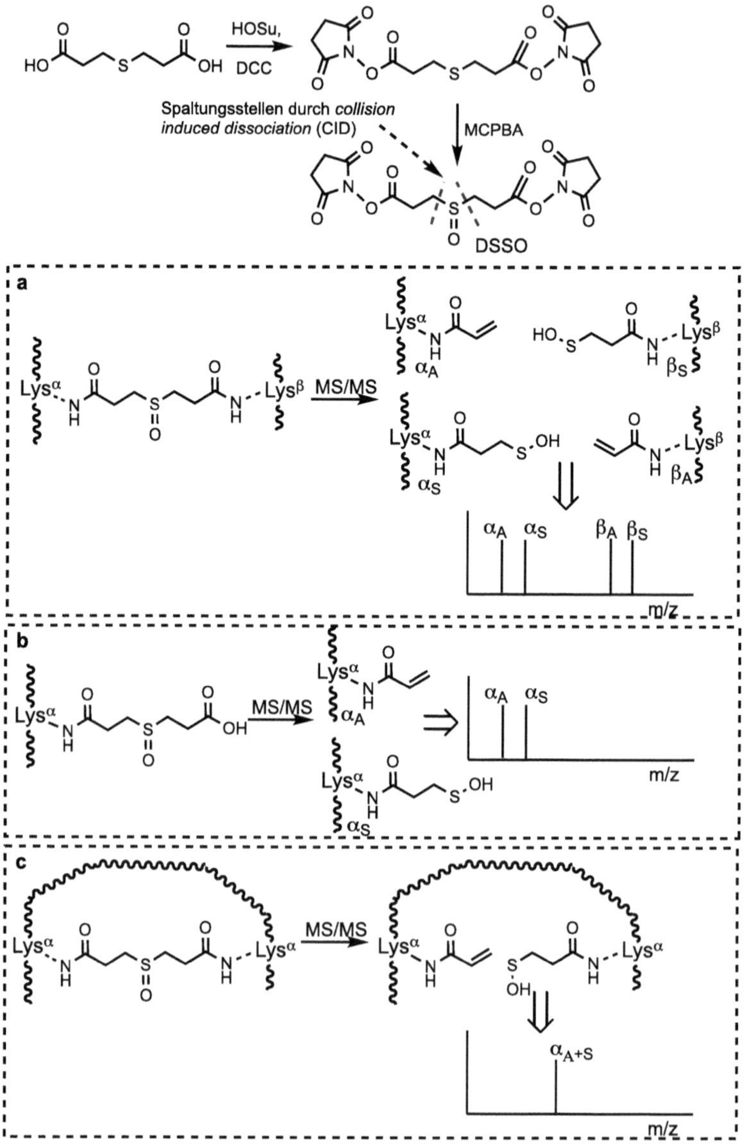

Fig. 7.15 Synthesis and use of the disuccinimidyl sulfoxide (DSSO) reagent for MS detection of protein interactions; **a** two proteins α and β have been linked; **b** no linkage; **c** intramolecular linkage within protein α [MCPBA…*m*-chloroperbenzoic acid]. (After [20])

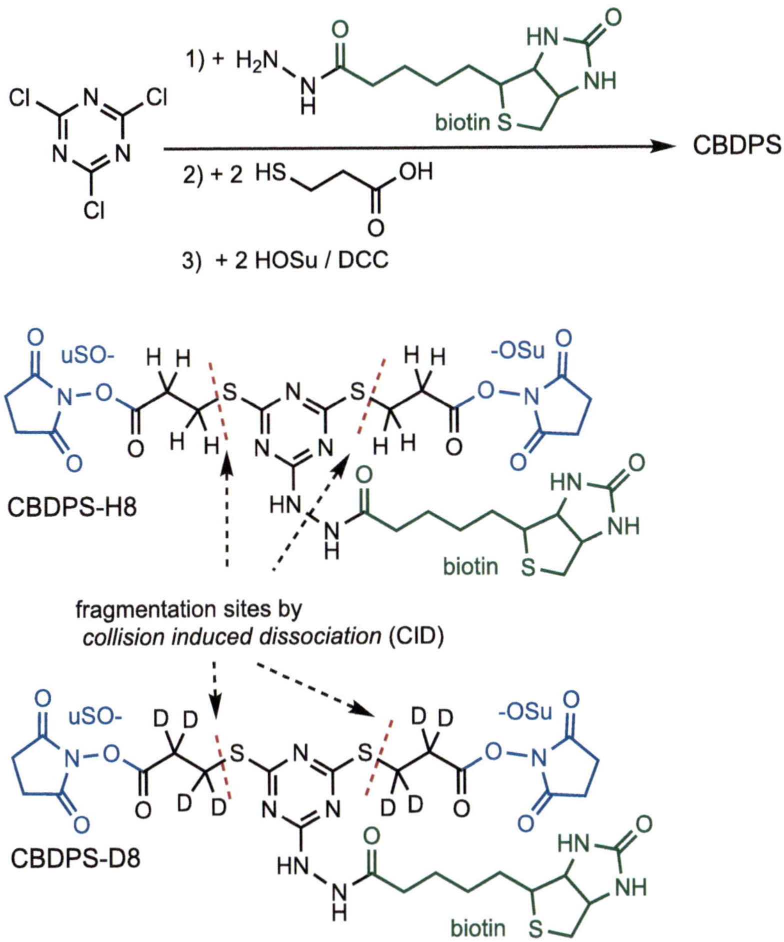

Fig. 7.16 Cyanurbiotindipropionylsuccinimide (CBDPS), a reagent for bridging spatially adjacent amino groups, for MS analysis of protein-protein interactions. (After [21])

Due to the extremely small dissociation constant of 10^{-15} M, one can assume that every biotin residue will “find” and bind its streptavidin. Thus, for biotin-labeled molecules, an entire arsenal of streptavidin conjugates can be employed:

- Conjugates with enzymes such as *horseradish peroxidase*, alkaline phosphatase, galactosidase
- Conjugates with fluorophores
- Conjugates with “NANOGOLD” particles or colloidal gold, especially for immunoblotting, light and electron microscopy

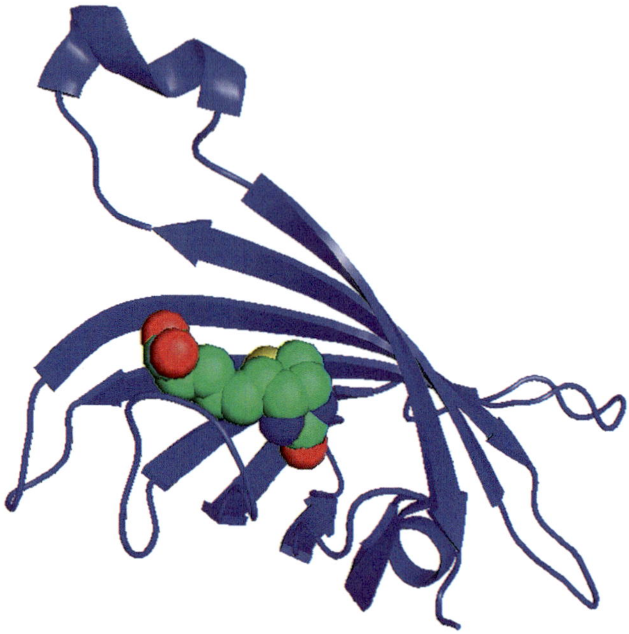

Fig. 7.17 Streptavidin–biotin complex. (According to Protein Data Bank, https://commons.wikimedia.org/w/index.php?curid=10383192 (Accessed on 01/10/2026))

- Conjugates with Dynabeads®: superparamagnetic particles with a monolayer of streptavidin for easy magnetic isolation of labeled proteins
- Conjugates with agarose for affinity chromatography
- Conjugates with *quantum dots* (*Qdot-nanocrystals*) for specialized fluorescence detection
- Conjugates with fluorescent polystyrene particles (*microspheres*)

Even if certain methods were not considered at the time of preparing the covalent biotin conjugate, one can later easily perform these analyses using the appropriate streptavidin derivative. However, one must consider the steric requirements of the streptavidin conjugate, which can naturally affect the properties of the substance under investigation. An example that exploits the increased mass of the conjugate with streptavidin will be discussed in Sect. 9.1.

There are also numerous efforts to modify the biotin–streptavidin (or –avidin) interaction, through variations on the small biotin molecule or mutations in the protein partner. As an example of rational design on biotin, see Ref. [27].

7.9 Examples and Summary

7.9.1 Labeling of Oligosaccharides at Neuraminic Acid (Sialic Acid) [28]

By mild periodate oxidation [recall: sodium periodate generally cleaves –CHOH–CHOH– groups to form the corresponding aldehydes], only the side chain of *N*-acetylneuraminic acid in oligosaccharides is attacked and degraded. As a result, in glycoproteins, for example, the usually terminal neuraminic acids are labeled. The specific reaction of aldehydes with hydroxylamine derivatives (NH_2O–R) to form oximes was already discussed in Sect. 7.3.2. In the publication mentioned here, a biotin modified with –O–NH_2 is specifically bound to the neuraminic acids in this way (Fig. 7.18). This procedure is an alternative to metabolic labeling (*metabolic labeling*), in which a benzaldehyde derivative of *N*-acetylneuraminic acid (BA-NeuAc) must be taken up by the cell. After the oligosaccharides synthesized in the cell are brought back to the surface, they can then be labeled with the same biotin-hydroxylamine reagent. However, incorporation of BA-NeuAc often works poorly, so periodate cleavage followed by oxime formation—which, according to this publication, proceeds much better in the presence of aniline—promises to be more successful. Through the highly specific and practically irreversible binding of the protein streptavidin to biotin, all kinds of labeled streptavidins (Sect. 7.8) can ultimately be used.

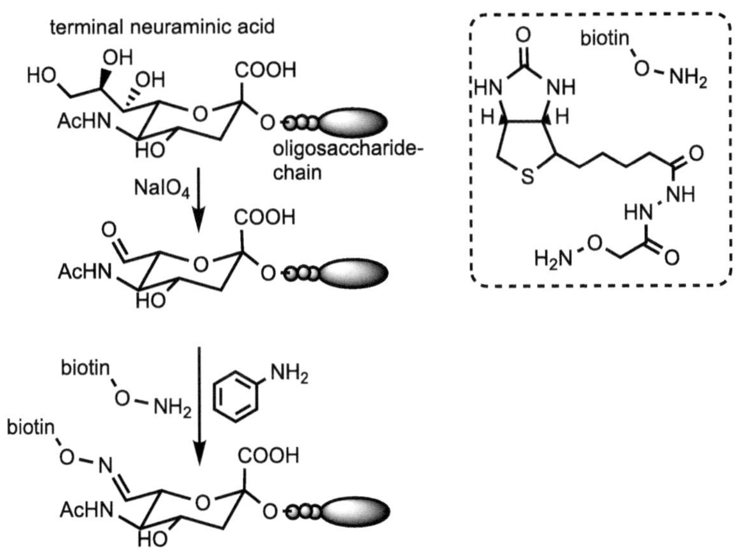

Fig. 7.18 Labeling method for terminal neuraminic acid on the cell surface. (According to Ref. [28])

7.9.2 Enzymatic Introduction of the Amino Acid Formylglycine (FGly) and Its Possibilities for Conjugate Formation [29]

Formylglycine is a non-coded amino acid that is introduced post-translationally from cysteine or serine by enzymatic means and is catalytically active in sulfatases. Two enzymes with different mechanisms have been isolated that can also perform this transformation *in vitro*: the oxygen-dependent *FGly-generating enzyme* (FGE) from eukaryotes and AtsB, an iron-sulfur protein that carries out this conversion from cysteine or serine independently of oxygen and is found only in prokaryotes. Both enzymes act only on a specific amino acid sequence following the cysteine residue. In Ref. [29], two orthogonal sequences were used: In the case of –Cys–Thr–Ala–Gly–Arg–, FGly is introduced by AtsB, and for the sequence –Cys–Thr–Pro–Ser–Arg–, the enzyme FGE is active.

But why is FGly of interest for bioconjugation methods? The aldehyde group does not otherwise occur freely in proteins and is therefore suitable for bioorthogonal linkages. We have already encountered some in Sects. 7.3.2 and 7.9.1. In the cited works, another very specific method is used: the hydrazino-*iso*-Pictet-Spengler (HIPS) reaction, which essentially also relies on a type of hydrazone formation, followed by an intramolecular electrophilic attack on the indole system (Fig. 7.19). By using the two differently acting enzymes, each followed by conjugation via HIPS two different payloads could be introduced into a protein. This methodology is of interest with regard to the production of ADCs (*antibody-drug conjugates*).

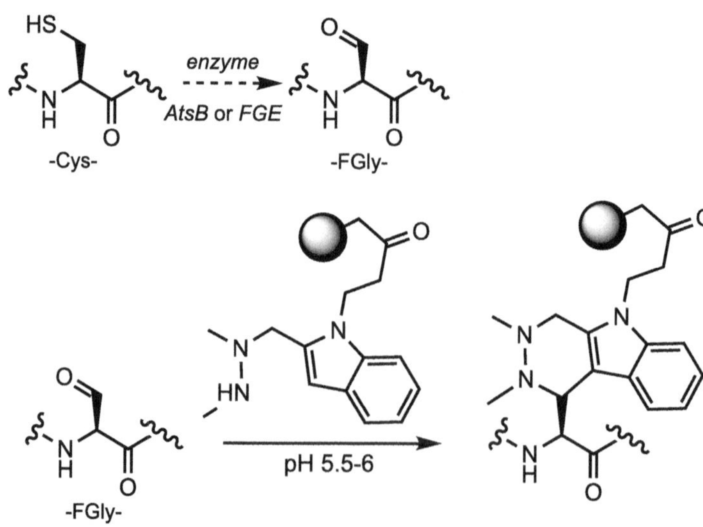

Fig. 7.19 Introduction of the amino acid formylglycine (FGly) and its derivatization by the hydrazino-iso-Pictet-Spengler (HIPS) reaction. (According to [29])

7.9.3 Two examples of Conjugation of Biomolecules to Ligands for Transition Metal Ions

In Ref. [30], the synthesis of a ligand for europium(III) is described, which can be bound to amino groups of proteins (e.g., antibodies) via an isothiocyanate group (Fig. 7.20a). In this complex, however, the rare earth ion shows only weak fluorescence. For the corresponding fluorescence detection, the europium is then complexed by another ligand in a detergent-containing solution, and now strong fluorescence can be measured. The high sensitivity results from the pronounced Stokes shift, i.e., excitation and emission wavelengths differ significantly, and in addition, the emission of europium is much more delayed in time than that of typical fluorophores, and can be detected during the pauses in irradiation when using a pulsed laser, so that noise is particularly minimized (see also [31]).

Radioactive tracers are often used in medical diagnostics. The ^{99m}Tc isotope is particularly easy to handle, as it can be eluted in the hospital itself from a ^{99}Mo column (which has a longer half-life) [32]. One possible low-molecular ligand for Tc is 6-hydrazinonicotinic acid. Its introduction was proposed via the *N*-hydroxysuccinimide ester (Fig. 7.20b). In one study [33], we attempted to use this reagent to label the cyclic peptide octreotide. This worked very poorly. Closer investigation showed that this *N*-hydroxysuccinimide ester was inactivated by the influence of the Boc-protected 6-hydrazinonicotinic acid. Thus, the activity of an active ester also depends on the carboxylic acid residue—especially with aromatic carboxylic acids. This Tc ligand also has the disadvantage that it offers at most two coordination sites for Tc. Therefore, binary or, better yet, ternary systems must be used [32].

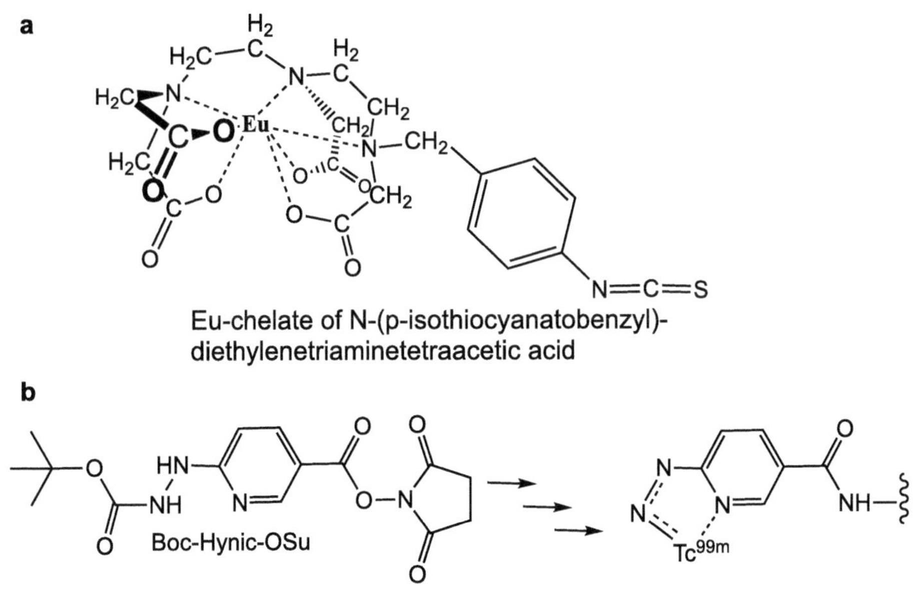

Fig. 7.20 Examples of conjugates for the introduction of transition metal ions

7.9.4 Labeling with Fluorocarbon Chains (*Fluorous Tags*)

We have already encountered this method in Sect. 4.2.3. When molecules are conjugated with C_nF_{2n+1} groups, they can subsequently be easily separated from all other substances by liquid-liquid extraction with fluorinated solvents or solid-liquid extraction with a fluorinated solid phase (FSPE). Ref. [34] shows that, after trypsin digestion of the entire proteome of cells, peptides with certain characteristics can be selectively derivatized with *fluorous tags* and then easily enriched by SPE for MS analysis. As an example, the peptide mixture was reduced (i.e., disulfide bridges were broken and free cysteine generated) and then reacted with *N*-[(3-perfluorooctyl)-propyl]iodoacetamide (see Sect. 7.2.1). In this way, all peptides with cysteine residues were captured. These could then be isolated by FSPE.

7.9.5 Summary of the Methods

Overview of the Methods Discussed

To conclude, the various conjugation methods discussed in this chapter are summarized again (Table 7.1).

Table 7.1 Summary of the most important methods described for the preparation of conjugates

Type of functional group	Active group of the reagent	Type of bond
Amino: $R–NH_2$	*N*-hydroxysuccinimide ester (–OSu)	Amide bond: –CO–NH–R
Amino: $R–NH_2$	Isothiocyanate (–N=C=S)	Thiourea: –NH–CS–NH–R
Amino: $R–NH_2$	Sulfonyl chloride ($–SO_2Cl$)	Sulfonamide: $–SO_2–NH–R$
Amino: $R–NH_2$	Aromatic aldehyde (Ar–CHO)	Imine: –Ph–CH=N–R
Thiol: R–SH	Maleimide (N*–CO-CH=CH–CO–N*)	Thioether: …–CH(S–R)–CH_2–…
Thiol: R–SH	Alkyl iodide ($–NH–CO–CH_2I$)	Thioether: $–NH–CO–CH_2–S–R$
Thiol: R–SH	Diethynylphosphinate R'O–P(=O)–$(C{\equiv}CH)_2$	Thioether: R'O–P(=O)–$(CH=C–S–R)_2$
Thiol: R–SH	Pyridyl disulfide (–S–S–Py)	Disulfide: –S–S–R (+ 2-TPy)
Alcohols: R–OH	Isocyanate: (–N=C=O)	Urethane: –NH–CO–O–R
H–Ser–	Periodate ($NaIO_4$)	Aldehyde: OHC–CO–
Aldehyde, ketone: RR'C=O	Various hydrazine derivatives ($–NH–NH_2$)	Various hydrazones: –NH–N=CRR'
Aldehyde, ketone: RR'C=O	1) Amine: $R–NH_2$, 2) $NaCNBH_3$	Amine: R–NH–CHRR'

(Continued)

Table 7.1 (Continued)

Type of functional group	Active group of the reagent	Type of bond
Aldehyde, ketone: RR'C=O	Hydroxylamine (-O–NH_2)	Oxime: –O–N=CRR'
Azide: R–N_3	Terminal alkyne (R'-C≡CH)	1,2,3-triazole derivative (TZ): R-TZ-R'
Terminal alkyne: R–C≡CH	Azide (R'–N_3)	1,2,3-triazole derivative (TZ): R'–TZ–R
Azide: R^1–N_3, + R_3P	Ester: R^2–COOR'	Amide: R^2–CO–NH–R^1 (+ R_3P=O)
Azide: R-N_3	hν, ··· H–C–	R–NH–C–
Diazirine: R–C$<N_2$	hν ··· H–C–	R–CH–C–
Benzophenone: Ph–CO–Ph	hν, ··· H–C–	Ph_2C(OH)–C–
Biotin: R—Biotin	Streptavidin conjugate	R—Biotin@Streptavidin conjugate

R = alkyl or aryl residues

7.10 Questions

Let's Test Your Understanding of this Chapter's Content

7.1 What requirements are placed on bioconjugation methods?

7.2 How are free amino groups generally conjugated? Give two examples plus reaction equations!

7.3 What is the advantage of cysteine as a target for conjugate formation?

7.4 What type of reaction is the reaction of thiols with maleimide derivatives?

7.5 Aldehydes form which conjugates with hydroxylamine derivatives?

7.6 Aldehydes are rare in biomolecules; how can they be generated?

7.7 How did K.B. Sharpless define click reactions?

7.8 Name three completely different applications of azides for conjugate formation!

7.9 Give examples of bifunctional conjugation reagents (with formulas) and explain the possible reactions!

7.10 Explain the purpose of bifunctional conjugation reagents in MS proteome analysis!

7.11 Why are thiols more easily oxidized by 2,2'-dithiodipyridine (Py–S–S–Py) than by other disulfides?

7.12 Why is biotin often used as a marker in conjugates?

References

1. Thermo Fisher Scientific, „The Molecular Probes Handbook". https://www.thermofisher.com/at/en/home/global/forms/mp-handbook-download-request-form-2014.html. Zugegriffen am 30.12.2025

2. You, J., Zhang, J., Wang, J., Jin, M.: Cysteine-based coupling: challenges and solutions. Bioconjugate. Chem. **32**(8), 1525–1534 (2021). https://doi.org/10.1021/acs.bioconjchem.1c00213; Ochtrop, P., Hackenberger, C.P.R.: Recent advances of thiol-selective bioconjugation reactions. Curr. Opinion. Chem. Biol. **58** 28–36 (2020). https://doi.org/10.1016/j.cbpa.2020.04.017
3. Grassetti, D.R., Murray Jr., J.F.: Determination of sulfhydryl groups with 2,2′- or 4,4′-dithiodipyridine. Arch. Biochem. Biophys. **119**(1), 41–49 (1967). https://doi.org/10.1016/0003-9861(67)90426-2
4. Stieger, C.E., Franz, L., Körlin, F., Hackenberger, C.P.R.: Diethynyl phosphinates for cysteine-selective protein labeling and disulfide rebridging. Angew. Chem. Int. Ed. **60**(28), 15359–15364 (2021). https://doi.org/10.1002/anie.202100683
5. Szijj, P.A., Kostadinova, K.A., Spears, R.J., Chudasama, V.: Tyrosine bioconjugation—an emergent alternative. Organic & Biomolecular Chemistry. **18**(44), 9018–9028 (2020). https://doi.org/10.1039/D0OB01912G
6. Kjærsgaard, N.L., Nielsen, T.B., Gothelf, K.V.: Chemical conjugation to less targeted proteinogenic amino acids. ChemBioChem. **23**(19), e202200245 (2022). https://doi.org/10.1002/cbic.202200245
7. Agten, S.M., Suylen, D., Ippel, H., …, Hackeng, T.M.: Chemoselective oxime reactions in proteins and peptides by using an optimized oxime strategy: the demise of levulinic acid. ChemBioChem. **14**(18), 2431–2434 (2013). https://doi.org/10.1002/cbic.201300598
8. Kolb, H.C., Finn, M.G., Sharpless, K.B.: Click-Chemie: diverse chemische Funktionalität mit einer Handvoll guter Reaktionen. Angew. Chem. **113** (11), 2056–2075 (2001). https://doi.org/10.1002/1521-3757(20010601)113:113.0.CO;2-W; Kolb, H.C., Finn, M.G., Sharpless, K.B.: Click chemistry: diverse chemical function from a few good reactions. Angew. Chem. Int. Ed. **40** (11), 2004–2021 (2001). https://doi.org/10.1002/1521-3773(20010601)40:113.0.CO;2-5
9. Stump, B.: Click bioconjugation: modifying proteins using click-like chemistry. ChemBioChem. **23**(16), e202200016 (2022). https://doi.org/10.1002/cbic.202200016
10. Bednarek, C., Wehl, I., Jung, N., Schepers, U., Bräse, S.: The Staudinger ligation. Chem. Rev. **120**(10), 4301–4354 (2020). https://doi.org/10.1021/acs.chemrev.9b00665
11. Poulou, E., Hackenberger, C.P.R.: Staudinger ligation and reactions—from bioorthogonal labeling to next-generation biopharmaceuticals. Israel J. Chem. **63**(1–2), e202200057 (2023). https://doi.org/10.1002/ijch.202200057
12. Dormàn, G., Prestwich, G.D.: Benzophenone photophores in biochemistry. Biochemistry. **33**(19), 5661–5673 (1994). https://doi-org.uaccess.univie.ac.at/10.1021/bi00185a001
13. Fancy, D.A., Kodadek, T.: Chemistry for the analysis of protein-protein interactions: rapid and efficient cross-linking triggered by long wavelength light. Proc. Natl. Acad. Sci. U.S.A. **96**(11), 6020–6024 (1999). https://doi.org/10.1073/pnas.96.11.6020; Denison, C., Kodadek, T.: Toward a general chemical method for rapidly mapping multi-protein complexes. J. Proteome Res. **3**(3), 417–425 (2004). https://doi-org.uaccess.univie.ac.at/10.1021/pr034071j
14. Dankbar, D., Gauglitz, G.N.: A study on photolinkers used for biomolecule attachment to polymer surfaces. Anal. Bioanal. Chem. **386**(7), 1967–1974 (2006). https://doi.org/10.1007/s00216-006-0871-x
15. Patnaik, S., Swami, A., Sethi, D., …, Kumar, P.: N-(Iodoacetyl)- N'-(anthraquinon-2-oyl)-ethylenediamine (IAED): A new heterobifunctional reagent for the preparation of biochips. Bioconjugate Chem. **18**(1), 8–12 (2007). https://doi-org.uaccess.univie.ac.at/10.1021/bc0602634
16. Valverde, I.E., Delmas, A.F., Aucagne, V.: Click à la carte: Robust semi-orthogonal alkyne protecting groups for multiple successive azide/alkyne cycloadditions. Tetrahedron. **65**(36), 7597–7602 (2009). https://doi.org/10.1016/j.tet.2009.06.093

17. Clave, G., Volland, H., Flaender, M., …, Renard, P.-Y.: A universal and ready-to-use heterotrifunctional cross-linking reagent for facile synthetic access to sophisticated bioconjugates. Org. Biomol. Chem. **8**(19), 4329–4345 (2010). https://doi.org/10.1039/C0OB00133C
18. Beal, D.M., Jones, L.H.: Molecular scaffolds using multiple orthogonal conjugations: applications in chemical biology and drug discovery. Angew. Chem. Int. Ed. **51**(26), 6320–6326 (2012). https://doi.org/10.1002/anie.201200002; Beal, D.M., Jones, L.H.: Mehrfache orthogonale Konjugationen mit Molekülgerüsten: Anwendung in der chemischen Biologie und Wirkstoff-Forschung. Angew. Chem. **124**(26), 6426–6432 (2012). https://doi.org/10.1002/ange.201200002
19. Liu, F., Rijkers, D.T.S., Post, H., Heck, A.J.R.: Proteome-wide profi ling of protein assemblies by cross-linking mass spectrometry. Nat. Methods. **12**(12), 1179–1184 (2015). https://doi.org/10.1038/nmeth.3603; Walzthoeni, T., Joachimiak, L.A., Rosenberger, G. ,…, Aebersold, R.: xTract: Software for characterizing conformational changes of protein complexes by quantitative cross-linking mass spectrometry. Nat. Methods. **12**(12), 1185–1190 (2015). https://doi.org/10.1038/nmeth.3631
20. Kao, A., Chiu, C.-l., Vellucci, D., …, Huang, L.: Development of a novel cross-linking strategy for fast and accurate identification of cross-linked peptides of protein complexes. Mol. Cell. Proteomics. **10**(1), M110.002170 (2011). https://doi.org/10.1074/mcp.M110.002212
21. Petrotchenko, E.V., Serpa, J.J., Borchers, C.H.: An isotopically coded CID-cleavable biotinylated cross-linker for structural proteomics. Mol. Cell. Proteomics. **10**(2), S1–S8 (2011). https://doi.org/10.1074/mcp.M110.001420
22. Soderblom, E.J., Goshe, M.B.: Collision-induced dissociative chemical cross-linking reagents and methodology: applications to protein structural characterization using tandem mass spectrometry analysis. Anal. Chem. **78**(23), 8059–8068 (2006). https://doi.org/10.1021/ac0613840
23. Müller, M.Q., Dreiocker, F., Ihling, C.H., Schäfer, M., Sinz, A.: Cleavable cross-linker for protein structure analysis: reliable identification of cross-linking products by tandem MS. Anal. Chem. **82**(16), 6958–6968 (2010). https://doi.org/10.1021/ac101241t
24. Highlight, de Souza, N.: Chemical biology: a multi-purpose cross-linker. Nat. Methods. **13**(5), 392–393 (2016). https://doi.org/10.1038/nmeth.3860; Tan, D., Li, Q., Zhang, M.-J., …, Lei, X.: Trifunctional cross-linker for mapping protein-protein interaction networks and comparing protein conformational states. eLife. **5**, e12509 (2016). https://doi.org/10.7554/eLife.12509
25. Stadlmeier, M., Runtsch, L.S., Streshnev, F., Wühr, M., Carell, T.: A click-chemistry-based enrichable crosslinker for structural and protein interaction analysis by mass spectrometry. ChemBioChem. **21**(1-2), 103–107 (2020). https://doi.org/10.1002/cbic.201900611
26. Thermo Fisher Scientific, „The Molecular Probes Handbook“: Streptavidin. https://www.thermofisher.com/search/results?query=streptavidin&persona=Catalog&navId=&resultPage=1. Zugegriffen am 30.12.2025
27. Terai, T., Maki, E., Sugiyama, S., …, Nagano, T.: Rational development of caged-biotin protein-labeling agents and some applications in live cells. Chem Biol. **18**(10), 1261–1272 (2011). https://doi.org/10.1016/j.chembiol.2011.09.007
28. Zeng, Y., Ramya, T.N.C., Dirksen, A., …, Paulson, J.C.: High-efficiency labeling of sialylated glycoproteins on living cells. Nat. Methods. **6**(3), 207–209 (2009). https://doi.org/10.1038/nmeth.1305
29. Krüger, T., Dierks, T., Sewald, N.: Formylglycine-generating enzymes for site-specific bioconjugation. Biological Chem. **400**(3), 289–297 (2019). https://doi.org/10.1515/hsz-2018-0358; Krüger, T., Weiland, S., Falck, G., …, Sewald, N.: Two-fold bioorthogonal derivatization by different formylglycine-generating enzymes. Angew. Chem. Int. Ed. **57**(24), 7245–7249 (2018). https://doi.org/10.1002/anie.201803183

30. Mukkala, V.M., Mikola, H., Hemmilä, I.: The synthesis and use of activated N-benzyl derivatives of diethylenetriaminetetraacetic acids: alternative reagents for labeling of antibodies with metalions. Anal. Biochem. **176**(2), 319–325 (1989). https://doi.org/10.1016/0003-2697(89)90316-3
31. Takalo, H., Mukkala, V.-M., Mikola, H., Liitti, P., Hemmila, I.: Synthesis of europium(III) chelates suitable for labeling of bioactive molecules. Bioconjugate Chem. **5**(3), 278–282 (1994). https://doi.org/10.1021/bc00027a015; Mukkala, V.-M., Helenius, M., Hemmilä, I., Kankare, J., Takalo, H.: Development of luminescent europium(III) and terbium(III) chelates of 2,2′:6′,2″- terpyridine derivatives for protein labelling. Helv. Chim. Acta. **76**(3), 1361–1378 (1993). https://doi.org/10.1002/hlca.19930760323
32. Liu, S., Edwards, D.S.: 99m Tc-labeled small peptides as diagnostic radiopharmaceuticals. Chem. Rev. **99**(9), 2235–2269 (1999). https://doi.org/10.1021/cr980436l
33. Krois, D., Riedel, C., Kalchhauser, H., …, Virgolini, I.: Synthesis of N-α-(6-Hydrazinonicotinoyl)-Octreotide: a precursor of a [99m Tc] complex. Liebigs Annalen. **1996**(9), 1463–1469 (1996). https://doi.org/10.1002/jlac.199619960919
34. Brittain, S.M., Ficarro, S.B., …, Peters, E.C.: Enrichment and analysis of peptide subsets using fluorous affinity tags and mass spectrometry. Nat. Biotechnol. **23**(4), 463–468 (2005). https://doi.org/10.1038/nbt1076

Ligation Methods for the Chemical Synthesis or Semisynthesis of Proteins

8

Abstract

The chemical total synthesis and semisynthesis of proteins is of great importance for obtaining homogeneous, fully characterized proteins with defined post-translational modifications (PTMs). Contributing to this are methods for native chemical ligation (NCL), which allow peptide fragments to be joined via natural amide bonds without the need for additional side-chain protecting groups. In this chapter, we will become acquainted with these methods and also discuss in detail two examples of such total syntheses.

Apart from purely molecular biological methods for protein synthesis, chemical synthesis and semisynthesis of proteins (especially with defined, homogeneous post-translational modifications (PTMs)) are becoming increasingly important [2, 3]. Homogeneous PTMs are often not possible at all when expressing a protein in prokaryotes, and even in yeast or mammalian cells, these are generally very heterogeneous. Until recently, direct linear solid-phase protein synthesis seemed impossible, and even today it is rarely practiced, although in principle it should be possible through *automated flow-based SPPS* (AFPS) [4]. The main problem is certainly that only the finished full-length protein could be purified from the many side products that occur. Achieving the desired homogeneity and purity is therefore difficult. For this reason, efforts have been made to assemble proteins from several fragments that can be purified individually. Smaller proteins can thus be formed from two or more peptides (with 30–50 amino acids, synthesized on the solid phase), larger ones in a semisynthesis from one or two peptides (carrying the PTMs) and a truncated protein that has been expressed molecular biologically. The prerequisite for this is a chemical methodology that allows the formation of peptide bonds without protecting groups on the amino acid side chains.

D. Krois, *Organic Chemical Methods*, https://doi.org/10.1007/978-3-662-73573-2_8

8.1 Ligation Methods for Protein Synthesis

Here, we will learn about methods that allow the synthesis of proteins from fragments via "natural" amide bonds. This is therefore referred to as *native chemical ligation* (NCL).

8.1.1 Ligation via Sulfur or Selenium Compounds

The first such reaction is the coupling of a C-terminal thioester with a peptide fragment bearing an N-terminal cysteine (Fig. 8.1a) [2, 3]. Carboxylic acid thioesters behave like activated esters (Section 3.1.2). Under more or less neutral pH conditions, the thioester reacts with a nucleophilic thiol group of a cysteine: initially, another thioester is formed, but if this cysteine has a free amino group, it is in such a favorable position that an *S*→*N* acyl transfer occurs—resulting in a native peptide bond. However, this method requires a cysteine at the corresponding coupling site, and cysteines occur in proteins at less than 2%. Alanine, for example, is much more common (about 8%), and cysteine can be converted into alanine ("*desulfurization*"). There are several ways to achieve this [5], the most common variant is shown in Fig. 8.1b. With the help of a radical initiator (often an azo compound), a radical R is generated, which abstracts an H-radical from the thiol, forming a thiyl radical. This then reacts with a phosphine, and the phosphoranyl radical spontaneously eliminates $R_3P{=}S$. The $–CH_2–$ radical stabilizes itself by H-abstraction from an added thiol, and the transformation to alanine is complete. Instead of sulfur compounds, NCL also works very well, sometimes even better, with the corresponding selenium compounds. Selenocysteine (Sec), which does occur naturally in some proteins, will be present at the coupling site, but can also be converted to alanine [5]. However, if amino acids other than Cys, Sec, or Ala are to be present at the N-terminus of the C-terminal fragment, unnatural synthetically produced amino acids with a β- or γ-thiol function can be used, which can then be desulfurized to the corresponding natural amino acids after NCL [6], or one can also use cleavable thiol-containing linkers (*auxiliaries*) [7].

How can the active thioesters at the C-terminus of the N-terminal fragment be obtained? If Merrifield's Boc-SPPS is used, it is possible to use a thioester linker so that a thioester is obtained directly after cleavage. This is not possible with the Fmoc strategy, as the basic cleavage of the Fmoc protecting groups would immediately cleave the thioester as well. Options for indirect introduction after SPPS are shown in Fig. 8.1c, d, e. With an appropriate polymer anchor, a hydrazide is generated directly upon deprotection and cleavage (Fig. 8.1c), which can be converted to the azide by an oxidation step with sodium nitrite. Substitution of the azide by a thiol yields a thioester. This method also has the advantage that hydrazides are not reactive in NCL itself, essentially serving as a C-terminal protecting group and are also much more stable than thioesters [2, 3, 8]. Another attractive route is shown in Fig. 8.1d. With an appropriate linker on the polymer,

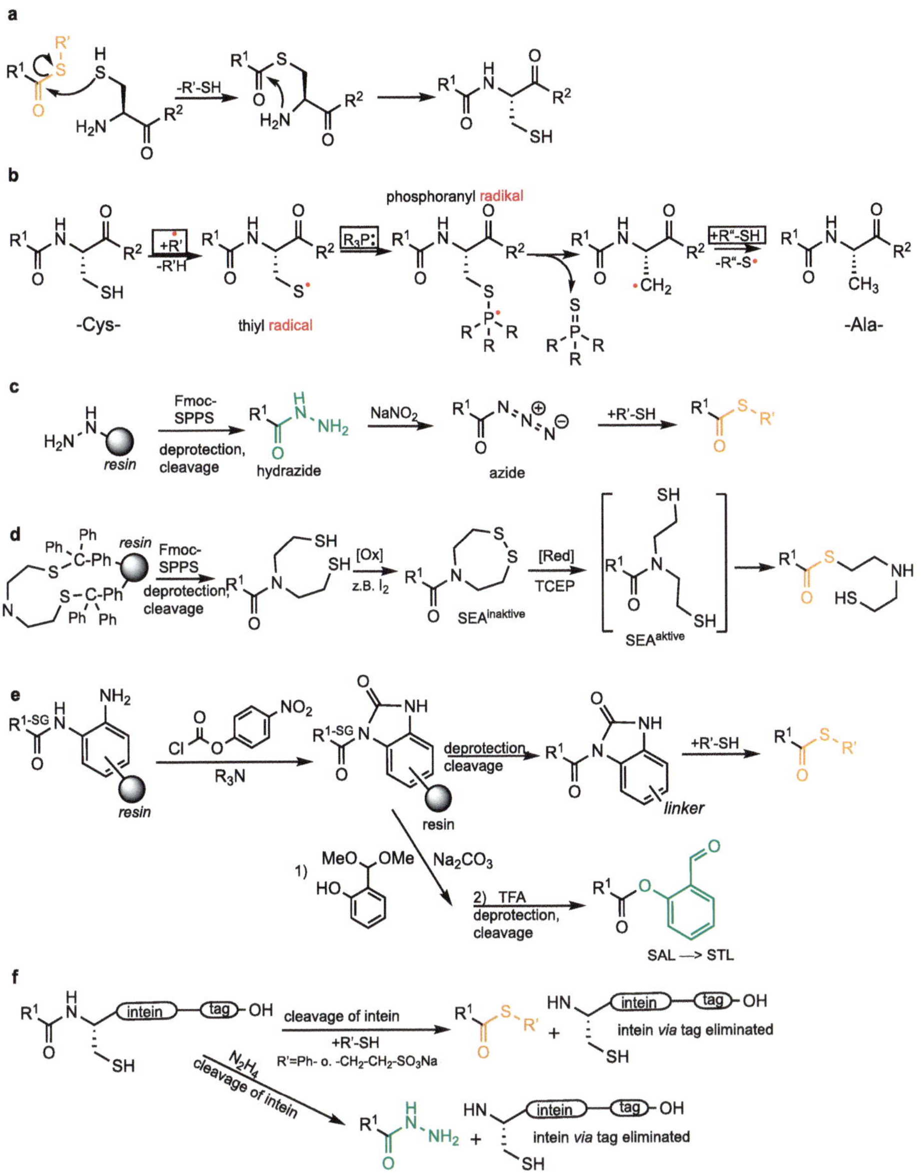

Fig. 8.1 Ligation methods for chemical protein synthesis or semisynthesis. **a** NCL via thioester and cysteine; **b** Conversion Cys➔Ala ("desulfurization"); **c**, **d**, **e** and **f** Methods for introducing the thioester (R^1, R^2 … peptide fragments)

the bis(2-sulfanylethyl)amide group is introduced at the C-terminus of the peptide. Oxidation with, for example, iodine produces the cyclic disulfide, which is not reactive and can thus be used as a protecting group. Reduction with the phosphine TCEP (Section 7.2.2) regenerates the dithiol, which now has very special properties: the proximity and double concentration of –SH favors an *N*→*S* acyl transfer, which is further facilitated by the weakening of the amide bond through the S–H…N hydrogen bond, so that the desired thioester is formed [9]. In Fig. 8.1e, a C-terminal amide on the polymer is sufficiently activated by formation of a cyclic aromatic urea so that after cleavage and deprotection, addition of a thiol yields the thioester. The activated intermediate also enables the formation of a salicylaldehyde ester (SAL), which can be used for another ligation method (see Section 8.1.2).

These are the chemical prerequisites for peptide fragment couplings by NCL, but there are still some problems to address: The concentrations would have to be relatively high for the partners to react with each other, and thioesters that are stable enough to be isolated are often not reactive enough for NCL. There are catalytic methods to increase reactivity [10], or the local concentration of the coupling partners can be increased by template formation. For this, PNA helix formation has been used or cleavable anchors (linkers) via click reactions, etc. [11, 12]. The solubility of the fragments can also sometimes be a problem [13]. This is of particular importance in the synthesis of lipid-modified proteins and membrane proteins [14, 15].

8.1.2 Ligation to Serine or Threonine as N-terminal Amino Acid

A ligation method has been developed that does not require sulfur or selenium and enables coupling to serine or threonine. It is based on salicylaldehyde (SAL) and is referred to as serine/threonine ligation (STL). The mechanism of this reaction is shown in Fig. 8.2. A C-terminal ester of salicylaldehyde (for preparation see e.g. Figure. 8.1e) reacts in an acetal-like reaction with the free $-NH_2$ and –OH groups of an N-terminal serine or threonine to form a cyclic aminal, which then spontaneously stabilizes via an

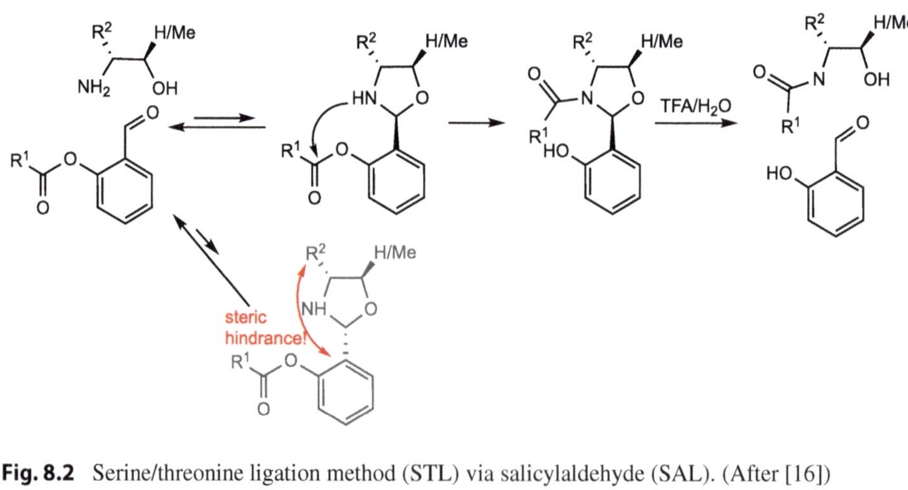

Fig. 8.2 Serine/threonine ligation method (STL) via salicylaldehyde (SAL). (After [16])

O→*N* acyl shift. Acid-catalyzed opening of the aminal then releases the salicylaldehyde again [16]. If the aldehyde group is protected as a thioacetal, this method can also be used for several consecutive ligation steps [17].

8.1.3 Ligation Using Inteins for Protein Semisynthesis

As already mentioned, in the semisynthesis of larger proteins, peptide fragments carrying modifications are linked to a truncated expressed protein. For this purpose, fusion proteins with a modified intein (inteins are responsible for the *splicing* of proteins) and a tag sequence (*tag*), such as a chitin-binding domain, have become established. The *tag* is used to purify the fusion protein, and then the desired truncated protein is cleaved off either as a thioester with a thiol or as a hydrazide with hydrazine (see Fig. 8.1f) [2, 3].

8.2 Examples of Synthesized Proteins

8.2.1 Synthesis of Alpha-Synuclein

Alpha-synuclein (aSyn) is a protein associated with neurotransmitter disorders such as Parkinson's disease. Point mutations as well as certain post-translational modifications (PTMs) play a role, which is why synthetic approaches have been and remain of interest [18]. The synthesis from three peptide fragments described in Ref. [19] is suitable for synthesizing various homogeneous variants and studying their properties (especially aggregation and fibril formation). Using this example, we will look at the chemical total synthesis of a protein (Fig. 8.3b). aSyn was synthesized from three peptide fragments via NCL, with Gly–Ala chosen as the ligation sites for NCL. As we have seen, alanine can be obtained from cysteine by "desulfurization," and the glycine thioester reacts faster in NCL for steric reasons. Fmoc-SPPS was performed on hydrazide-modified trityl polymer. The hydrazides of fragments 69–106 and 1–68 were not converted to the thioester *via* azide, but via a pyrazole intermediate (Fig. 8.3a) [20]. Since the synthesis was planned from C➔N-terminus, cysteine-69 had to be protected. Thus, at the end of SPPS, not Fmoc-Cys but a Boc-protected thiazolidine carboxylic acid was introduced. After NCL with fragment 107–140 (A107C), the thiazolidine was cleaved with a hydroxylamine. What reaction is this? The thiazolidine can also be regarded as a thioaminal of formaldehyde, which is how it is formed, and formaldehyde is then trapped as an oxime during cleavage under mildly acidic conditions. Finally, fragment 1–68 as a thioester was ligated to the now free cysteine-106 to form the complete protein by NCL. At the end, the native sequence was generated by radical "desulfurization" (Fig. 8.1). The same strategy also enabled the production of modifications: a mutant A53T, which has been attributed physiological significance, a phosphorylated serine-129, and, using an N-terminally acetylated fragment 1–68, various acetylated variants.

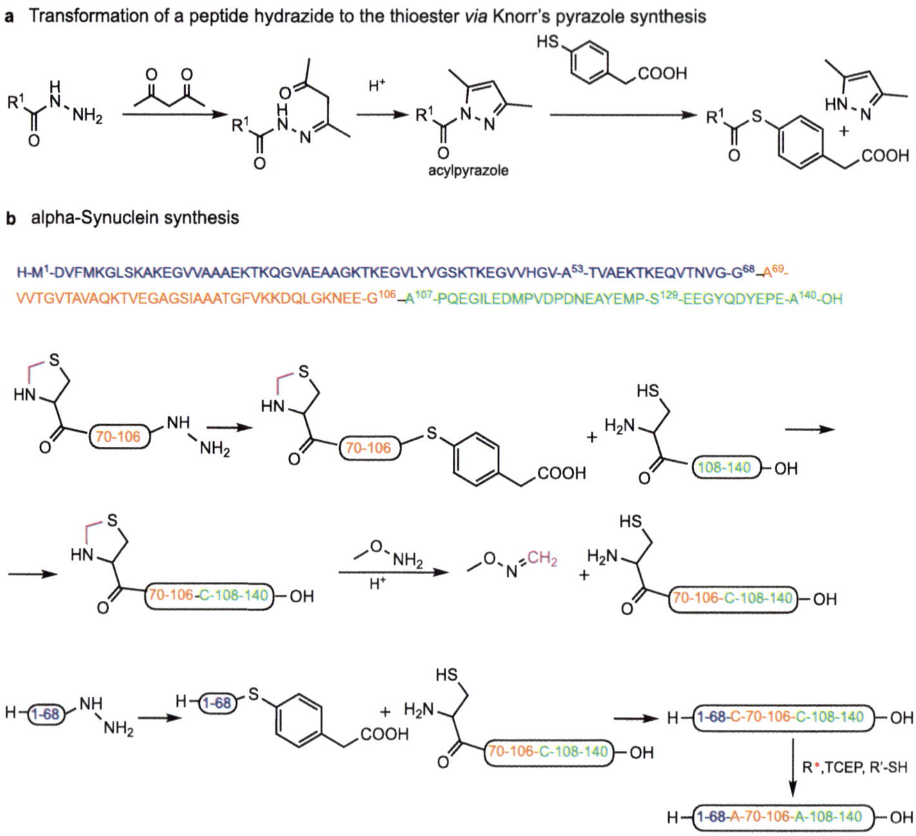

Fig. 8.3 **a** Transformation of peptide hydrazides into thioesters via acylpyrazoles [20]; **b** Synthesis of alpha-synuclein. (After Ref. [19])

8.2.2 Chemoenzymatic Synthesis of Glycoproteins

As already discussed in Section 5.2.3, we have touched on the problems in the synthesis of glycopeptides and glycoproteins and mentioned some examples. Now we want to look at such syntheses in more detail. The chemical or chemoenzymatic synthesis of these biopolymers is particularly interesting and in demand due to the microheterogeneity of these proteins from natural sources. Here are some general recent references on this topic: [21–23]. Especially when it comes to producing larger quantities for therapeutic applications, solid-phase synthesis of oligosaccharides [24] is apparently still not an option today. The *N*-glycans described in recent syntheses are mostly isolated from natural sources, e.g., egg yolk, purified, and then further modified enzymatically.

Also in the synthesis of sialylated glycoforms of erythropoietin (Section 5.2.3) [25, 26], which we will discuss in more detail here, the glycan component (Fig. 8.4a) was obtained from chicken egg yolk [28] by enzymatic conversions and purification steps (see Supporting Information to [29]). The synthesis of the complete protein was finally achieved by NCL from three glycopeptide fragments and two peptide fragments. The choice of the right protecting groups was only successful after some less successful preliminary attempts. As in the example above, the corresponding hydrazides were isolated after Fmoc-SPPS, which were then converted to the thioesters via the azides before the respective NCL step. But how were the oligosaccharides attached to the respective asparagines? Incorporation of the glycosylated amino acid during SPPS is problematic for more complex oligosaccharide chains, so after cleavage from the polymer by Boc–NH–NH_2, the side-chain protecting group of the aspartic acid was removed and the amine-derivatized oligosaccharide was coupled using a coupling reagent (HATU). This method is also called *Lansbury aspartylation* [27]. The main problem with Asp–Aaa–sequences, and especially Asp–Gly–, is the formation of succinimides (see Section 3.2.4, Fig. 3.20). As we have learned (Section 5.2.4), biosynthetic *N*-glycans are post-translationally incorporated only at Asn-Xxx-Ser/Thr sequences. It has been shown—somewhat surprisingly—that masking the hydroxy amino acid as a pseudoproline significantly reduces aspartimide formation by conformational changes (Fig. 8.4b) [29]. Accordingly, this approach was also used in our example [25]. Figure. 8.4c gives an overview of the procedure for the total synthesis (without details of reaction conditions and all side-chain protecting groups). First, the two C-terminal peptides, which do not have oligosaccharide side chains, were ligated via NCL, with alanines 98 and 128 introduced as cysteines to enable NCL. Cysteine 98 was masked as a thiazolidine derivative during the first fragment coupling (Section 8.2.1). The water-soluble phosphine TCEP was used to release the thiol from the tBu–S–S– disulfide. MPAA acted as a catalyst for NCL. Then, the thiazolidine ring was opened with hydroxylamine CH_3–O–NH_2, releasing cysteine-98. In parallel, for fragment (29–67) (*N*-glycan at Asn^{38}), the hydrazide was converted to the thioester (first to the azide with $NaNO_2$, then to the thioester with the thiol MESNa) and ligated with fragment (68–97) (*N*-glycan at Asn^{83}) after treatment with TCEP under MPAA catalysis. The hydrazide of (29–97) was converted to the thioester as above and then coupled with [98–166] under MPAA catalysis. At this point, cysteines 68, 98, and 128 were converted to the desired alanines (of the EPO protein) by radical "desulfurization" (Section 8.1.1). After removal of the acetamidomethyl (Acm) protecting group from Cys^{29} with $PdCl_2$, the final fragment coupling with the glycopeptide (1-28) was performed. The aim of the publication was to enzymatically introduce sialic acid residues at the ends of the *N*-glycans; see Ref. [25].

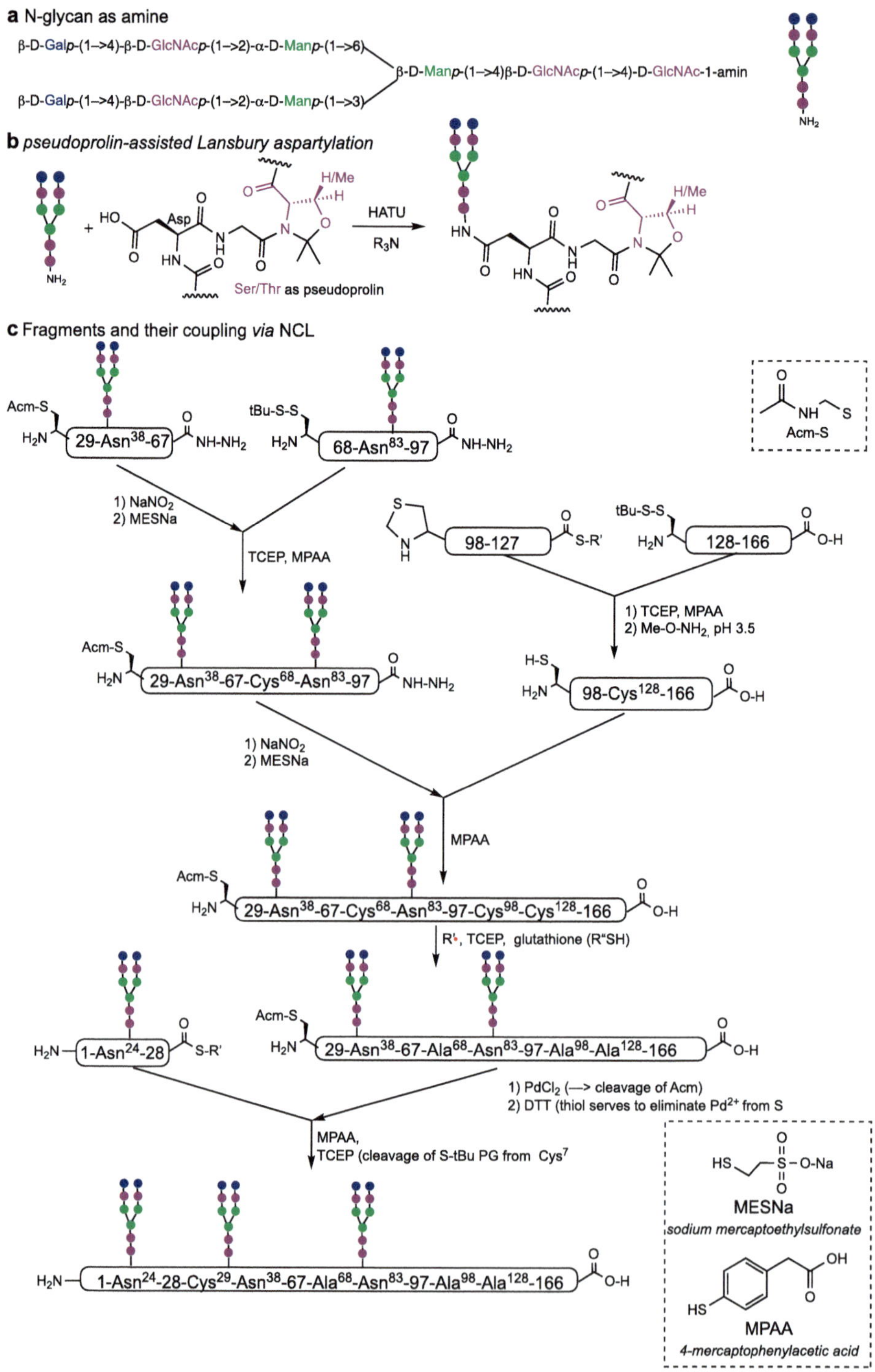

Fig. 8.4 Total synthesis of the glycoprotein erythropoietin (EPO): **a** Structure of the *N*-glycan; **b** Coupling of an aminoglycan according to Lansbury [27] using serine or threonine masked as pseudoproline; **c** Overview of the fragment couplings via NCL. (After Ref. [25])

8.3 Questions

Let us test our understanding of the content of this chapter

8.1 Why must special methods be used for the ligation of large peptide fragments (up to approx. 50 amino acids), rather than coupling reagents as described in Sect. 3.1.2?
8.2 What are the prerequisites for the application of NCL (in the strict sense)?
8.3 Why is it often difficult to determine the fragment ligation sites?
8.4 Why can alanine also be chosen as a ligation site?
8.5 Can the sulfur atom in NCL also be replaced by another element?
8.6 What is the basis of the serine/threonine ligation method (STL)?
8.7 How can C-terminal thioesters be introduced after SPPS?
8.8 How can a protected cysteine be introduced directly at the N-terminus?
8.9 What is meant by *Lansbury aspartylation*?
8.10 What are pseudoprolines and into which amino acids can they be converted?
8.11 Why is the chemical synthesis of glycoproteins important, but also particularly challenging?

References

1. Zu diesem Thema ist 2021 ein Buch erschienen, hier der Verweis auf die Inhaltsangabe: Total Chemical Synthesis of Proteins, Index, Brik, A., Dawson, P. and Liu, L. (Hrsg). WILEY-VCH (2021). https://onlinelibrary.wiley.com/doi/book/10.1002/9783527823567. Zugegriffen am 03.01.2026
2. Conibear, A.C., Watson, E.E., Payne, R.J., Becker, C.F.W.: Native chemical ligation in protein synthesis and semi-synthesis. Chem. Soc. Rev. **47**(24), 9046–9068 (2018). https://doi.org/10.1039/C8CS00573G
3. Thompson, R.E., Muir, T.W.: Chemoenzymatic semisynthesis of proteins. Chem. Rev. **120**(6), 3051–3126 (2020). https://doi.org/10.1021/acs.chemrev.9b00450
4. Gates, Z.P., Hartrampf, N.: Flow-based SPPS for protein synthesis: a perspective. Peptide Sci. **112**(6), e24198 (2020). https://doi.org/10.1002/pep2.24198
5. Diemer, V., Roy, E., Agouridas, V., Melnyk, O.: Protein desulfurization and deselenization. Chem. Soc. Rev. **53**(17), 8521–8545 (2024). https://doi.org/10.1039/D4CS00135D
6. Guan, I., Williams, K., Liu, J. S. T., Liu, X.: Synthetic Thiol and Selenol derived amino acids for expanding the scope of chemical protein synthesis. Front. Chem. **9**, 826764 (2021). https://doi.org/10.3389/fchem.2021.826764; Watson, E. E., Malins, L. R., Payne, R. J.: Expanding native chemical ligation methodology with synthetic amino acid derivatives. In: Brik, A., Dawson, P., Liu, L. (Hrsg.) Total Chemical Synthesis of Proteins, Kap: 5, S. 119–159. WILEY-VCH (2021). https://doi.org/10.1002/9783527823567.ch5
7. Loibl, S.F., Dallmann, A., Hennig, K., Juds, C., Seitz, O.: Features of auxiliaries that enable native chemical ligation beyond glycine and cleavage via radical fragmentation. Chem. Eur. J. **24**(14), 3623–3633 (2018). https://doi.org/10.1002/chem.201705927
8. Zuo, C., Tan, X., Tan, X., Liu, L.: Chemical synthesis of proteins through native chemical ligation of peptide hydrazides. In: Brik, A., Dawson, P. and Liu, L. (Hrsg.) Total

Chemical Synthesis of Proteins, Kap: 4, S. 87–118. WILEY-VCH (2021). https://doi.org/10.1002/9783527823567.ch4

9. Agouridas, V., Ollivier, N., Vicogne, J., Diemer, V., Melnyk, O.: Redox-controlled chemical protein synthesis: sundry shades of latency. Acc. Chem. Res. **55**(18), 2685–2697 (2022). https://doi.org/10.1021/acs.accounts.2c00436
10. Diemer, V., Firstova, O., Agouridas, V., Melnyk, O.: Pedal to the metal: the homogeneous catalysis of the native chemical ligation reaction. Chem. Eur. J. **28**(16), e202104229 (2022). https://doi.org/10.1002/chem.202104229
11. Vazquez, O., Seitz, O.: Templated native chemical ligation: peptide chemistry beyond protein synthesis. J. Pept. Sci. **20**(2), 78–86 (2014). https://doi.org/10.1002/psc.2602
12. Giesler, R.J., Erickson, P.W., Kay, M.S.: Enhancing native chemical ligation for challenging chemical protein syntheses. Curr. Opin. Chem. Biol. **58**, 37–44 (2020). https://doi.org/10.1016/j.cbpa.2020.04.003
13. Giesler, R. J., Fulcher, J. M., Jacobsen, M. T., Kay, M.S.: Controlling segment solubility in large protein synthesis. In: Brik, A., Dawson, P., Liu, L. (Hrsg.) Total chemical synthesis of proteins, Kap: 7, S. 185–209. WILEY-VCH (2021). https://doi.org/10.1002/9783527823567.ch7 (2021)
14. Hanna, C.C., Kriegesmann, J., Dowman, L.J., Becker, C.W.F., Payne, R.J.: Chemical synthesis and semisynthesis of lipidated proteins. Angew. Chem. Int. Ed. **61**(15), e202111266 (2022). https://doi.org/10.1002/anie.202111266
15. Schmid, A., Becker, C. F. W.: Chemical synthesis of membrane proteins. In: Brik, A., Dawson, P., Liu, L. (Hrsg.) Total chemical synthesis of proteins, Kap: 16, S. 437–462. WILEY-VCH (2021). https://doi.org/10.1002/9783527823567.ch16
16. Liu, H., Li, X.: Serine/Threonine ligation: origin, mechanistic aspects, and applications. Acc. Chem. Res. **51**(7), 1643–1655. https://doi.org/10.1021/acs.accounts.8b00151 (2018); Cheung, C. H. P., Li, X.: Ser/Thr ligation for protein chemical synthesis. In: Brik, A., Dawson, P., Liu, L. (Hrsg.) Total Chemical Synthesis of Proteins, Kap: 10, S. 285–306. WILEY-VCH (2021). https://doi.org/10.1002/9783527823567.ch10
17. Huang, D.-L., Li, Y., Liang, J., …, Zheng, J.-S.: The new salicylaldehyde S,S-Propanedithioacetal ester enables N-to-C sequential native chemical ligation and Ser/Thr ligation for chemical protein synthesis. J. Am. Chem. Soc. **142**(19), 8790–8799 (2020). https://doi.org/10.1021/jacs.0c01561
18. Gatzemeier, L.M., Meyer, F., Outeiro, T.F.: Synthesis and semi-synthesis of alpha-synuclein: insight into the chemical complexity of synucleinopathies. ChemBioChem. **25**(20), e202400253 (2024). https://doi.org/10.1002/cbic.202400253
19. Gatzemeier, L.M., Meyer, F., Diederichsen, U., Outeiro, T.F.: Chemical synthesis of alpha-synuclein proteins via solid-phase peptide synthesis and native chemical ligation. Chem. Eur. J. **29**(33), e202300649 (2023). https://doi.org/10.1002/chem.202300649
20. Flood, D.T., Hintzen, J.C.J., Bird, M.J., …, Dawson, M.E.: Leveraging the knorr pyrazole synthesis for the facile generation of thioester surrogates for use in native chemical ligation. Angew. Chem. Int. Ed. **57**(36), 11634–11639 (2018). https://doi.org/10.1002/anie.201805191
21. Ochiai, H., Elouali, S., Yamamoto, T., …, Nishiuchi, Y.: Chemical and chemoenzymatic synthesis of peptide and protein therapeutics conjugated with human N-Glycans. ChemMedChem. **19**(13), e202300692 (2024). https://doi.org/10.1002/cmdc.202300692
22. Roy, R. and Mousavifar, L.: Carrier diversity and chemical ligations in the toolbox for designing tumor-associated carbohydrate antigens (TACAs) as synthetic vaccine candidates. Chem. Soc. Rev. **52**(10), 3353–3396. https://doi.org/10.1039/D2CS01032A (2023); Nomura, K., Liu, Y., Kajihara, Y.: Chapter four – synthesis of homogeneous glycoproteins with diverse N-glycans. In: Baker, D.C. (Hrsg.) Adv. Carbohydrate. Chem. Biochem. **81**, 57–93, Academic Press (2022). https://doi.org/10.1016/bs.accb.2022.09.004

23. Lin, J.D., Liu, X.-W.: Recent development in ligation methods for glycopeptide and glycoprotein synthesis. Chem Asian J. **15**(17), 2548–2557 (2020). https://doi.org/10.1002/asia.202000566
24. Niggemeyer, G.B., Danglad-Flores, J.A., Seeberger, P.H.: Automated synthesis of C1-functionalized oligosaccharides. J. Am. Chem. Soc. **147**(2), 1649–1655 (2025). https://doi.org/10.1021/jacs.4c11798
25. Hessefort, H., Gross, A., Seeleithner, S., …, Unverzagt, C.: Chemical and enzymatic synthesis of sialylated glycoforms of human erythropoietin. Angew. Chem. Int. Ed. **60**(49), 25922–25932 (2021). https://doi.org/10.1002/anie.202110013
26. Hessefort, M., Hessefort, H., Seeleithner, S., …, Unverzagt, C.: Strategies for the highly efficient synthesis of erythropoietin N-glycopeptide hydrazides. J. Pept. Sci. **27**(1), e3283 (2021). https://doi.org/10.1002/psc.3283
27. Cohen-Anisfeld, S.T., Lansbury Jr., P.T.: A practical, convergent method for glycopeptide synthesis. J. Am. Chem. Soc. **115**(23), 10531–10537 (1993). https://doi.org/10.1021/ja00076a010
28. Seko, A., Koketsu, M., Nishizono, M., …, Yamamoto, T.: Occurrence of a sialylglycopeptide and free sialylglycans in hen's egg yolk. Biochim. Biophys. Acta General Subjects. **1335**(1), 23–32 (1997). https://doi.org/10.1016/S0304-4165(96)00118-3
29. Ullmann, V., Rädisch, M., Boos, I., …, Unverzagt, C.: Convergent solid-phase synthesis of N-Glycopeptides facilitated by pseudoprolines at consensus-sequence Ser/Thr residues. Angew. Chem. Int. Ed. **51**(46), 11566–11570 (2012). https://doi.org/10.1002/anie.201204272; Wang, P., Aussedat, B., Vohra, Y., Danishefsky, S.J.: An advance in the chemical synthesis of homogeneous N-linked glycopolypeptides by convergent aspartylation. Angew. Chem. Int. Ed. **51**(46), 11571–11575 (2012). https://doi.org/10.1002/anie.201205038

Part II

Selected Examples of the Application of Organic-Chemical Methods to Biologically Significant Questions

Synthesis of Lipopeptides and Protein Conjugates for the Function of the N-Ras Protein

9

Abstract

This chapter discusses in detail three studies from the research group of Prof. Herbert Waldmann, which focus on the biological topic of the "Ras protein." In the first two publications, lipopeptides corresponding to the C-terminus of this protein are synthesized using newly developed strategies for peptide synthesis of these sometimes highly labile compounds. They are appropriately labeled for biophysical investigative methods, such as *Surface Plasmon Resonance* or fluorescence microscopy studies. In the second study, these peptides are also assembled with a correspondingly truncated protein to form biologically active, fluorescence-labeled protein chimeras. The third publication addresses a different question in the field of Ras proteins. It concerns the synthesis of a natural inhibitor of farnesyl transferase, which catalyzes the first enzymatic step in the formation of the mature, active Ras protein. Although these studies are not of the most recent date, they particularly highlight the interesting aspects of biological-chemical research in a highly interdisciplinary field.

▶ In this chapter, we will follow how a leading biological-organic chemist takes up a topic from biology and, initially as a chemist, synthesizes "simple" model compounds using new synthetic strategies, tests them in a targeted manner, but then goes further and ultimately synthesizes defined protein conjugates. Keeping the topic in mind, he also pursues a completely different question—the elucidation and further development of an enzyme inhibitor—which, although not entirely fitting the chapter heading, we still want to consider. These works by Herbert Waldmann were certainly one reason for his rapid rise to director of a Max Planck Institute—an aspect that may also be of interest to ambitious students.

D. Krois, *Organic Chemical Methods*, https://doi.org/10.1007/978-3-662-73573-2_9

Important note on the use of this second part of the textbook As already mentioned, this second part of the book discusses selected publications. It is assumed that everyone will download these publications from the internet; this should not be a problem at any university, as they are all from very common journals. We will not reprint any parts of these publications in the textbook, but will only interpret them and provide important additional information.

9.1 Synthesis of Lipopeptides of the N-Ras Protein

The basis of this section is the publication Chemoenzymatic Synthesis of Biotinylated Ras Peptides and Their Use in Membrane Binding Studies of Lipidated Model Proteins by Surface Plasmon Resonance, Schelhaas, M., Nägele, E., Kuder, N., Bader, B., Kuhlmann, J., Wittinghofer, A., Waldmann, H., *Chem. Eur. J.* **5** (4), 1239–1252, *doi*: 10.1002/(SICI)1521-3765(19990401)5:4<1239::AID-CHEM1239>3.0.CO;2-N (1999) [1].

If you search the literature, you will often encounter the Ras protein. There are a wide variety of studies dealing with it. Why is this protein so interesting? To answer this, we need to take a closer look at its biological role. It is one of the most important control points in the Ras/MAP kinase signal transduction cascade (for a general review on signal transduction and organic synthesis, see Ref. [2]); MAP stands for *mitosis activating protein*. Figure 9.1 shows this circuit, which is so important for the regulation of cell growth.

When the corresponding receptor tyrosine kinase (RTK) is activated by an extracellular signal, a growth factor (GF), it autophosphorylates its intracellular domain. The adaptor molecule Grb2 can now bind (Sect. 12.2.2), which localizes the protein SOS, thereby activating the Ras protein by converting GDP to GTP. The signal is then relayed via phosphorylations all the way to the nucleus, where it influences gene transcription. Why is the Ras protein such a sensitive link in this chain? It possesses intrinsic GTPase activity, meaning it normally deactivates itself immediately. However, often a single mutation in the *ras* gene is enough to abolish this enzymatic activity—it becomes an oncogene. If the signal remains constantly switched on, this leads to uncontrolled cell division, i.e., cancer. In fact, in 30–40% of all human cancers, a point mutation in the *ras* gene is found, and in certain types even more than 80%. No wonder, then, that it is such an interesting research object.

As is often the case with gene products of crucial efficacy, the protein is inactive when it leaves the ribosome. For the N-Ras protein to acquire its active structure, four enzymes must act on it (Fig. 9.2). The original C-terminus (–Cys–Val–Leu–Ser–OH) adopts a β-*turn* conformation. First, Ras-farnesyltransferase transfers a farnesyl group (a sesquiterpene with 15 C atoms) from farnesyl pyrophosphate to the last cysteine. Then, a specific protease cleaves off the last three amino acids, and the farnesylated, now terminal cysteine is methylated to the methyl ester by a methyltransferase. Finally, a palmitoyl

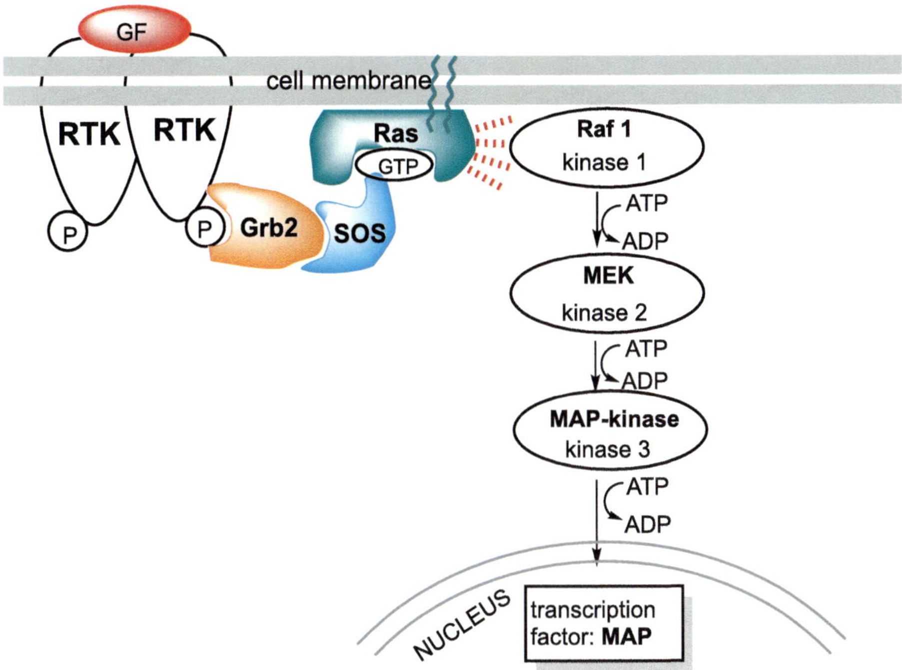

Fig. 9.1 The Ras/MAP kinase signal transduction cascade; GF … *growth factor*, RTK … receptor tyrosine kinase (in the phosphorylated active form); Grb2 … adaptor protein; SOS … Ras activating protein; Ras (in its active form with GTP); Raf 1 … a kinase; MEK (also known as MAP kinase kinase) and MAP kinase relay the phosphorylation signal further into the nucleus

(hexadecanoic acid) group is transferred to the penultimate cysteine. Thus, the protein is equipped at the C-terminal region with two lipid residues, which evidently serve to anchor it in the cell membrane. Further information on the regulation of Ras proteins and their significance in cancer development can be found in the reviews [2–4].

Investigating the precise function and significance of these lipid anchors was the focus of Herbert Waldmann's work on this topic. In the 1990s, the synthesis of lipopeptides was a challenge that would have appealed to a chemist with a strong background in enzymatic reactions. This was the starting point for the present publication [1].

Scheme 3 [1] presents the target compounds, and Scheme 4 the retrosynthetic analysis. Biotin was chosen as the label for the peptides (Sect. 7.8), and we will see later why. For the following reactions, I will not comment further on the steps already familiar from Sect. 3.1. However, you should try to think through these steps—as an exercise for Chap. 3 and "peptide synthesis."

Scheme 5 shows the synthesis of the C-terminal tripeptide **12**. Here, the formation of the choline ester from the 2-bromoethyl ester and its cleavage by the enzyme butyrylcholinesterase is tested. At this point, this would not have been necessary, but for the

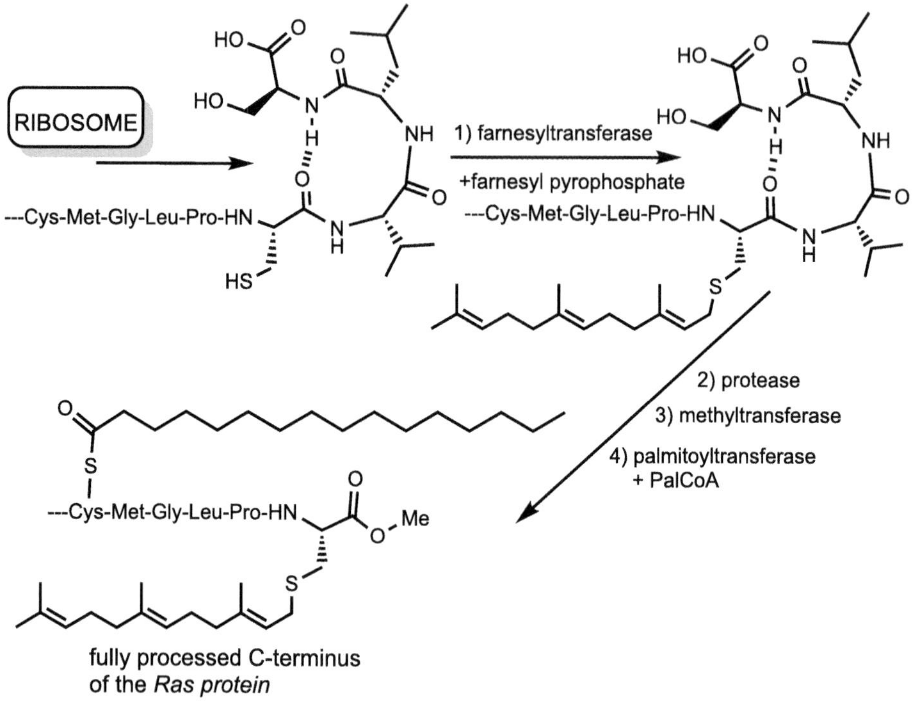

Fig. 9.2 Posttranslational modifications at the C-terminus of the N-Ras protein [PalCoA = palmitoyl coenzyme A]

cleavage of the ester protecting group from the palmitoylated tripeptide, the reaction was to be used. Scheme 6 shows the introduction of palmitic acid to an Aloc- and TMS-protected cysteine (TMS = trimethylsilyl-) via the acid chloride. In Scheme 7, the N-terminal tripeptide is then assembled, but cleavage of the choline ester initially proved difficult, as the enzyme required an aqueous solvent mixture in which the lipopeptide was not soluble. Therefore, dimethyl-β-cyclodextrin was added for solubilization. Ultimately, the N-terminally protected (Aloc-) lipidated hexapeptide was obtained.

Cleavage of Allyl Protecting Groups

Allyl groups are activated for nucleophilic attack by binding to Pd(0) complexes, which then leads to cleavage of the allyl group. Morpholine or the non-basic dimedone can be used as the nucleophile [5] (Fig. 9.3a).

Mode of Action of the Coupling Reagent EEDQ (1-Ethoxycarbonyl-2-ethoxy-1,2-dihydroquinoline)

The acyl group to be activated first substitutes the 2-ethoxy group of EEDQ. Then, either quinoline is immediately eliminated and a mixed anhydride is formed (see Sect. 3.1.2.2), which then reacts with the already present amine component to form the amide, or quinoline elimination and amide formation occur in a single step (Fig. 9.3b).

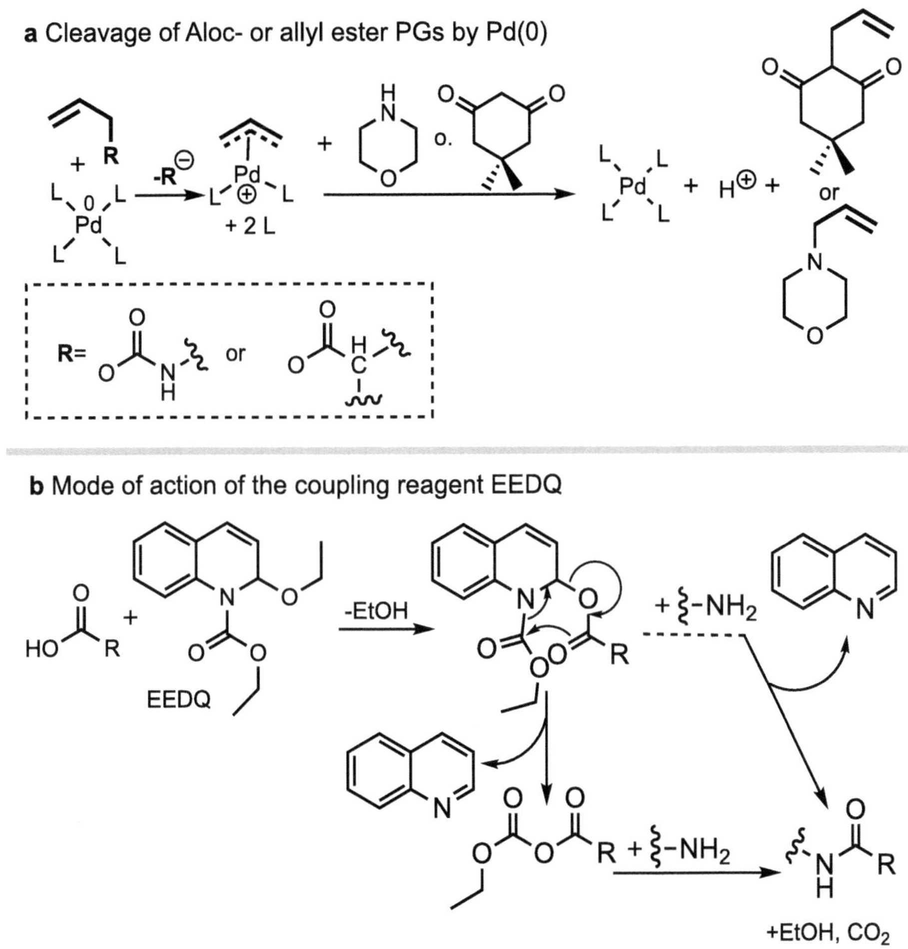

Fig. 9.3 **a** For the cleavage of allyl protecting groups (PG) and **b** for the mechanism of peptide coupling via EEDQ

However, it is noted that cleavage of the Aloc protecting group could lead to migration of the palmitoyl group to the now free N-terminus. Therefore, a different route was taken for the synthesis of the biotinylated lipopeptides. The C-terminal farnesylated pentapeptide was synthesized (Scheme 8) using similar reactions as before. For the N-terminal part (Scheme 9), cystine was protected as the allyl ester, and after coupling with Boc–Gly–OH and reduction with dithiothreitol (DTT), the palmitoyl or hexadecyl group was introduced. After Boc cleavage with TFA, coupling with biotinylaminocaproic acid (Bio-tAcaOH) was performed. Although the secondary amine morpholine was used as the nucleophile for cleavage of the allyl ester [5], no elimination reaction occurred at the cysteine under these conditions. Finally, the C-terminal pentapeptide (30) was coupled with BiotAcaOH to give the solely farnesylated biotinylated product **2**, and with the palmitoylated or hexadecylated biotinylated dipeptides to give **3** and **4** (Scheme 10).

What was the purpose of these biotin-containing lipopeptides? The aim was to investigate the significance of the second lipid residue biophysically. For this, a then completely new method, *surface plasmon resonance* (SPR) [6] (Fig. 9.4), was to be used.

Surface Plasmon Resonance (SPR) [6]

The use of the SPR effect for biophysical measurements began around 1990. The company Biacore probably developed the first commercial instruments. Production is now part of Cytiva.

Let us consider Fig. 9.4: A light beam passes through a prism and is totally reflected. On the underside is the sensor chip, generally an ultrathin gold film onto which various biomaterials can be applied. At a certain angle α, a fraction of the incident light intensity penetrates the interface (as an evanescent wave), and the intensity of the reflected light at this angle is reduced. This is measured by a diode array detector. The angle is now extremely sensitive to the mass (refractive index) in the thin layer on the sensor chip. If you want to test interactions with a membrane, for example, the HPA chip is used, which carries an alkanethiol layer bound to the gold surface. Liposomes can be adsorbed onto this. A liquid stream is then passed over this surface, and if molecules adsorb to the membrane layer, the angle changes; when rinsed with pure liquid, some molecules desorb again, while others may remain adsorbed. The kinetics of adsorption and dissociation (or desorption) can be tracked in the diagram.

Now the significance of biotinylation (Sect. 7.8) in this case becomes clear: If biotinylated lipopeptides are immobilized on the membrane-like surface of the SPR chip, the protein streptavidin, which binds to them, is also captured. This results in a strong

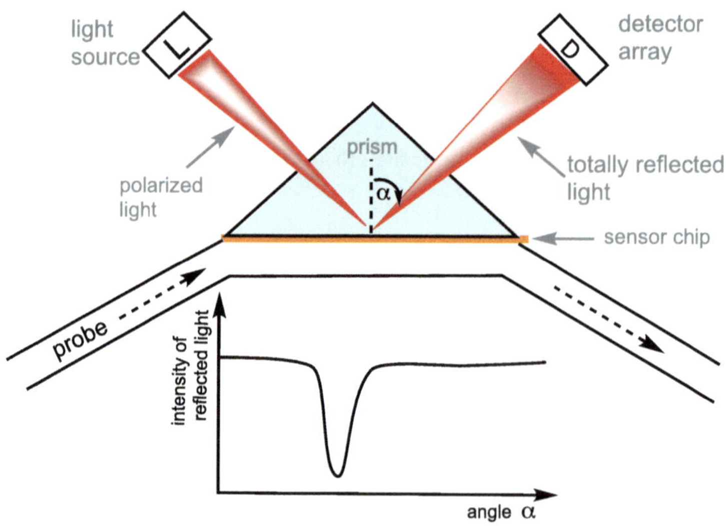

Fig. 9.4 Principle of surface plasmon resonance (SPR) [L … light source, D … diode array detector]

change in the adsorbed mass, i.e., a pronounced SPR signal. This only decreases when the lipopeptide is washed out again.

Figures 3 and 4 (see [1]) show the results of these experiments. Figure 3 shows the entire procedure of the SPR measurement. The surface of the HPA chip was loaded with dimyristoylphosphatidylcholine (DMPC) liposomes and the corresponding lipopeptide (5% w/w). After a brief, vigorous wash with NaOH, the remaining free surface was blocked with BSA (*bovine serum albumin*). This was followed by incubation with streptavidin. The desorption of streptavidin during rinsing with buffer solution was then recorded. The original alkanethiol surface could be regenerated each time by washing with a detergent.

Figure 4 then compares the results for the solely farnesylated pentapeptide and the additionally palmitoylated heptapeptide. As references, both the pure DMPC surface without peptide ("null signal," i.e., no streptavidin is adsorbed) and a mixture of DMPC with a biotinylated detergent (B-DHPE, biotinylated dihexadecanoyl-phosphatidylethanolamine) (≫ irreversible binding of streptavidin) were investigated. The result is clear: While the monolipidated pentapeptide **2** is slowly washed out (together with streptavidin) during buffer washing, the SPR signal for the heptapeptide **3** with two lipid chains remains almost unchanged, i.e., **3** remains irreversibly anchored in the membrane layer. This experiment proves the importance of the double lipid modification of the N-Ras protein for stable anchoring to the cell membrane, which in turn is essential for biological activity.

The results are consistent with a previous publication from the group [7], in which fluorescence labeling of such lipopeptides showed that the doubly lipid-modified heptapeptide (fluorophore–Gly–Cys–Met–Gly–Leu–Pro–Cys–OMe) is localized in the membrane, whereas the monolipo pentapeptide is not. It was also shown that a farnesylated heptapeptide with a free cysteine at position 2 is apparently palmitoylated in a membrane by a membrane-bound protein *S*-acyl transferase and then becomes membrane-localized, whereas this does not occur with an analogue containing a Ser^2.

For a thorough understanding of the synthetic steps, the experimental section of Ref. [1] is also well worth reading.

9.2 Semisynthesis of Fluorescently Labeled N-Ras Proteins

▶ **The basis of this section is the publication** Synthesis and application of fluorescent Ras proteins for live-cell imaging, Reents, R., Wagner, M., Schlummer, S., Kuhlmann, J., Waldmann, H., *ChemBioChem* 6, 86–94; doi: https://doi-org.uaccess.univie.ac.at/10.1002/cbic.200400233 (2005) [8].

These recently discussed studies practically led directly to the challenge of obtaining N-Ras **proteins** with variations in the lipid chains at the C-terminus. Herbert Waldmann pursued a semisynthetic approach for this. A correspondingly truncated oncogenic N-Ras protein was to be produced in pure form by molecular biological expression and then linked to synthetic C-terminal lipopeptides. However, the two parts were not joined by a native peptide bond—the methods for this were not as advanced at the time as they are today (Chap. 8)—but rather, a Ras chimera was to be generated by conjugating a cysteine at the C-terminus of the protein with a maleimide group on the peptides (Sect. 7.2). Of course, the success of this strategy was by no means guaranteed from the outset. In Ref. [9], it was demonstrated that such protein chimeras actually retain biological activity.

The main goal of the present work was to obtain fluorescently labeled Ras proteins. Previously, such proteins could only be obtained by conjugation with *green fluorescent protein* (GFP) (Sect. 14.1) or other fluorescent proteins. From a biophysical perspective, this is problematic when comparing the masses: 21 kDa (Ras) and 26 kDa (GFP).

In the meantime, the group had adapted solid-phase peptide synthesis so that it could also be applied to such sensitive lipopeptides (especially the *S*-palmitoylated ones). The main problem was the anchor (i.e., the polymer-bound C-terminal protecting group). In publication [10], a hydrazide linkage is used, which, by mild oxidation with Cu(II) salts in the presence of oxygen and methanol as nucleophile, yields the corresponding methyl esters. Under these conditions, farnesyl- and palmitoyl-containing peptides can also be cleaved from the solid phase. In the case of a 4-chlorotrityl anchor [11] or 4-methyltrityl anchor [8], the peptide can be released with 10% TFA or 1% TFA in dichloromethane, respectively. However, in these cases, the palmitoyl group was not introduced on the solid phase.

Solid-Phase Synthesis of Lipopeptides

Due to the acid sensitivity of terpene residues (farnesyl or geranylgeranyl) and the extreme base lability of acylated cysteines (= thioesters), other anchors for the solid phase became necessary (Fig. 9.5). This also had to be considered for the temporary protecting group and the permanent side-chain protecting groups. The methods for this were accordingly adapted in Refs. [10, 11].

The fluorophore was first attached to the farnesyl or geranylgeranyl group. The corresponding heptapeptide with Cys(StBU) and a maleimide group at the N-terminus formed the thioether bond to the C-terminal cysteine of the truncated Ras protein in buffer solution, and the lipidated proteins could be obtained in pure form and characterized after cleavage of the *tert*-butyldisulfide protecting group with dithioerythritol (DTE) (Schemes 1 and 2).

However, as the paper puts it, frustration followed: The protein labeled with the fluorophore NBD (7-nitro-4-benzoxadiazolyl or 7-nitrobenzofurazanyl group) was clearly visible, but was not localized to the membrane and also showed no oncogenic activity. It was therefore apparently not palmitoylated. The Bodipy-FL chromophore showed the best fluorescence properties, but the protein chimera behaved just as in the case of NBD. Only the dansyl-labeled protein conjugate showed the expected biological activity, but no usable fluorescence (for fluorescence and fluorophores, see Sect. 14.1). It was concluded from this that Bodipy FL would be the fluorophore of choice, but the more polar moiety at the terpene unit apparently interfered with the palmitoylation step. Analogous lipopeptides were then synthesized in which the fluorescence label was attached via an ethylenediamine unit to the **carboxyl group** of the terminal farnesylated cysteine. These fluorescent lipopeptides were then linked to the protein part as above (Scheme 3, Fig. 2). This protein now met expectations: It showed a comparable differentiation efficiency to the wild-type oncogenic Ras and could also be excellently visualized in the plasma membrane, about 1 h after microinjection into the cells (Ref. [8] Figs. 5 and 6).

Summary and Outlook

Of course, these protein–lipopeptide conjugates were then also used for a wide variety of biophysical, spectroscopic, or biological studies [12, 13], but we will not pursue this topic further here. It is certainly a key aspect of the path to success that one not only does not shy away from collaboration with all kinds of research groups, but actively fosters it. And finally, the field of biological chemistry, as it might be called, is a particularly interdisciplinary field. This is how this book is intended to work: It aims to strengthen mutual understanding between chemists, biologists, and physicians, and thus promote collaboration.

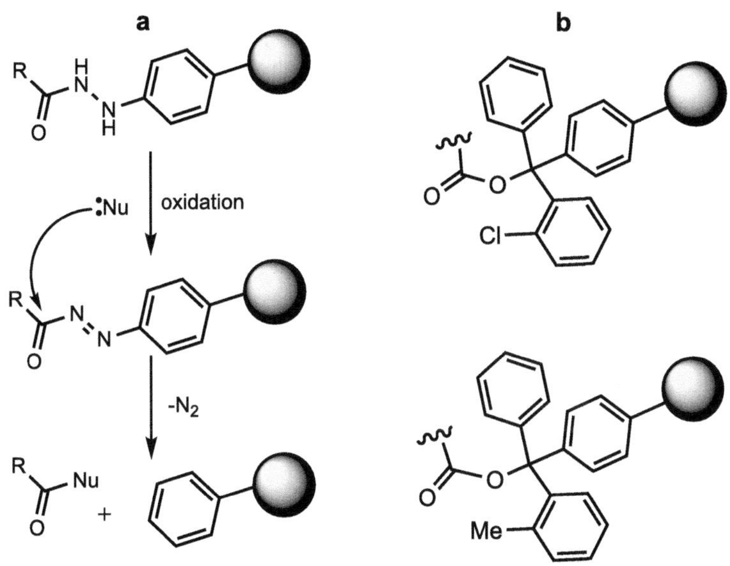

Fig. 9.5 Anchors for the solid-phase synthesis of lipopeptides: **a** Mechanism for pH-neutral cleavage of the hydrazide anchor by mild oxidation with Cu(II) salts to the acyldiazene, followed by nitrogen elimination under nucleophilic attack. **b** The two trityl linkages can be cleaved under mildly acidic conditions (10% or 1% TFA in dichloromethane)

9.3 Pepticinnamin E: A Natural Inhibitor of Ras Farnesyltransferase

> **This section is based on the publication** Synthesis and In Vitro Evaluation of the Farnesyltransferase Inhibitor Pepticinnamin E, Hinterding, K., Hagenbuch, P., Rétey, J., Waldmann, H., *Chem. Eur. J.* **5**, 227–236, https://doi.org/10.1002/(SICI)1521-3765(19990104)5:1<227::AID-CHEM227>3.0.CO;2-X (1999) [14]

When I first put together the lecture, the article [1] in Sect. 9.1 was the interesting starting point for me. But inevitably, I could not pass by this title [14]: The same research group, the same topic "Ras," but a completely different aspect of organic-chemical work on biologically relevant topics. This concerns the posttranslational modifications of the Ras protein, which we have already depicted in Fig. 9.2, specifically the first step, farnesylation. If this step is blocked, there is no active Ras protein—even in the case of a dangerous mutant—thus representing a possible therapeutic target. Therefore, there was great interest in inhibitors of this farnesyltransferase (Fig. 9.6).

A potent inhibitor discovered only shortly before, pepticinnamin E, isolated from a *Streptomyces* species, seemed particularly interesting for a peptide chemist, as it consists of several, though mostly non-proteinogenic, amino acids. In addition, the configuration of the natural product at the α-carbon atom of the highly substituted aromatic amino acid had not yet been elucidated. This work now describes the synthesis of pepticinnamin E. The authors also found the compound interesting because it has the structure of a bisubstrate inhibitor. The peptide portion could mimic the C-terminus of the unprocessed Ras, while the lipophilic N-terminal moiety likely imitates farnesylpyrophosphate (see Fig. 3). Inhibitors that replace both substrates of an enzyme in the active site should always be particularly effective.

The synthesis is described in Schemes 3, 4, and 5 (Ref. [14]). The previously unknown aromatic amino acid was prepared in both configurations from the corresponding benzyl bromide using the Schöllkopf method (see also Sect. 13.1.5). Both enantiomers were then converted to the final product in several steps. Comparison of the data showed that the natural pepticinnamin E contains this amino acid in the (*S*)-configuration, while the product with the (*R*)-configuration was called epi-pepticinnamin E.

To quantitatively describe the inhibition, an assay was established. The C-terminal tetrapeptide of the unprocessed Ras protein (–Cys–Val–Leu–Ser–OH, Fig. 9.2) was linked via a glycine residue at the N-terminus to the fluorescent dansyl group (Scheme 6). If the farnesyltransferase is active, the farnesyl group is transferred to the cysteine and the fluorescence of the dansyl chromophore shifts to significantly shorter wavelengths. As is well known, fluorescence is highly sensitive to the lipophilicity or hydrophilicity of its environment. During the course of the enzymatic reaction, fluorescence at the lower wavelength thus increases. If this time-dependent fluorescence

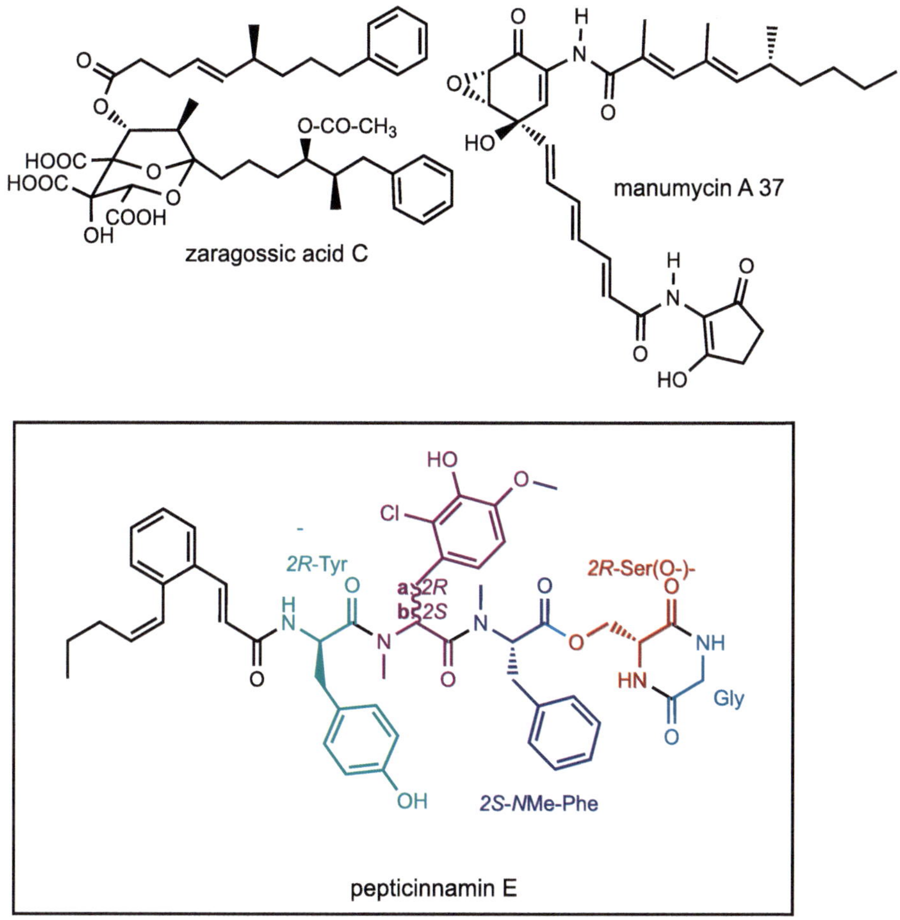

Fig. 9.6 Naturally occurring inhibitors of farnesyltransferase

intensity (sigmoidal curve) is evaluated kinetically, one obtains the IC_{50} values (see Sect. 13.1.2), which indicate the concentration at which only half of the enzyme activity is achieved. Pepticinnamin E shows the best effect, while the diastereomer epi-pepticinnamin E has an IC_{50} value six times higher. The central tripeptide with the correct configuration also shows quite good inhibition (Table 2).

This result provided a good starting point for the systematic development of analogs. In two subsequent studies [15], a library of variants was synthesized, some of which showed significantly lower IC_{50} values. However, only one compound was able to induce the death of cancer cells. Further studies are needed to understand the mechanism of inhibition and induced apoptosis.

Summary and Outlook

If one follows the literature in this area further, it appears that to this day there is still no truly effective cancer therapy through inhibition of Ras proteins. In certain types of cancer, the H-Ras protein can be treated with a farnesyltransferase inhibitor, but for the apparently more dangerous K-Ras mutants, there are still no suitable inhibitors of prenylation at the terminal cysteine. This is mainly because, in the case of farnesyltransferase inhibition, geranylgeranyltransferase takes over, so the oncogenic K-Ras protein still becomes active. Therefore, inhibitors for both enzymes are being developed [16].

Furthermore, one must of course also ask whether impaired prenylation might have negative side effects on completely different proteins. The most targeted approach to combating cancer caused by Ras mutants would be the specific inhibition of the *oncogenic* Ras protein. There have already been some successes in this direction: For a specific mutant of K-Ras (G12C), a compound has been developed that is now being tested in clinical trials. However, the goal of these developments is a therapy that works generally for oncogenic mutants [17, 18].

9.4 Questions

Let's test our understanding of the content of this chapter!

9.1 Explain the biological significance of the Ras protein!

9.2 Describe how the Ras protein is activated after translation! Which enzymes are involved?

9.3 What is the purpose of the lipid modifications on the N-Ras protein?

9.4 Why were allyl and choline ester protecting groups, among others, used in the synthesis of the lipopeptides?

9.5 Try to understand and explain the reaction schemes in Ref. [1]—especially as an exercise for understanding Chap. 3.

9.6 Where have we already encountered the reaction of maleimide with cysteines, which was used for the formation of the protein chimera in Ref. [8]?

9.7 What should be considered when incorporating fluorophores into biologically active molecules?

9.8 What is meant by a bisubstrate inhibitor?

9.9 Explain the mode of action of the assay for testing farnesyltransferase inhibition! See Ref. [14].

9.10 Explain the method of *surface plasmon resonance*!

References

1. Schelhaas, M., Nägele, E., Kuder, N., …, Waldmann, H.: Chemoenzymatic synthesis of biotinylated Ras peptides and their use in membrane binding studies of lipidated model proteins by surface plasmon resonance. Chem. Europ. J. **5**(4), 1239–1252 (1999). https://doi.org/10.1002/(SICI)1521-3765(19990401)5:4<1239::AID-CHEM1239>3.0.CO;2-N
2. Hinterding, K., Alonso-Díaz, D., Waldmann, H.: Organic Synthesis and Biological Signal Transduction. Angew. Chem. Int. Ed. **37**(6), 688–749 (1998). https://doi.org/10.1002/(SICI)1521-3773(19980403)37:6<688::AID-ANIE688>3.0.CO;2-B; Hinterding, K., Alonso-Dìaz, D., Waldmann, H.: Organische Synthese und biologische Signaltransduktion. Angew. Chem. **110**, 716–780 (1998). https://doi.org/10.1002/(SICI)1521-3757(19980316)110:6<716::AID-ANGE716>3.0.CO;2-8
3. Simanshu, D.K., Nissley, D.V., McCormick, F.: RAS proteins and their regulators in human disease. Cell. **170**(1), 17–33 (2017). https://doi.org/10.1016/j.cell.2017.06.009; Zinatizadeh, M. R., Zarandi, P. K., Keshavarz-Fathi, M., …, Rezaei, N.: The role of ral signaling and post translational modifications (PTMs) of Ras in cancer. Genome. Instab. Dis. **3**(1), 22–32 (2022). https://doi.org/10.1007/s42764-022-00059-0; Osaka, N., Hirota, Y., Ito, D., …, Sasaki, A.T.: Divergent mechanisms activating RAS and small GTPases through post-translational modification. Front. Molecular Biosci. **8**, 707439 (2021). https://doi.org/10.3389/fmolb.2021.707439; Campbell, S.L., Philips, M.R.: Post-translational modification of RAS proteins. Curr. Opinion. Struct. Biol. **71**, 180–192 (2021). https://doi.org/10.1016/j.sbi.2021.06.015
4. Lu, S., Jang, H., Muratcioglu, S., …, Zhang, J.: Ras conformational ensembles, allostery, and signaling. Chem. Rev. **116**(11), 6607–6665 (2016). https://doi.org/10.1021/acs.chemrev.5b00542 Dai, W., Xie, S., Chen, C., Choi, B.H.: Ras sumoylation in cell signaling and transformation. Seminars Cancer Biol. **76**, 301–309 (2021). https://doi.org/10.1016/j.semcancer.2021.03.033; Simão, S., Agostinho, R.R., Martínez-Ruiz, A., Araújo, I.M.: Regulation of Ras signaling by S-Nitrosylation. Antioxidants. **12**(8), 1562 (2023). https://doi.org/10.3390/antiox12081562
5. Kunz, H., Unverzagt, C.: The Allyloxycarbonyl (Aloc) Moiety–conversion of an unsuitable into a valuable amino protecting group for peptide synthesis. Angew. Chem. Int. Ed. **23**23(6), 436–437 (1984). https://doi.org/10.1002/anie.198404361; Kunz, H., Unverzagt, C.: Der Allyloxycarbonyl(aloc)-Rest – die Verwandlung einer untauglichen in eine wertvolle Aminoschutzgruppe für die Peptidsynthese. Angew. Chem. **96**(6), 426–427 (1984). https://doi.org/10.1002/ange.19840960612; Friedrich-Bochnitschek, S., Waldmann, H., Kunz, H.: Allyl esters as carboxy protecting groups in the synthesis of O-glycopeptides. J. Org. Chem. **54**(4), 751–756 (1989). https://doi.org/10.1021/jo00265a007; Trost, B.M.: New rules of selectivity: Allylic Alkylations catalyzed by Palladium. Acc. Chem. Res. **13**(11), 385–393 (1980). https://doi.org/10.1021/ar50155a001
6. Weimar, T.: Recent trends in the application of evanescent wave biosensors. Angew. Chem. Int. Ed. **39**(7), 1219–1221 (2000). https://doi.org/10.1002/(SICI)1521-3773(20000403)39:7<1219::AID-ANIE1219>3.0.CO;2-5; Weimar, T.: Neue Trends bei der Verwendung von Biosensoren auf der Basis evaneszierender Wellen. Angew. Chem. **112**(7), 1271–1274. (2000). https://doi.org/10.1002/(SICI)1521-3757(20000403)112:7<1271::AID-ANGE1271>3.0.CO;2-Q; Mozsolits, H., Aguilar, M.-I.: Surface plasmon resonance spectroscopy: an emerging tool for the study of peptide – membrane interactions. Biopolymers (Pept. Sci.) **66**(1), 3–18 (2002). https://doi.org/10.1002/bip.10200; Mozsolits, H., Thomas, W.G., Aguilar, M.-I.: Surface plasmon resonance spectroscopy in the study of membrane-mediated cell signalling. J. Pept. Sci. **9**9(2), 77–89 (2003). https://doi.org/10.1002/psc.439

7. Waldmann, H., Schelhaas, M., Nägele, E., …, Silvius, J.R.: Chemoenzymatic synthesis of fluorescent N-Ras lipopeptides and their use in membrane localization studies in Vivo. Angew. Chem. Int. Ed. **36**(20), 2238–2241 (1997). https://doi.org/10.1002/anie.199722381 (1997); Waldmann, H., Schelhaas, M., Nägele, E., …, Silvius, J.R.: Chemoenzymatische Synthese fluoreszierender N-Ras-Lipopeptide und ihre Verwendung bei In-vivo-Studien zur Membranlokalisierung. Angew. Chem. **109**(20), 2334–2337 (1997). https://doi.org/10.1002/ange.19971092028
8. Reents, R., Wagner, M., Schlummer, S., Kuhlmann, J., Waldmann, H.: Synthesis and application of fluorescent ras proteins for live-cell imaging. ChemBioChem. **6**(1), 86–94 (2005). https://doi-org.uaccess.univie.ac.at/10.1002/cbic.200400233 s. auch: Reents, R., Wagner, M., Kuhlmann, J., Waldmann, H.: Synthesis and Application of Fluorescence-Labeled Ras-Proteins for Live-Cell Imaging. Angew. Chem. Int. Ed. **43**(20), 2711–2714 (2004). https://doi.org/10.1002/anie.200353265; Reents, R., Wagner, M., Kuhlmann, J., Waldmann, H.: Synthese und Anwendung fluoreszenzmarkierter Ras-Proteine in der Bildgebung lebender Zellen. Angew. Chem. **116**(20), 2765–2768 (2004). https://doi.org/10.1002/ange.200353265
9. Kuhn, K., Owen, D.J., Bader, B., Wittinghofer, A., Kuhlmann, J., Waldmann, H.: Synthesis of functional Ras lipoproteins and fluorescent derivatives. J. Amer. Chem. Soc. **123**(6), 1023–1035 (2001). https://doi.org/10.1021/ja002723o
10. Ludolph, B., Waldmann, H.: The synthesis of acid- and base-labile lipopeptides on solid support. Chem. Eur. **9**(15), 3683–3691 (2003). https://doi.org/10.1002/chem.200304822
11. Völkert, M., Uwai, K., Tebbe, A., …, Waldmann, H.: Synthesis and biological activity of photoactivatable N-Ras peptides and proteins. J. Am. Chem. Soc. **125**(42), 12749–12758 (2003).. https://doi.org/10.1021/ja036178d
12. Triola, G., Waldmann, H., Hedberg, C.: Chemical biology of lipidated proteins. ACS Chem. Biol. **7** (1), 87–99 (2012). https://doi.org/10.1021/cb200460u (2012); Goermer, K., Buerger, M., Kruijtzer, J.A.W., …, Waldmann, H.: Chemical-biological exploration of the limits of the Ras de- and repalmitoylating machinery. ChemBioChem. **13**13(7), 1017–1023 (2012). https://doi.org/10.1002/cbic.201200078; Mejuch, T., Waldmann, H.: Synthesis of lipidated proteins. Bioconjugate Chem. **27** (8), 1771–1783 (2016). https://doi.org/10.1021/acs.bioconjchem.6b00261
13. Erwin, N., Patra, S., Dwivedi, M., …, Winter, R.: Influence of isoform-specific Ras lipidation motifs on protein partitioning and dynamics in model membrane systems of various complexity. Biol. Chem. **398**(5–6), 547–563 (2017). https://doi.org/10.1515/hsz-2016-0289 (2017); Hu, J., Zhu, P., Li, Y., Chen, Y.: Synthesis of Ras proteins and their application in biofunctional studies. Chin. Chem. Lett. **29**(7), 1043–1050 (2018). https://doi.org/10.1016/j.cclet.2018.05.035
14. Hinterding, K., Hagenbuch, P., Rétey, J., Waldmann, H.: Synthesis and In Vitro evaluation of the Farnesyltransferase Inhibitor Pepticinnamin E. Chem. Eur. J. **5**(1), 227–236 (1999). https://doi.org/10.1002/(SICI)1521-3765(19990104)5:1<227::AIDCHEM227>3.0.CO;2-X
15. Thutewohl, M., Kissau, L., Popkirova, B., …, Waldmann, H.: Solid-phase synthesis and biological evaluation of a pepticinnamin E library. Angew. Chem. Int. Ed. **41**(19), 3616–3620 (2002). https://doi.org/10.1002/1521-3773(20021004)41:19<3616::AID-ANIE3616>3.0.CO;2-F; Thutewohl, M., Kissau, L., Popkirova, B., …, Waldmann, H.:Festphasensynthese und biologische Evaluierung einer Pepticinnamin-E-Bibliothek. Angew. Chem. **114**(19), 3768–3772 (2002). https://doi.org/10.1002/1521-3757(20021004)114:19<3768::AID-ANGE3768>3.0.CO;2-5; Thutewohl, M., Kissau, L., Popkirova, B., …, Waldmann, H.: Identification of mono- and bisubstrate inhibitors of protein farnesyltransferase and inducers of apoptosis from a pepticinnamin E library. Bioorg. Med. Chem. **11**(12), 2617–2626 (2003). https://doi.org/10.1016/S0968-0896(03)00160-3

16. Dai, X., Sun Y., Zhang, T., …, Hongwei, G.: An overview on natural farnesyltransferase inhibitors for efficient cancer therapy. J. Enzyme. Inhibition. Med. Chem. **35**(1), 1027–1044 (2020). https://doi.org/10.1080/14756366.2020.1732366; Xia, M.-t., Yu, L., Yan, Z.-w., …, Li, S.: FTase inhibitors and cancer: prospects for use in targeted therapies. Med. Chem. Res. **33**(1), 21–35 (2024). https://doi.org/10.1007/s00044-023-03171-2; Horiuchi, N., Omer, R., Sugino, F., …, Ohkanda, J.: Design and evaluation of bivalent K-Ras inhibitors that target the CAAX binding site and the acidic surface of Farnesyltransferase and Geranylgeranyltransferase I. Chem. Eur. J. **31** (27), e202500306 (2025). https://doi.org/10.1002/chem.202500306
17. Khan, I., Rhett, J.M., O'Bryan, J.P.: Therapeutic targeting of RAS: New hope for drugging the "undruggable". Biochimica et Biophysica Acta (BBA) Mol Cell Res. **1867**(2), 118570 (2020). https://doi.org/10.1016/j.bbamcr.2019.118570
18. Qin, W., Liu, Z., Huang, M., …, Wei, X.: Recent advances in peptide inhibitors targeting wild-type Ras protein interactions in cancer therapy. Internat. J. Mol. Sci. **26**(4), 1425 (2025). https://doi.org/10.3390/ijms26041425

Development of Synthetic Peptide-based Vaccines

10

Abstract

This chapter addresses the topic of "Synthetic Peptide-Based Vaccines" from the field of immunology, using the example of efforts to develop anti-malaria vaccines. Following a brief introduction clarifying some immunological terms, the life cycle of the malaria parasite, *Plasmodium falciparum*, is described. Subsequently, publications dealing with the development of peptide-based vaccines targeting the sporozoite and merozoite stages of the parasite in humans are discussed in detail, and the current status of malaria control using vaccines and therapeutics is mentioned. In addition, various techniques for determining the efficacy of antibodies—including a publication on malaria—are briefly discussed.

10.1 Brief Immunological Introduction

In this chapter, we chemists are treading on thin ice. By this I mean myself and also the readers with a primarily chemical background. But immunology is an interesting field, especially for bioorganic chemists. And antibodies (Ab)have often been and continue to be used in biochemistry—so one should have at least some understanding of them.

Antibodies are protein molecules (immunoglobulins) produced by the so-called B cells of the immune system after presentation of an antigen (Ag) (usually a protein molecule from the pathogen). Some recognize a short, specific amino acid sequence, while others recognize a spatial arrangement of amino acids within a protein structure—this is referred to as a continuous or discontinuous epitope. Usually, several different antibodies (which also recognize different epitopes) are produced against an antigen; these are called polyclonal antibodies in a serum. Through special methods, monoclonal

D. Krois, *Organic Chemical Methods*, https://doi.org/10.1007/978-3-662-73573-2_10

antibodies (mAb) can be obtained and further propagated—these are then a single type of immunoglobulin molecule.

Most vaccines consist of dead or attenuated pathogens, and our immune system ultimately responds by producing antibodies that will then recognize and neutralize the actual pathogen. However, this reaction only occurs if the B and T cells (which ultimately mediate the cellular immune response) of the immune system are activated. The use of whole pathogens or at least large parts of them, however, carries some disadvantages: many antibodies are also produced against the more variable parts, which may be ineffective in a later infection if that infection is caused by a slightly different strain. This is why, for example, the flu vaccine has to be reformulated every year and in some years provides only partial or even no protection at all.

Thus, the idea of synthetic vaccines has been around for a long time. These should covalently combine only specific immunogenic and antigenic regions—ideally those that are little or not at all variable—into a single molecule. Ideally, they should also contain another molecular component that further enhances the immune response, i.e., has adjuvant properties (for externally added adjuvants, see Table 10.1). Despite the attractiveness of this idea, no such vaccine has yet reached the market. One problem seems to be that especially the immunogenic regions intended to activate T cells are not perceived equally strongly by every individual. On the other hand, with a huge protein molecule from the pathogen, this is less of an issue, as there is likely a suitable T-cell epitope for everyone somewhere in the molecule. In fact, immunogenicity depends on whether the MHC molecules (*major histocompatibility complex*), which vary between individuals and in humans constitute the HLA genotype (*histocompatibility* or *human leukocyte antigens*), can form a complex with the peptide and present it to the TCR (*T cell receptor*).

10.2 Malaria

10.2.1 Plasmodium Falciparum—The Causative Agent of Malaria

Malaria is one of the most widespread and also most dangerous diseases. In *Nature*, a special themed section titled *Outlook Malaria* appeared in 2004, 2012, and 2023 [1]. Each year, well over 100 million people contract the disease, and nearly a million die annually, mainly in Africa. Only in recent years have two vaccines finally been approved by the WHO, and this is partly due to the complex life cycle of this parasite (Fig. 10.1).

The primary infection is caused by sporozoites, which are transmitted via the bite of the *Anopheles* mosquito—the two vaccines approved so far are directed against the sporozoites. These enter the liver via the bloodstream and infect hepatocytes (liver cells). Various sporozoite-like stages are passed through in the liver. Ultimately, merozoites are produced and released into the bloodstream. These then infect red blood cells (erythrocytes) and multiply massively. This leads to the actual severe disease. Upon another mosquito bite, some sexual gametocytes, produced in smaller quantities, are taken up by the

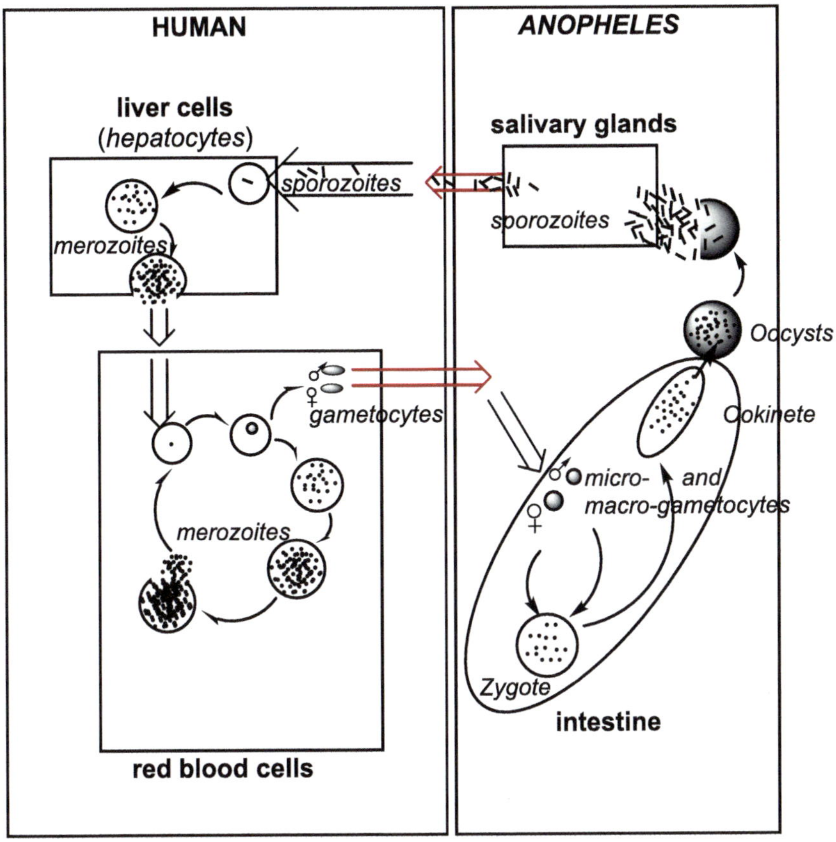

Fig. 10.1 The life cycle of the parasite Plasmodium falciparum, shown schematically

mosquito, where they develop into a zygote in its gut after fertilization. Through further stages, the oocysts reach the salivary gland, where sporozoites are released again, and the cycle begins anew. For a human vaccine, the sporozoite and merozoite stages are thus suitable targets, but they display very different surface proteins. The literature shows that it is much easier to induce effective antibodies against the sporozoites. The merozoite proteins, on the other hand, have highly variable regions that differ from strain to strain, and conserved regions that are not immunogenic. Manuel Elkin Patarroyo, who leads a special research institute on this topic in Colombia, is convinced that peptide-based synthetic vaccines combining several antigenic regions would be best suited for preventive vaccination. The challenge is, on the one hand, to identify the appropriate conserved (not highly variable) regions of the proteins [2] and, on the other hand, to make the corresponding peptides immunogenic, since these parts often are not immunogenic at all. In this context, he often refers to the *code of immunological silence* [3, 4].

10.2.2 Synthetic Peptide-Based Vaccines

There are several approaches to developing malaria vaccines. Here, we will discuss only a few that deliberately pursue the path toward a synthetic, peptide-based vaccine. The goal is to induce effective, "neutralizing" (i.e., those that recognize and disable the "enemy") anti-peptide antibodies. Much has been published regarding the antigen–antibody relationship. If the transition state of a chemical reaction is mimicked in a stable molecule and antibodies are produced against this molecule, these antibodies should and can have catalytic activity, essentially acting as artificial enzymes [5]. When peptides are to be used as effective vaccines, not only is the optimal antigen–antibody relationship important, but various components of the immune system come into play, and for immunogenicity (interaction with T cells), processed antigens are relevant, which is difficult to predict. This challenge was generally highlighted in the articles in Ref. [6] in 2001.

The publications we wish to discuss here also deal precisely with these problems.

10.3 Vaccines Against the Circumsporozoite Protein (CSP)

10.3.1 A Fully Synthetic Vaccine Based on Dendrimers

▶ **This section is based on the publication** A totally synthetic polyoxime malaria vaccine containing ***Plasmodium falciparum*** B cell and universal T cell epitopes elicits immune responses in volunteers of diverse HLA types, Nardin, E.H., Calvo-Calle, J.M., Oliveira, G.A., Nussenzweig, R.S., Schneider, M., Tiercy, J.-M., Loutan, L., Hochstrasser, D., Rose, K., *The Journal of Immunology*, **166** (1), 481–489, https://doi.org/10.4049/jimmunol.166.1.481 (2001) [7].

In this study, a fully synthetic vaccine against the envelope protein of sporozoites (CSP) was actually produced and tested in human volunteers. Two T-cell epitopes and one B-cell epitope were combined in the peptide component. The circumsporozoite protein CSP has a repetitive region with the sequence –Asn–Ala–Asn–Pro– (sometimes also written as –Asn–Pro–Asn–Ala–, which is the same), and it is known that antibodies against this sequence also have a neutralizing effect on sporozoites. However, and this is a main point of this publication, this region is not immunogenic. Therefore, two identified T-cell epitopes were also integrated into the peptide component. In earlier experiments, only the T1 epitope had been included, and only individuals with a certain HLA class II genotype responded well (about 25–35% of the population). By additionally incorporating the universal T* epitope, a broader range of effectiveness was hoped for.

Since linear peptides are generally not very immunogenic, they are usually either polymerized (e.g., via disulfide bridges), or—preferably—synthesized as dendrimers, known as *multiple antigenic peptides* (MAPs) [8] (Fig. 10.2). The method of oxime

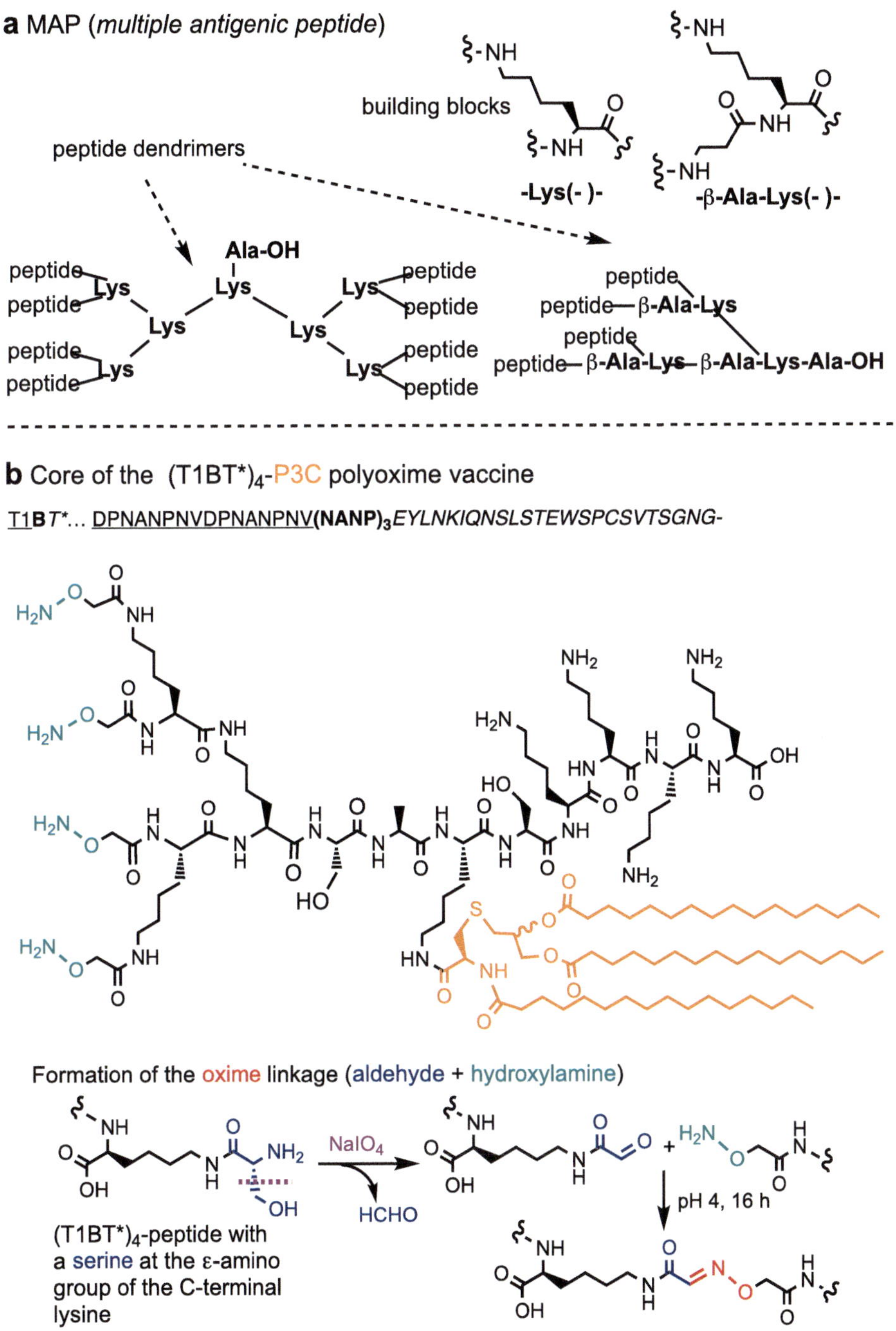

Fig. 10.2 **a** Multiple antigenic peptides (MAPs) in general (after [8]) and **b** Structure of the synthetic polyoxime malaria vaccine (after [7])

linkage (see Section 7.3) for the synthesis of MAPs had already been identified as particularly suitable by one of the authors in 1994 [9]. The fully synthetic vaccine, consisting of four 48-mer peptides at the core along with the endogenous Pam3Cys adjuvant (three palmitoyl residues on cysteine) [10] (Fig. 10.2), was shown by HPLC, MALDI-MS, and SDS-PAGE to be homogeneous and of high purity.

This study then focuses primarily on the immune response of the volunteers depending on their HLA genotype. Of the ten individuals, only three showed little or no response. All others showed good antibody titers after the three vaccinations, and these antibodies were also able to immobilize (and thus neutralize) sporozoites in a CSP test. The question is then discussed as to why a T-cell epitope from the pathogen's proteins should be better than a well-known, highly immunogenic structure from another organism (e.g., tetanus toxin, diphtheria toxin, influenza hemagglutinin, or hepatitis B surface protein … HBsAg). It is plausibly argued that using "**own**" T-cell epitopes would mean that immunity would be boosted with every contact with the pathogen (via a mosquito bite); this would not happen with a "foreign" immunogen.

This work was published in 2001. But what became of it afterwards? Ten years later, the corresponding author Elizabeth Nardin published a review [11] in which she addresses this. It turned out that synthesizing the polyoxime vaccine on the scale required for clinical trials was too difficult. In animal experiments, the 48-mer peptide *T1***B***T** alone with an external adjuvant (Table 10.1) produced good antibody titers. Thus, the MAP structure does not seem to be necessary. However, the adjuvants commonly used in animal experiments are usually not approved for human use. The review does mention a licensed topical ointment containing the TLR-7 agonist imiquimod (Aldara ointment, by 3M™) as a very effective adjuvant when applied immediately after vaccination. In a 2021 publication [12], the effects of various monoclonal antibodies are compared, and it is concluded that the *T1* epitope, which is a less common repeat unit on CSP, is also the most effective antigen for generating neutralizing antibodies.

Adjuvants for Vaccinations

For strong antibody production and robust immune responses, adjuvants are usually used, which, however, can be more or less painful.

However, the concept was not completely abandoned: Recently, a method was presented [13] in which the *T1***B***T** peptide covalently linked to the Pam3Cys adjuvant is incorporated into a specially prepared microparticle (MP). This synthetic vaccine has already been tested in primates and is said to have excellent properties.

Table 10.1 Adjuvants to Enhance the Immune Response

Name	Composition	Mode of Action
Alum [$Al(OH)_3$] *)	Aluminum hydroxide gel	Delayed release of the antigen Enhanced uptake by macrophages
Incomplete Freund's adjuvant	Mineral oil-in-water emulsion	Delayed release of the antigen Enhanced uptake by macrophages
Complete Freund's adjuvant	Mineral oil-in-water emulsion + dead mycobacteria	Delayed release of the antigen Enhanced uptake by macrophages, costimulation in macrophages
Freund's adjuvant with MDP	Mineral oil-in-water emulsion + muramyl dipeptide (component of mycobacteria)	Like complete Freund's adjuvant
Montanide ISA 51 VG	Mineral oil-in-water emulsion + detergent	Similar to incomplete Freund's adjuvant
Imiquimod (Aldara, 3M)*)	4-Amino-1-isobutyl-1*H*-imidazo[4,5-*c*]quinoline	Induces an inflammatory skin reaction by binding to TLR-7

*) Approved for use in humans

On the Preparation of Microparticles (MP)

First, $CaCO_3$ cores are produced by rapidly mixing a 0.33 M Na_2CO_3 solution (with 1.2 g/L polyglutamic acid (PGA)) with a 0.33 M $CaCl_2$ solution. The spherical, porous $CaCO_3$ microparticles have a PGA surface layer. Using dynamic light scattering (DLS), particles of suitable size are selected (*flow process*). Subsequently, alternating layers of polylysine (PLL, positively charged) and PGA (negatively charged) are applied to the microparticles (7 layers) (LbL … *layer-by-layer* assembly) and crosslinked, with an outermost layer of the design peptide (DP … *T1***B***T** -K_{20}+Pam3Cys). This results in the following structure: $CaCO_3$:PGA:PLL:PGA:PLL:PGA:PLL:PGA:DP (the ":" stands for the crosslinked layers) [13]. Apparently a complicated process, but one that is evidently automatable and in any case leads to highly reproducible synthetic vaccines *without* the use of virosomes or surface structures from other pathogens (VLPs, *virus-like particles*).

10.3.2 Approved CSP-Based Vaccines

Mosquirix™ or RTS,S/AS01 Vaccine (Manufacturer: Glaxo Smith Kline)

In the past four years, the first vaccines against malaria have been approved by the WHO, specifically for children. In 2021, Mosquirix™ was generally approved for vaccination, especially in Africa. The malaria vaccine under the name RTS,S/AS01 was developed over many years (since 1987) by the Walter Reed Army Institute of Research and Glaxo Smith Kline. The RT peptide consists of the repeat units of CSP Asn–Ala–Asn–Pro (here … R, in Section 10.3.1 this was the **B** part) and an immunodominant part (T). It is co-expressed with the N-terminal surface antigen of the hepatitis B virus (HBsAg, 226 amino acids) (S) in yeast, so that after combination with additional HBsAg, virus-like particles (VLP) with CSP and S on the surface are formed. AS01 refers to the specific adjuvant of the vaccine. In the phase III clinical trial, success rates of 25–50% were achieved in children, with decreasing efficacy over time [14, 15] and [16].

R21/Matrix-M^{TM} Vaccine (Manufacturer: Serum Institute of India PVT Ltd.)
This vaccine, approved by the WHO in 2023, was developed by the University of Oxford. It also consists of a virus-like particle (VLP) made from the fusion protein CSP-HBsAg, but without the addition of extra HBsAg, so that more CSP is exposed on the surface. The adjuvant is Matrix-M^{TM} (MM), a formulation of the *SARS-CoV-2 spike* (S) *glycoprotein* (NVX-CoC2373) and MM-adjuvant (Novamax, Gaithersburg, MD, USA). Clinical studies demonstrated higher efficacy (up to over 70%) in protection against malaria, but here too, efficacy wanes over time [15, 16] and [17].

Review [18] provides a general overview of vaccine development and opens a perspective toward mRNA vaccines. However, the development presented there toward self-replicating mRNA (in the vaccine) will certainly trigger even greater fears among vaccine-hesitant individuals. The development and production of RNA vaccines may be faster and cheaper, but for routine vaccines, synthetic vaccines would probably be safer and more widely accepted. In emergencies such as the COVID epidemic, however, this was certainly the best solution.

We should also keep in mind: *Plasmodium falciparum* is only one of the parasites that cause malaria. Others include *P. vivax*, *P. ovale*, *P. malariae*, and also *P. knowlesi*. In particular, *P. vivax* is also responsible for severe cases of malaria, and no vaccine has yet been approved against it, although there are approaches in development (e.g., Ref. [15, 19])

10.4 Peptide-based Vaccines Against Merozoites

10.4.1 Merozoite Proteins and Their Function

Proteins of the merozoites involved in the invasion of red blood cells (erythrocytes) have been identified as suitable targets for a vaccine against the blood-stage forms of the parasite. If these are blocked, infection should be prevented. The invasion process is quite complex. This is well summarized, for example, in the introduction of a dissertation [20]. Initially, the attachment involves surface proteins (*merozoite surface proteins* MSP), of which there are about ten; MSP-1 is probably the best studied. It is essential for the survival of the parasite. This is followed by a reorientation, during which the apical pole of the pear-shaped merozoite is aligned toward the host. This process involves the *apical membrane antigen*-1 (AMA-1), without which the parasites are not viable. Antibodies against AMA-1 prevent invasion despite successful attachment. Next, a tight junction is formed between the parasite and the host cell. This involves other proteins localized at the apical pole, in the micronemes; these include various EBA proteins and EBL-1. The merozoite then penetrates deeper into the invaginating erythrocyte membrane (with the help of an actin-myosin motor), and lipids and proteins from the rhoptries are released, including the RH and RAP complexes. In the end, the merozoite is located within the erythrocyte in a parasitophorous vacuole membrane. The *dense granules* migrate to the surface of the merozoite and release their contents (RESA proteins), which are likely involved in the transformation of blood cells. For an illustration of the merozoite, see Refs. [4] and [20].

10.4.2 Mechanisms of Immune Evasion by Merozoites

Plasmodium falciparum has developed a number of techniques to continue proliferating despite immune responses. First, the process of erythrocyte invasion is very rapid (about 1 min), leaving little time for the immune system to react. In addition, there are likely several independent invasion pathways, meaning that if an essential protein of one pathway is blocked by antibodies, infection can still occur via other factors.

Some proteins exhibit pronounced amino acid dimorphism [21]: In the MSP-1 protein, at positions 44, 47, and 52, there are two or three different amino acids in an otherwise highly conserved region (Table 10.2). However, monoclonal antibodies against one variant usually show little cross-reactivity with other variants (Sect. 10.5).

In all essential merozoite proteins, highly conserved and variable regions alternate. While the variable regions are usually immunogenic, antibodies cannot be generated against the conserved regions. There is a *code of immunological silence* [3, 4].

10.4.3 The Attempt to Make Conserved Essential Regions of Merozoite Proteins Immunogenic

> **This section is based on the publication** The molecular basis for peptide-based antimalarial vaccine development targeting erythrocyte invasion by P. falciparum. Aza-Conde, J., Reyes, C., Suárez, C. F., Patarroyo, M. A., Patarroyo, M. E.: *Biochem. Biophys. Res. Commun.*, **534,** 86–93, https://doi.org/10.1016/j.bbrc.2020.11.090 (2021) [22].

Over many years, the research group led by Manuel Elkin Patarroyo in Colombia has attempted to break the aforementioned *code of immunological silence*. First, peptides about 20 amino acids long each, spanning the entire length of the essential merozoite

Tab 10.2 Sequence variants of the sequence region 38 to 55 of the MSP-1 protein [21]

	38	39	40	41	42	43	**44**	45	46	**47**	48	49	50	51	**52**	53	54	55
1)	Ala	Val	Leu	Thr	Gly	Tyr	**Ser**	Leu	Phe	**Gln**	Lys	Glu	Lys	Met	**Val**	Leu	Asn	Glu
2)	Ala	Val	Leu	Thr	Gly	Tyr	**Gly**	Leu	Phe	**Gln**	Lys	Glu	Lys	Met	**Val**	Leu	Asn	Glu
3)	Ala	Val	Leu	Thr	Gly	Tyr	**Gly**	Leu	Phe	**His**	Lys	Glu	Lys	Met	**Val**	Leu	Asn	Glu
4)	Ala	Val	Leu	Thr	Gly	Tyr	**Gly**	Leu	Phe	**His**	Lys	Glu	Lys	Met	**Leu**	Leu	Asn	Glu
5)	Ala	Val	Leu	Thr	Gly	Tyr	**Gly**	Leu	Phe	**His**	Lys	Glu	Lys	Met	**Ile**	Leu	Asn	Glu
6)	Ala	Val	Leu	Thr	Gly	Tyr	**Ser**	Leu	Phe	**His**	Lys	Glu	Lys	Met	**Ile**	Leu	Asn	Glu

Designations of the sequence variants: 1) MAD 20, 2) NF 54, 3) IFA 9,2, 4) Brazil 608, 5) RO 71, 6) K 1 proteins, were synthesized using the Boc strategy of solid-phase peptide synthesis, and those that bind to erythrocytes with high affinity were identified, the so-called *conserved high activity binding peptides* (cHABP). The argument that peptides are unsuitable because they do not adopt a structure could be refuted, since the HABPs, with a length between 18 and 25 amino acids, according to NMR, exhibited structures largely identical to the corresponding parts of the protein itself [4]. Thus, such peptides would be suitable as vaccine components. However, even though these conserved and apparently functionally essential regions had now been identified and synthesized, they were useless as antigens, as they did not elicit an immune response. Individual amino acids were then replaced by others, i.e., small mutations were introduced, and thus for each cHABP a whole array of modified peptides (mHABP) was generated (see Fig. 2 in Ref. [22] and [4, 23]). Many of these were as unsuitable as the parent peptide, i.e., no antibodies against the protein were produced, but some were indeed effective! Whether immunization provides protection against infection with *Plasmodium falciparum* was tested in Aotus monkeys, as they are immunologically quite similar to humans and can be infected by *P. falciparum*. If, 15 days after an experimentally induced infection, no pathogens could be detected in the blood, the animals were considered protected. Only a few peptides showed both antibody titers and protection against malaria in the Aotus monkeys (see Fig. 2 in Ref. [22]). It may be disappointing that even in these cases only about 20–30% of the monkeys showed a response. Figure 2 also shows the correlation with the corresponding HLA genotypes and how the respective mHABPs fit into the binding pockets of these HLA molecules. Thus, in addition to immunological experiments and chemical peptide synthesis, a wealth of structural data was used both experimentally and computationally to achieve optimal interaction with HLA molecules and thus immune responses. Overall, a very elaborate and lengthy process, see Refs. [3, 4, 21–24]. Nevertheless, the research group is convinced that ultimately a fully developed synthetic vaccine, which by combining short peptides (mHABP) covers many different epitopes from several stages of the pathogen, will be the optimal solution.

10.5 Methods for Determining the Immunoreactivity of Antisera and Antibodies

In the previous sections, we have discussed examples of vaccine development. Now we will look at some general immunological methods for determining antigen-antibody interactions and the cross-reactivity of an antibody with modified antigens [25]. In this publication the specificity of monoclonal antibodies against a *Plasmodium falciparum* merozoite antigen is addressed, making this a very fitting example for the previous chapter.

> **This section is based on the publication** Comparison of analytical methods for the evaluation of antibody responses against epitopes of polymorphic protein antigens, Helg, A., Müller, M.S., Joss, A., Poltl-Frank, F., Stuart, F., Robinson, J.A., Pluschke, G., J. Immunol. Methods, **276**, 19–31, https://doi.org/10.1016/S0022-1759(03)00075-9 (2003) [25].

The techniques **IFA** (*immunofluorescence assay*), **Western Blot**, **ELISA** (*enzyme-linked immunosorbent assay*), **competitive ELISA**, **SPR** (*Surface Plasmon Resonance*, SPR, Surface Plasmon Resonance see Sect. 9.1) and **ITC** (*isothermal titration calorimetry*) are compared.

Two monoclonal mouse antibodies (mAB 7.19 and mAB 7.27), obtained by immunization against the MSP-1 variant MAD-20 (Table 10.2), were tested using the mentioned techniques for their affinity to *Plasmodium falciparum* schizonts (merozoite stage) of various strains (MAD-20, RO 71, NF 54, IFA 9.2, and K1, Table 10.2). The experimental section also describes all the conditions of the individual tests in detail.

Execution of the Individual Tests

In the IFA, erythrocytes infected with parasites of different strains are incubated with the monoclonal antibodies (mAB) and then treated with anti-mouse antibodies and a fluorescent dye. Detection is performed by fluorescence microscopy (Ref. [25], Fig. 2).

For the Western blot, the infected blood cells are lysed, the proteins separated by gel electrophoresis, and transferred to nitrocellulose filters (*blotting*). Strips of these are incubated with the mAB, and after washing, the remaining antibodies are stained using anti-mouse antibody-enzyme conjugates and an enzyme substrate (Ref. [25], Fig. 3).

For ELISA, microtiter plates are coated with polymerized peptides corresponding to the individual sequence variants, and after washing, incubated with dilutions of the mAB, washed, and then treated as in the Western blot with anti-mouse antibody-enzyme conjugates and finally with enzyme substrate. After a set time, the enzyme reaction is stopped. Absorbance is measured as a function of antibody dilution, typically yielding a sigmoidal curve; the antiserum dilution at the inflection point (half saturation) is referred to as the antibody titer. Strong affinity of the antibody to the antigen means strong absorbance and a high dilution at half saturation (Fig. 10.3). The interaction of the *bound immobilized* antigen with the antibody is measured (Ref. [25], Fig. 4a).

In the competitive ELISA, the plate is coated with the antigen used for immunization, and then a mixture of the mAB at the previously determined dilution (half saturation) with the various peptides (proteins) of the sequence variants to be tested in different dilutions is added. The procedure then continues as above. Here, the

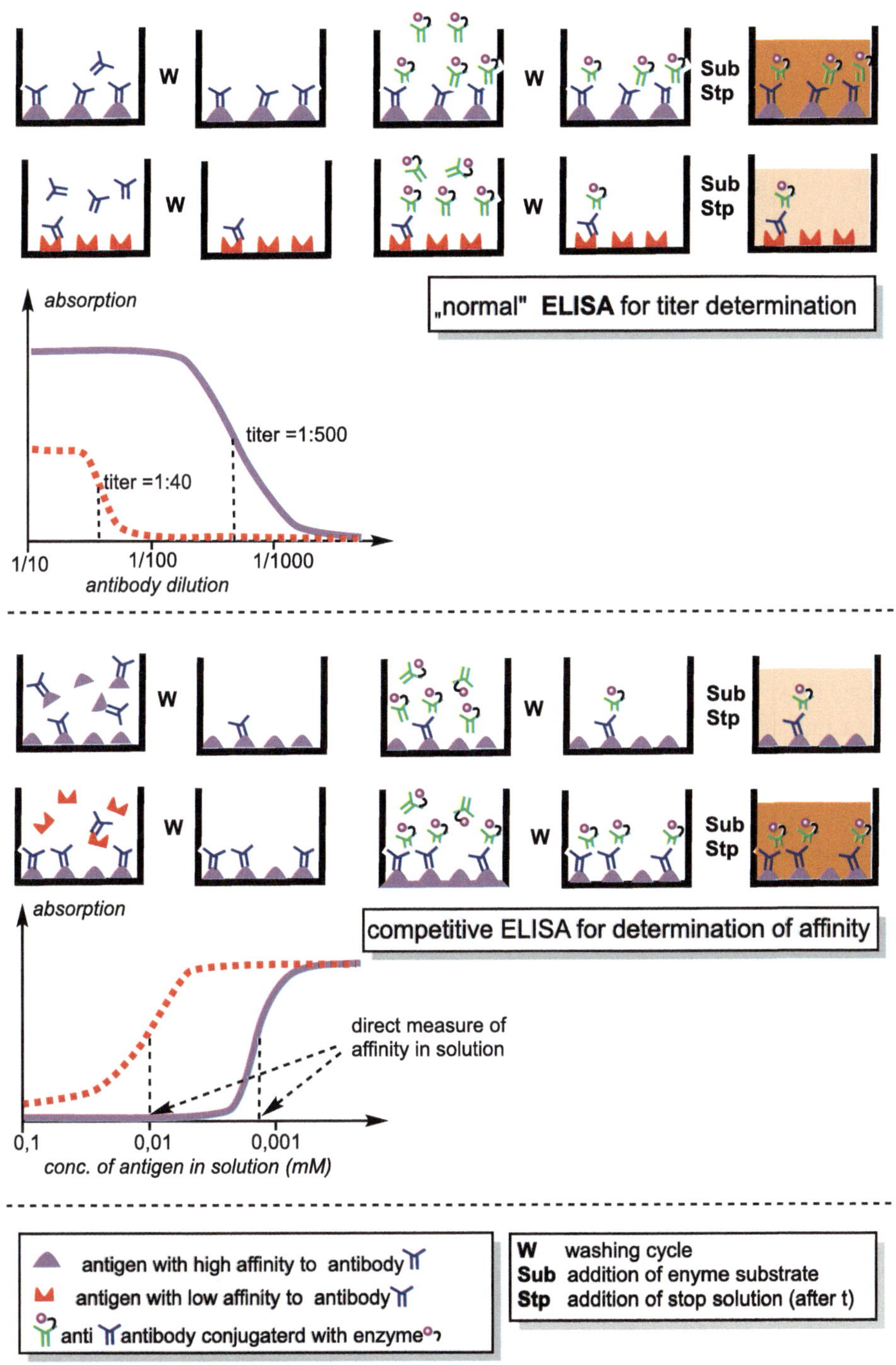

Fig. 10.3 Schematic representation of "normal" and competitive ELISA (*enzyme-linked immunosorbent assay*)

interactions of antibodies with antigens *in solution* are compared (Fig. 10.3) (see Ref. [25], Fig. 4b).

For SPR, the CM5 chip is activated as an *N*-hydroxysuccinimide ester and then conjugated with the antigen (SPf66 peptide) used for immunization. Dilutions of the mAB are passed over the chip, and the concentration corresponding to about half the maximum readout is selected. mAB at this dilution are preincubated with the various antigens to be tested and then passed over the microsensor chip in flow. The procedure thus corresponds roughly to the competitive ELISA. (Ref. [25], Fig. 4c).

ITC measurements directly determine the thermodynamic data of an interaction. In the sample cell, a solution of the mAB is placed, and then small amounts of the antigen to be tested are repeatedly added with a syringe, and the amount of heat released is measured (Ref. [25], Fig. 5).

The results discussed in the study showed that IFA and Western blot falsely suggested significant cross-reactivity of the mAB with all sequence variants of MSP-1. However, the much more sensitive ELISA (Fig. 10.3), SPR, and ITC measurements contradicted this, sometimes completely. The conclusion is that the simple tests often used in immunology (IFA, *Western blot*) do not distinguish between weak and strong antigen-antibody interactions and thus often yield misleading cross-reactivity results.

10.6 Therapeutics against Malaria

As we have seen, approved vaccines against malaria infection have only been available for a few years (and actually only for children). For a long time, drugs have been used to treat the disease, but these have not been sufficiently accessible to large segments of the population in affected countries [26]. After the malaria pathogen developed increasing resistance first to quinine and then to chloroquine (Fig. 10.4), the natural product artemisinin, isolated from *Artemisia annua*, was recognized in 1972 by the Chinese scientist Tu Youyou as a highly effective therapeutic, for which she was awarded the Nobel Prize in Physiology or Medicine in 2015 [27]. To improve bioavailability, derivatives (of the first and second generation) were then developed. Meanwhile, resistance to these artemisinin derivatives has also emerged, and the WHO has recommended combination therapies of artemisinin derivatives with, for example, mefloquine, amodiaquine, or lumefantrine (Fig. 10.4) [28]. There are also attempts to use chimeric therapeutics, i.e., those in which two different active substances are covalently linked [27].

Whereas in the previous examples effective natural products (quinine, artemisinin) served as starting points, from which analogs were then possibly developed, the compounds BRD3444 and BRD7929 shown in Fig. 10.4 were identified as hits in the context

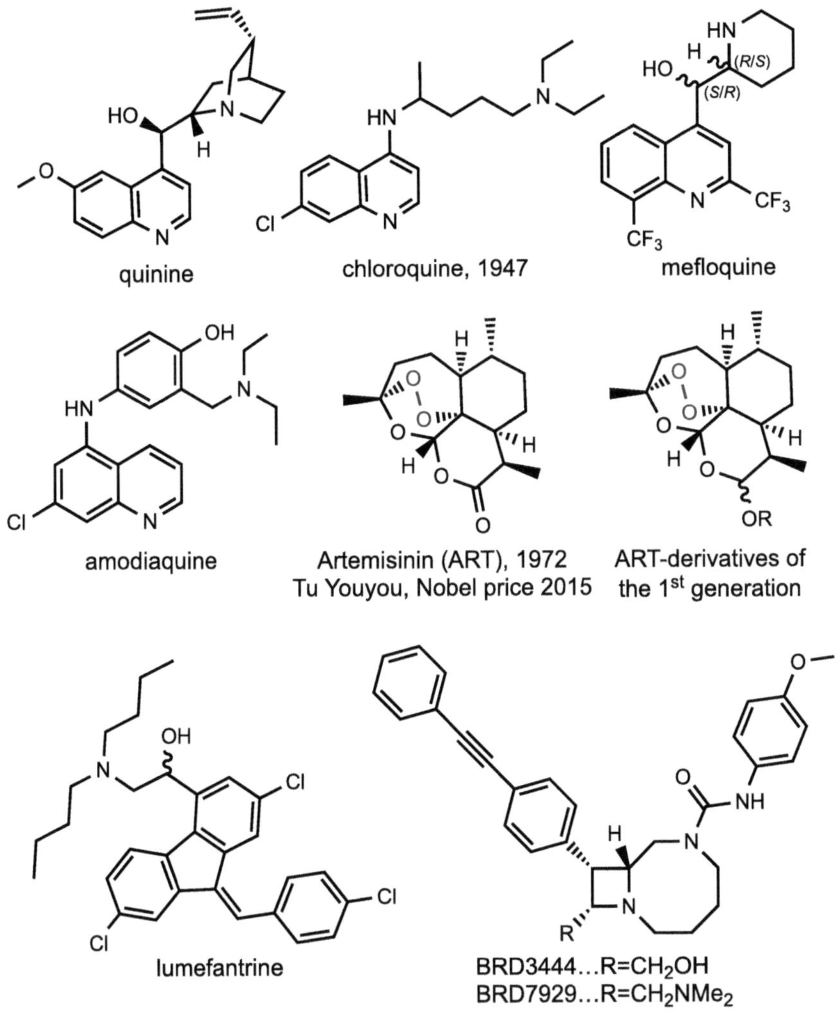

Fig. 10.4 Structures of some antimalarial therapeutics. (According to Ref. [27])

of a *diversity-oriented synthesis* (DOS) [29]. Starting from these substances, the "target" was then identified as the phenylalanine-tRNA synthetase of *Plasmodium falciparum* (PfcFRS). Recently, an X-ray crystal structure analysis of the complex of BRD7929 with the enzyme was also achieved [30]. This structural information can now serve as a basis for developing even more effective and, if possible, synthetically less complex analogs.

Summary and Outlook

In this chapter, using the example of the widespread disease malaria, we have primarily followed the development of vaccines. We have seen that, in addition to the conventional method of immunization with proteins from the pathogen, there are also approaches for synthetic vaccines. These would offer the advantage of standardized, analytically well-characterized components as well as the possibility of combining different antigenic regions in a single formulation. Especially for the complex developmental stages of the parasite, this would be the best solution. For a highly mutating pathogen such as the influenza virus or SARS-CoV-2, a synthetic vaccine would also be an advance. It should combine several essential conserved regions—in a slightly modified form to make them immunogenic, as outlined in Sect. 10.4.3. The same research group led by M. E. Patarroyo has indeed started such a project for SARS-CoV-2 [31].

10.7 Questions

Let's test our understanding of the content of this chapter!

10.1 What type of pathogen causes the disease malaria?

10.2 Can malaria normally be transmitted from person to person?

10.3 What are antibodies and which cells produce them?

10.4 The surface proteins of a pathogen often have highly conserved and highly variable regions. Why should antibodies be generated against the conserved regions?

10.5 What are MAPs?

10.6 In the example of the synthetic vaccine from Sect. 10.3.1, how are the antigenic peptides attached to the dendrimeric *core* ?

10.7 What is meant by dimorphism in the surface protein MSP-1 of merozoites?

10.8 How is a "normal" ELISA performed and what can be determined with it?

10.9 What is the principal difference between a "normal" and a competitive ELISA?

10.10 In the approved malaria vaccines, how is the antigen, the CSP protein, presented to the immune system?

10.11 Look at Fig. 10.4! Which structural element is common to the antimalarial drugs quinine, chloroquine, amodiaquine, lumefantrine, as well as BRD3444 and BRD7929?

References

1. Nature: Outlook malaria. *Nature*. **430**(7002), 923–946 (2004).; Nature: Outlook Malaria, *Nature*, **484**(7395), sup. S13–S59 (2012); Nature: Outlook Malaria. *Nature*, **618**(7967) (2023)
2. Patarroyo, M.E., Arévalo-Pinzón, G., Reyes, C., Moreno-Vranich, A., Patarroyo, M.A.: Malaria parasite survival depends on conserved binding peptides' critical biological functions. Curr. Issues Mol. Biol. **18**(1), 57–78 (2016). https://doi.org/10.21775/cimb.018.057

3. Patarroyo, M.E., Bermudez, A., Patarroyo, M.A.: Structural and immunological principles leading to chemically synthesized, multiantigenic, multistage, minimal subunit-based vaccine development. Chem. Rev. **111**(5), 3459–3507 (2011). https://doi.org/10.1021/cr100223m
4. Patarroyo, M.E., Patarroyo, M.A.: Emerging rules for subunit-based, multiantigenic, multistage chemically synthesized vaccines. Acc. Chem. Res. **41**(3), 377–386 (2008). https://doi.org/10.1021/ar700120t
5. Schultz, P.G.: Catalytic antibodies. Acc. Chem. Res. **22**(8), 287–294 (1989). https://doi.org/10.1021/ar00164a005; Schultz, P.G., Yin, J., Lerner, R.A.: The chemistry of the antibody molecule. Angew. Chem. Int. Ed. **41**(23), 4427–4437 (2002). https://doi.org/10.1002/1521-3773(20021202)41:23<4427::AID-ANIE4427>3.0.CO;2-K; Schultz, P.G., Yin, J., Lerner, R.A.: Die Chemie von Antikörpern. Angew. Chem. **114** (23), 4607–4618 (2002). https://doi.org/10.1002/1521-3757(20021202)114:23<4607::AID-ANGE4607>3.0.CO;2-U
6. Van Regenmortel, M.H.V.: The recognition of proteins and peptides by antibodies. J. Immunoassay. **21**(2-3), 85–108 (2000). https://doi.org/10.1080/01971520009349530; Van Regenmortel, M.H.V.: Pitfalls of reductionism in the design of peptide-based vaccines. Vaccine **19**(17–19), 2369–2374 (2001). https://doi.org/10.1016/S0264-410X(00)00456-4; Van Regenmortel, M.H.V.: Antigenicity and immunogenicity of synthetic peptides. Biologicals, **29** (3–4), 209–213, (2001). https://doi.org/10.1006/biol.2001.0308
7. Nardin, E.H., Calvo-Calle, J.M., Oliveira, G.A., …, Rose, K.: A totally synthetic polyoxime malaria vaccine containing Plasmodium falciparum B cell and universal T cell epitopes elicits immune responses in volunteers of diverse HLA types. J. Immunol. **166**(1), 481–489 (2001). https://doi.org/10.4049/jimmunol.166.1.481
8. Tam, J.P., Spetzler, J.C.: Multiple antigen peptide system. Methods Enzymol. **289**, 612–637., Academic Press (1997). https://doi.org/10.1016/S0076-6879(97)89067-2
9. Rose, K.: Facile synthesis of homogeneous artificial proteins. J. Am. Chem. Soc. **116**(1), 30–33 (1994). https://doi.org/10.1021/ja00080a004
10. Beekman, N.J.C.M., Schaaper, W.M.M., Tesser, G.I., …, Meloen, R.H.: Synthetic peptide vaccines: palmitoylation of peptide antigens by a thioester bond increases immunogenicity. J. Peptide Res. **50**(5), 357–364 (1997). https://doi.org/10.1111/j.1399-3011.1997.tb01195.x
11. Nardin, E.: The past decade in malaria synthetic peptide vaccine clinical trials. Hum. Vaccin. **6**(1), 27–38 (2010). https://doi.org/10.4161/hv.6.1.9601
12. Calvo-Calle, J.M., Mitchell, R., Altszuler, R., Othoro, C, Nardin, E.: Identification of a neutralizing epitope within minor repeat region of Plasmodium falciparum CS protein. npj Vaccines. **6**(1), 10 (2021). https://doi.org/10.1038/s41541-020-00272-6
13. Powell, T.J., Tang, J., Mitchell, R., …, Nardin, E.: Immunogenicity, efficacy, and safety of a novel synthetic microparticle pre-erythrocytic malaria vaccine in multiple host species. Vaccines. **11**(12), 1789 (2023). https://doi.org/10.3390/vaccines11121789
14. Nadeem, A.Y., Shehzad, A., Islam, S.U., …, Lee, Y.S.: Mosquirix™ RTS, S/AS01 vaccine development, immunogenicity, and efficacy. Vaccines. **10**(5), 713 (2022). https://doi.org/10.3390/vaccines10050713
15. Chen, J., Wang, Q., He, X., Yang, B.: Malaria vaccines: current achievements and path forward. Vaccines. **13**(5), 542 (2025). https://doi.org/10.3390/vaccines13050542
16. Duffy, P.E., Gorres, J.P., Healy, S.A., Fried, M.: Malaria vaccines: a new era of prevention and control. Nat. Rev. Miicrobiol. **22**(12), 756–772 (2024). https://doi.org/10.1038/s41579-024-01065-7
17. Aderinto, N., Olatunji, G., Kokori, E., …, Ojabo, R.E.: A perspective on Oxford's R21/Matrix-M™ malaria vaccine and the future of global eradication efforts. Malaria J. **23**(1), 16 (2024). https://doi.org/10.1186/s12936-024-04846-w
18. Tsoumani, M.E., Voyiatzaki, C., Efstathiou, A.: Malaria vaccines: from the past towards the mRNA vaccine era. Vaccines. **11**(9), 1452 (2023). https://doi.org/10.3390/vaccines11091452

19. De, S.L., Ntumngia, F.B., Nicholas, J., Adams, J.H.: Progress towards the development of a P. vivax vaccine. Expert Rev. Vacc. **20**(2), 97–112 (2021). https://doi.org/10.1080/14760584.2021.1880898
20. Viebig, S.: Charakterisierung von Membranproteinen der invasiven Stadien des Malariaerregers Plasmodium falciparum. Dissertation, Univ. Kiel (2011). http://d-nb.info/1012919889/34. Zugegriffen am 04.01.2026
21. Daubenberger, C.A., Nickel, B., Ciatto, C., …, Pluschke, G.: Amino acid dimorphism and parasite immune evasion: cellular immune responses to a promiscuous epitope of Plasmodium falciparum merozoite surface protein 1 displaying dimorphic amino acid polymorphism are highly constrained. Eur. J. Immunol. **32**(12), 3667–3677 (2002). https://doi.org/10.1002/1521-4141(200212)32:12<3667::AID-IMMU3667>3.0.CO;2-C
22. Aza-Conde, J., Reyes, C., Suárez, C.F., Patarroyo, M.A., Patarroyo, M.E.: The molecular basis for peptide-based antimalarial vaccine development targeting erythrocyte invasion by P. falciparum. Biochem. Biophys. Res. Commun. **534**, 86–93 (2021). https://doi.org/10.1016/j.bbrc.2020.11.090
23. Patarroyo, M.E., Bermudez, A., Alba, M.P., …, Vanegas, M.: Stereo electronic principles for selecting fully-protective, chemically-synthesised malaria vaccines. Front. Immunol. **13**, 926680 (2022). https://doi.org/10.3389/fimmu.2022.926680
24. Patarroyo, M.E., Bermúdez, A., Alba, M.P., …, Patarroyo, M.A.: IMPIPS: the immune protection-inducing protein structure concept in the search for steric-electron and topochemical principles for complete fully-protective chemically synthesised vaccine development. PLoS One. **10**(4), e0123249/0123241–e0123249/0123232 (2015). https://doi.org/10.1371/journal.pone.0123249
25. Helg, A., Mueller, M.S., Joss, A., …, Pluschke, G.: Comparison of analytical methods for the evaluation of antibody responses against epitopes of polymorphic protein antigens. J. Immunol. Methods. **276**(1-2), 19–31 (2003). https://doi.org/10.1016/S0022-1759(03)00075-9
26. Rathmes, G., Rumisha, S.F., Lucas, T.C.D., …, Weiss, D.J.: Global estimation of anti-malarial drug effectiveness for the treatment of uncomplicated Plasmodium falciparum malaria 1991–2019. Malaria J. **19**(1), 374 (2020). https://doi.org/10.1186/s12936-020-03446-8
27. Sharma, B., Singh, P., Singh, A.K., Awasthi, S.K.: Advancement of chimeric hybrid drugs to cure malaria infection: an overview with special emphasis on endoperoxide pharmacophores. Eur. J. Med. Chem. **219**, 113408 (2021). https://doi.org/10.1016/j.ejmech.2021.113408
28. Nguyen, T.D., Gao, B., Amaratunga, C., …, Aguas, R.: Preventing antimalarial drug resistance with triple artemisinin-based combination therapies. Nat. Commun. **14**(1), 4568 (2023). https://doi.org/10.1038/s41467-023-39914-3
29. Kato, N., Comer, E., Sakata-Kato, T., …, Schreiber, S.L.: Diversity-oriented synthesis yields novel multistage antimalarial inhibitors. Nature **538**(7625), 344–349 (2016). https://doi.org/10.1038/nature19804; Maetani, M., Kato, N., Jabor, V.A.P., …, Schreiber, S.L.: Discovery of antimalarial azetidine-2-carbonitriles that inhibit P. falciparum dihydroorotate dehydrogenase. ACS Med. Chem. Lett. **8**(4), 438–442 (2017). https://doi.org/10.1021/acsmedchemlett.7b00030; Maetani, M., Zoller, J., Melillo, B., …, Schreiber, S.L.: Synthesis of a bicyclic azetidine with in vivo antimalarial activity enabled by stereospecific, directed C(sp3)-H arylation. J. Am. Chem. Soc. **139**(32), 11300–11306 (2017). https://doi.org/10.1021/jacs.7b06994
30. Sharma, M., Mutharasappan, N., Manickam, Y., …, Sharma, A.: Inhibition of Plasmodium falciparum phenylalanine tRNA synthetase provides opportunity for antimalarial drug development. Structure. **30**(7), 962–972.e963 (2022). https://doi.org/10.1016/j.str.2022.03.017
31. Patarroyo, M.A., Patarroyo, M.E., Pabón, L., …, Avendano, C.: SM-COLSARSPROT: highly immunogenic supramutational synthetic peptides covering the world's population. Front. Immunol. **13**, 859905 (2022). https://doi.org/10.3389/fimmu.2022.859905

Applications of Modified Oligonucleotides

11

Abstract

Various applications of oligonucleotides with modified nucleobases are discussed. The first part addresses the significance of hydrogen bonds for the stability and replication of DNA. The replacement of the nucleobase thymine with an isosteric analogue that cannot form H-bonds with adenine yields particularly intriguing results regarding the selectivity of the polymerase reaction. The second section deals with the use of fluorophores and quenchers for the detection of specific RNA (or DNA) sequences. The focus is especially on innovative approaches that exploit chemical reactions for such assays. The third part presents a study in which targetedly modified oligonucleotides enabled the introduction of an electron at a specified site, and the effect on more distant nucleobases was investigated. Certainly, this is only a limited and arbitrarily selected excerpt from this fascinating and inspiring field of nucleic acid chemistry.

▶ In Sect. 6.3 we became acquainted with various types of nucleic acid modifications. Now we want to discuss some concrete applications based on publications that introduced new ideas and marked the beginning of an interesting development.

D. Krois, *Organic Chemical Methods*, https://doi.org/10.1007/978-3-662-73573-2_11

11.1 Targeted Modifications of Nucleobases for a Better Understanding of the DNA Double Helix According to the Watson and Crick Model

Our starting point is the following review article by Eric T. Kool Insights into the Stability and Replication of DNA by Mimicking Its Structure and Function, Kool, E.T., Morales, J.C., Guckian, K.M., *Angew. Chem.* **112** (6), 1046–1068, https://doi.org/10.1002/(SICI)1521-3757(20000317)112:6<1046::AID-ANGE1046>3.0.CO;2-9 (2000) [1].

The research group led by Prof. Eric T. Kool wanted clear evidence that it is indeed the hydrogen bonds that are crucial for the stability of the DNA double helix and responsible for selectivity during replication in the polymerase reaction. Certainly, others had asked these questions before, but to answer them, one would actually need to find model nucleobases that are completely identical in structure (i.e., isosteric) but cannot form H-bonds. By replacing the carbonyl oxygen atoms in thymine (**T**) with fluorine atoms and the –N–H group with –C–H, these requirements were met for a base. In Scheme 1 of the paper, you can see the space-filling models of **T** and **F** (this is the name given to this difluorotoluene derivative). There are practically no differences in spatial occupancy (see also Ref. [1], Fig. 2), but there are significant differences in charge distribution (Ref. [1], Scheme 3). Although the fluorine atom is more electronegative than O, in a C–F single bond it has a significantly lower partial charge. It was also known that F is a poor hydrogen bond acceptor. Accordingly, in the liquid-liquid partitioning of the nucleosides, **T** showed a partition coefficient $< 1/10$ between octanol and water, whereas the nucleoside **F** had a value >10. The analogous base pairings are shown in Fig. 11.1.

For the base adenine (**A**), it is not possible to construct a nonpolar, ideally isosteric base. The analog **Z** used here shows a significant difference in spatial occupancy, since the heterocyclic basic =N atom could only be replaced by a =CH group, with the H atom leading to unfavorable steric interactions with H–N (from **T**) (Fig. 11.1).

DNA Structure and Melting Temperature

As is well known, double-stranded DNA exists as an antiparallel double helix (Watson-Crick), *in vivo* as B-DNA. The backbone of 2-deoxyribose and phosphate is located at the outer periphery of the helix, while the base pairs fill the interior. Between stacked base pairs, π-π interactions (base stacking) occur, which naturally also affect the UV spectra of the heterocyclic bases. When the double helix separates, i.e., is denatured, the UV absorption at certain wavelengths changes significantly. This produces so-called melting curves, which provide information about the stability of the double helix (Fig. 11.2).

As a measure of stability during thermal denaturation, the melting temperature is defined, i.e., the temperature at which 50% is denatured, that is, the temperature at the inflection point of the curve. From this value, the thermodynamic stabilization energy of the double helix (ΔG) for the corresponding oligonucleotide can also be calculated.

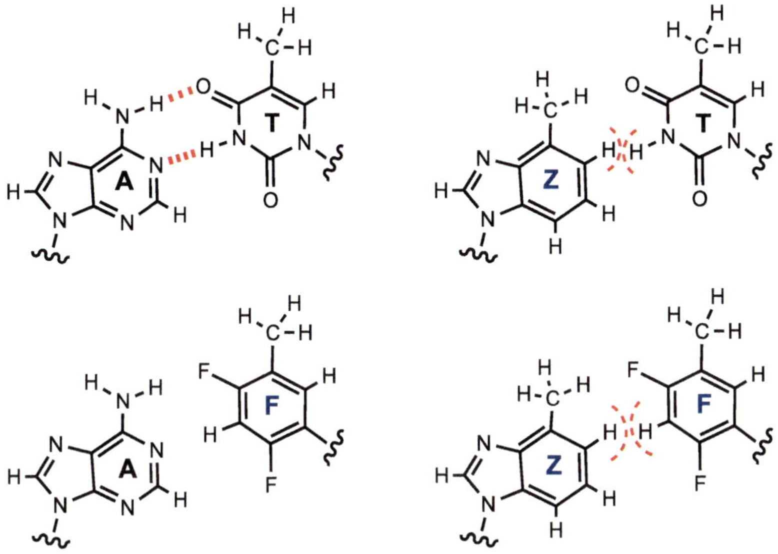

Fig. 11.1 Base pairings of **A–T** and the nonpolar analogs **Z** and **F**

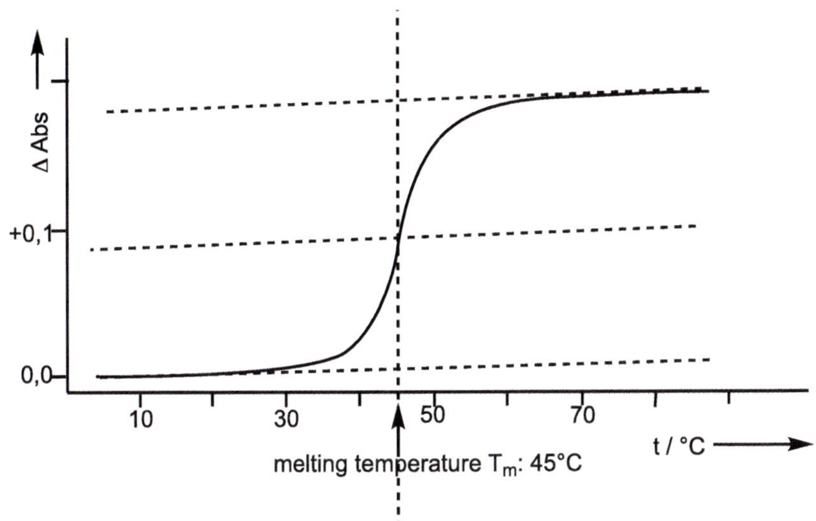

Fig. 11.2 Representative melting curve of a DNA double helix with determination of the melting temperature, measured at a wavelength of $\lambda = 260$ nm

To test the stability of the double helix, complementary dodecadeoxyribonucleotides with a single varied base in the middle were synthesized d(CTTTTC**X**TTCTT)–d(AAGAA**Y**GAAAAG), and the melting temperatures were determined (see Ref. [1], Table 2). All pairings **X–Y** with the base **F** show a clearly reduced melting temperature—thus lower stability, with no preference for pairing with **A** over the other pairings. The same is observed for **Z**, the nonpolar adenine analog. Only for the pairing **Z–F** (and also **F–F**) is a significantly smaller destabilization of the twelve-membered double helix measured.

This result is therefore as expected: it is the hydrogen bonds that stabilize the DNA double helix. Reference should be made to a publication from the same period that analyzes the nature of these H-bonds in more detail using theoretical chemical calculations [2]. However, the absence of H-bonds is probably not the only reason. Changes in the H_2O solvation shells and increased flexibility due to the absence of the hydrogen bond, and thus a disturbance in the orientation of the other base pairs, are also likely to play a role. What accounts for the certain stability of the nonpolar base pairings **Z–F** (and also **F–F**)? – This supports the theory that, in addition to direct H-bonds, bound water molecules also contribute to stabilization or destabilization. In a purely nonpolar pairing, H_2O is removed from the region, which, for entropic reasons, provides a certain thermodynamic stabilization effect. The structural consequences of a **Z–F** pairing in another dodecanucleotide were investigated in Ref. [3]. Despite destabilization, the overall structure of the DNA double helix changed very little. It was concluded that the targeted incorporation of such nonpolar bases could be used, for example, for targeted studies of protein-DNA interactions.

Nonpolar nucleobases were also incorporated at the unpaired ends of a double strand to estimate base stacking forces. Here, significantly increased stabilization could be achieved (see Ref. [1], Table 3).

Now, however, we come in the review article to the point that caused a great stir and apparently also considerable resistance [4]. The issue now was to clarify whether the selectivity of the polymerase reaction is also determined by the H-bonding patterns, as was widely assumed. For this purpose, two different experiments were conducted with respect to the ideal isostere **F**. A 23/28 nucleotide-long primer-template duplex was prepared, and *in vitro*, using the Klenow fragment of *E. coli* DNA polymerase I, the incorporation of each of the four natural deoxynucleoside triphosphates (Y) (dNTP) opposite the bases **T** or **F** at position X was investigated (Fig. 11.3).

To great surprise, in both cases dATP was incorporated, and in the case of **F** with almost unchanged selectivity and only about four times slower (Ref. [1], Figs. 5 and 6). Then the reverse experiment was performed: the incorporation of dTTP or dFTP (in general: dYTP) opposite X = **A** was monitored (Ref. [1], Fig. 7). The incorporation of dFTP occurred somewhat more slowly (about a factor of 10), but also with almost unchanged selectivity. The corresponding experiments with the sterically less perfect adenine analog **Z**, on the other hand, did not yield similarly high selectivity. The incorporation of dFTP opposite **Z** or dZTP opposite **F** was again quite selective (Ref. [1], Figs. 8 and 9). From

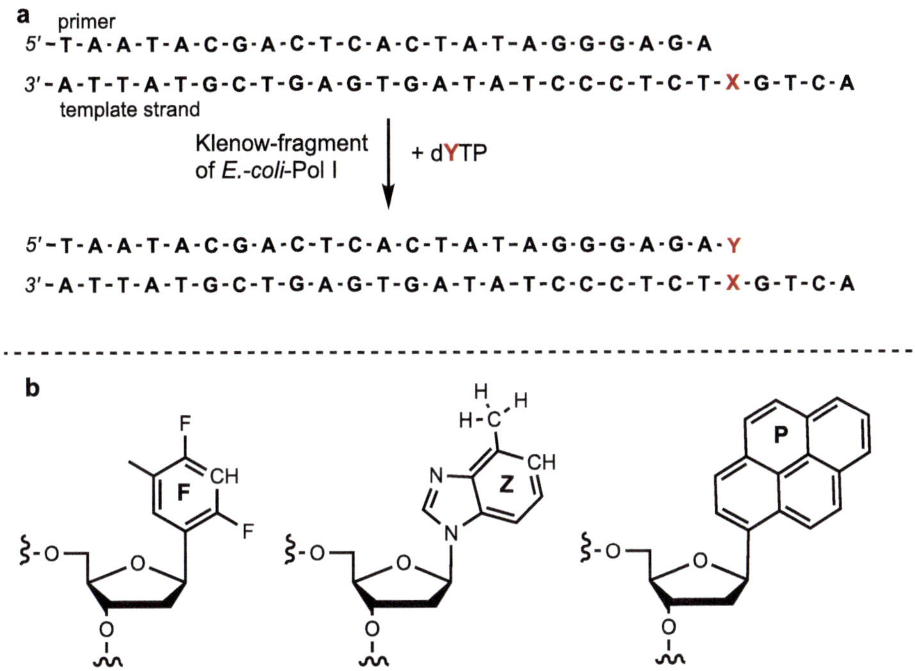

Fig. 11.3 **a** Testing of the difluorotoluene derivative **F** in the polymerase reaction; **b** Structural formulas of the nucleoside building blocks **F**, **Z**, and **P**

all these results, a model for the selectivity of the polymerase reaction based purely on steric criteria was developed (Ref. [1] Fig. 10). Only in the case of correct base pairing would the space in the active site be just right, with the influence of the hydration shell also playing a significant role here.

With any hypothesis or theory proposed in science, new experiments should arise from it that can then serve to test the thesis. This was also done in this case. Assuming that this concept of selection based on spatial occupancy and independent of H-bonds in the active site of the polymerase is correct, a nucleobase was chosen that should fit optimally opposite an *abasic site* (Φ, deoxyribose without nucleobase): the polycondensed aromatic pyrene as a nucleoside (**P**) fulfilled this requirement. The corresponding nucleoside triphosphate dPTP (Fig. 11.3b) was synthesized and tested as in the case of **F**. Whereas it was previously accepted that dATP is preferentially incorporated opposite abasic sites (A-rule), the incorporation of dPTP occurred 100 times faster and also selectively (factor >100) only opposite Φ (see Ref. [1], Figs. 13 and 14). This result excellently supported the theory.

In 2006, Prof. Eric T. Kool addressed discussions and counterarguments to these sensational results (the first article appeared in 1996) in Ref. [4] under the title: *The difluorotoluene debate – a decade later*. Further studies are also discussed, in which analogs of the base **F** with Cl, Br, or I atoms in place of the F atoms were tested. As expected, nucleobases with excessive spatial occupancy were not accepted by the polymerase, while the optimum selectivity was found with the 2-bromo-4-chlorotoluene derivative.

With regard to the nature of the polymerase, significant differences were subsequently observed. While the Klenow fragment of *E. coli* DNA polymerase I, Taq polymerase, and T7 DNA polymerase (class A polymerases), as well as class B polymerases and reverse transcriptases, accept the nonpolar base **F** with high efficiency and selectivity, there are other polymerases for which the formation of hydrogen bonds with the incoming dNTP is apparently also crucial [5].

Building on the insights from these studies, new base pairs have been and are being designed that could be of interest for genetic manipulations [6]. The concept of isosteric, nonpolar nucleobases was later revisited and expanded by E. T. Kool in another context: The functionality of enzymes that recognize and repair errors or missing nucleobases in DNA has become an important topic today, because low or absent activity of these enzymes can be a reason for the development of tumors. In review [7], methods are developed using specially modified nucleobases to elucidate the mechanisms of these enzymes, measure their activity, and also develop inhibitors. We thus see that the work from 1996–2006 continues to provide new impetus for research in the 2020s.

Results of These Studies

First, the stabilization of the DNA double helix by hydrogen bonds between the bases was clearly proven. Second, it was convincingly shown that for the selectivity of nucleotide incorporation by polymerases, either only the steric complementarity of the bases is decisive, or, in certain polymerases, at least plays a role alongside the importance of hydrogen bonds. Like any groundbreaking discovery, these studies have continued to exert their influence in subsequent years.

Although we have not considered the synthetic aspects in this chapter, we should not overlook the fact that the preparation of nucleotide building blocks with unnatural bases each required a separate organic-chemical synthesis.

11.2 Detection of RNA (and DNA) Using Specially Modified Oligonucleotides (*Molecular Beacons* and Others)

11.2.1 Hybridization with Fluorescent Reagents

We now turn to an analytical aspect from the field of nucleic acids. It is often of interest in research, as well as in medical diagnostics, to quickly detect the presence of certain mRNA molecules or point mutations in RNA (or DNA) using simple assays.

> As a starting point for this chapter, we will again refer to a publication by Eric T. Kool: Quenched probes for highly specific detection of cellular RNAs, Silverman, A. P., Kool, E. T., *Trends in Biotechnology*, **23** (5), 225–230, https://doi.org/10.1021/jacs.9b09812 (2005) [8]. This review article discusses various techniques for detecting specific RNA strands.

Initially, only the FISH technique (*fluorescence in situ hybridization*) was used for this purpose. This involves using relatively long (about 30 nucleotides) complementary oligonucleotides that are directly linked to a fluorophore at one end. The procedure typically includes the following steps: (i) fixation and permeabilization of the cells, (ii) hybridization, (iii) thorough washing to remove unbound fluorescent probe, and (iv) detection by fluorescence microscopy, etc.

Figure 11.4 shows a somewhat more complex variant: the primary probe is non-fluorescent and carries only recognition tags. After hybridization, washing can be performed, followed by the addition of antibodies or streptavidin conjugated to a fluorophore, and then another wash step. A disadvantage of the FISH technique is that the cells must be fixed, and nonspecific binding of fluorescent molecules can lead to a strong background signal. The oligonucleotides must be relatively long and are therefore not very selective—a single point mutation cannot be detected this way.

Around the year 2000, the first methods emerged that relied on the fluorescence of a fluorophore that was previously non-fluorescent due to quenching; the English umbrella term for this is *quenched probes*. The fluorescence quenching is based on quenching by direct contact or resonance energy transfer (RET) between the fluorophore and quencher. In the so-called *molecular beacons* [9], hybridization with the template—the target RNA sequence—induces a conformational change that separates the fluorophore and quencher (Fig. 11.5).

Due to a short complementary region at the 3′ and 5′ ends of the oligonucleotide, it adopts a loop structure, which causes the fluorescence of the fluorophore at one end to be quenched by the quencher at the other end. In the *loop* region, there is a sequence complementary to the template (green bases), which is significantly longer. When the *molecular beacon* finds the template, it hybridizes with the complementary red bases, causing the oligonucleotide to straighten. As a result, the fluorophore begins to fluoresce—but only as long as the double strand remains intact. This method has significant advantages over FISH. The washing steps can be omitted, and if the *molecular beacons* can be introduced into cells (by microinjection or permeabilization methods), experiments can also be performed in living or at least non-lysed cells. A problem is posed by nucleases, which cleave the oligonucleotide and thus lead to background fluorescence. Therefore, PNA-based probes (see Sect. 6.3.2) would be of interest for this [10]. More recent applications and developments based on the principle of *molecular beacons* can be found in the following publications: [10, 11].

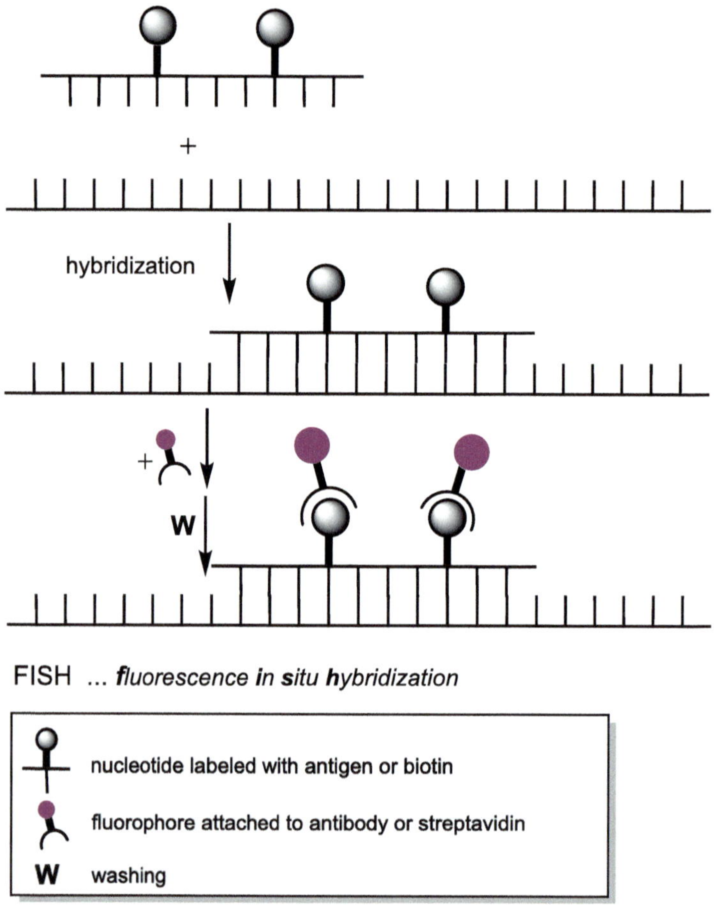

Fig. 11.4 Schematic representation of the FISH method for detecting RNA molecules—the nucleotide length is only symbolic

11.2.2 Template-Guided Reaction Activates Fluorophore

In the article we are discussing here [8], Prof. Eric T. Kool naturally focuses primarily on the *quenched probes* he developed. These are of particular interest to us as well, because the signal is produced by a chemical reaction. In addition to the always necessary chemical synthesis of the conjugated oligonucleotides, these techniques center on an organic-chemical transformation. He calls these reagents *QUAL probes* (*quenched autoligation probes*). The primary publication appeared in 2001 under the title *Quencher as Leaving Group* [12]. This title precisely describes the methodology: the fluorescence-quenching chromophore also serves as the leaving group in an S_N2 reaction that links two short oligonucleotides when they—bound to the template—are in close proximity to each other.

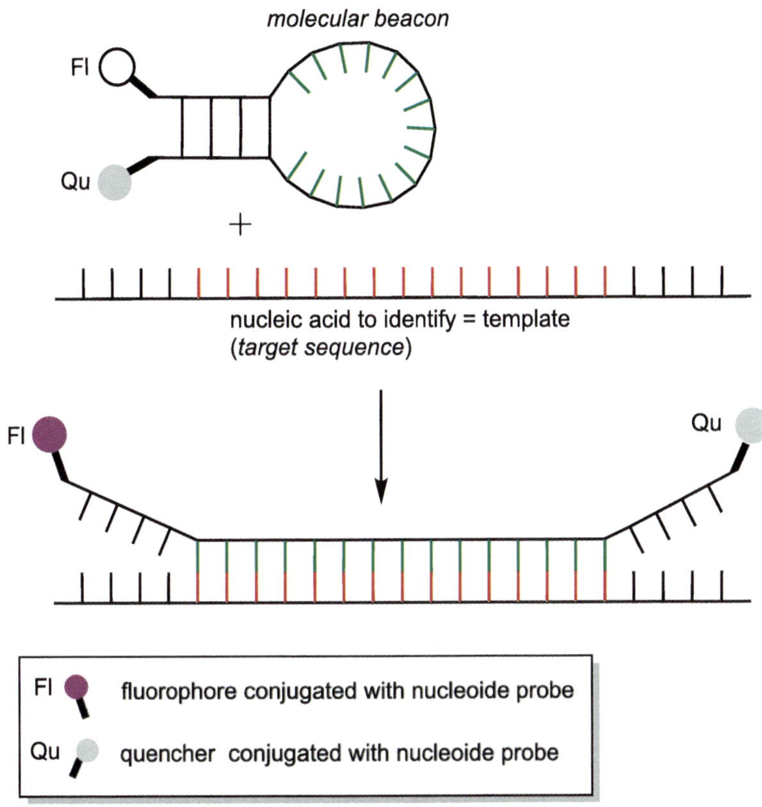

Fig. 11.5 Principle of molecular beacons

Figure 11.6 shows the method both schematically and with specific formulas. The quencher, the dabsyl group, is a sulfonic acid derivative that, as a sulfonate, activates the 5' position for nucleophilic attack. About two nucleotides further along, the fluorophore is attached to a thymine base (at the methyl group position). This oligonucleotide is somewhat longer (in the example from Ref. [12] 13 bases long). As a second reagent, a short oligonucleotide (about 2–3-fold excess) is used, which carries a phosphorothioate group at the 3′ end. As is well known, sulfur is much more nucleophilic than a corresponding oxygen atom (see Sect. 1.1). The synthesis of this oligonucleotide is performed on a solid phase with phosphoramidites, but in reverse, with the solid support at the 5′ end (see also Sect. 6.2.2). Then, and only then, when it binds to the template sequence (hybridization), does the S_N2 reaction occur, releasing the quencher and thus causing the now-linked oligonucleotide to fluoresce. In contrast to *molecular beacons*, this fluorescence persists even if the autoligated reagents dissociate from the template. In principle, this can lead to an amplified signal, allowing the detection of very low template concentrations. If, in addition, a possible point mutation site is chosen within the short oligonucleotide, the reaction will only occur if the sequence matches exactly.

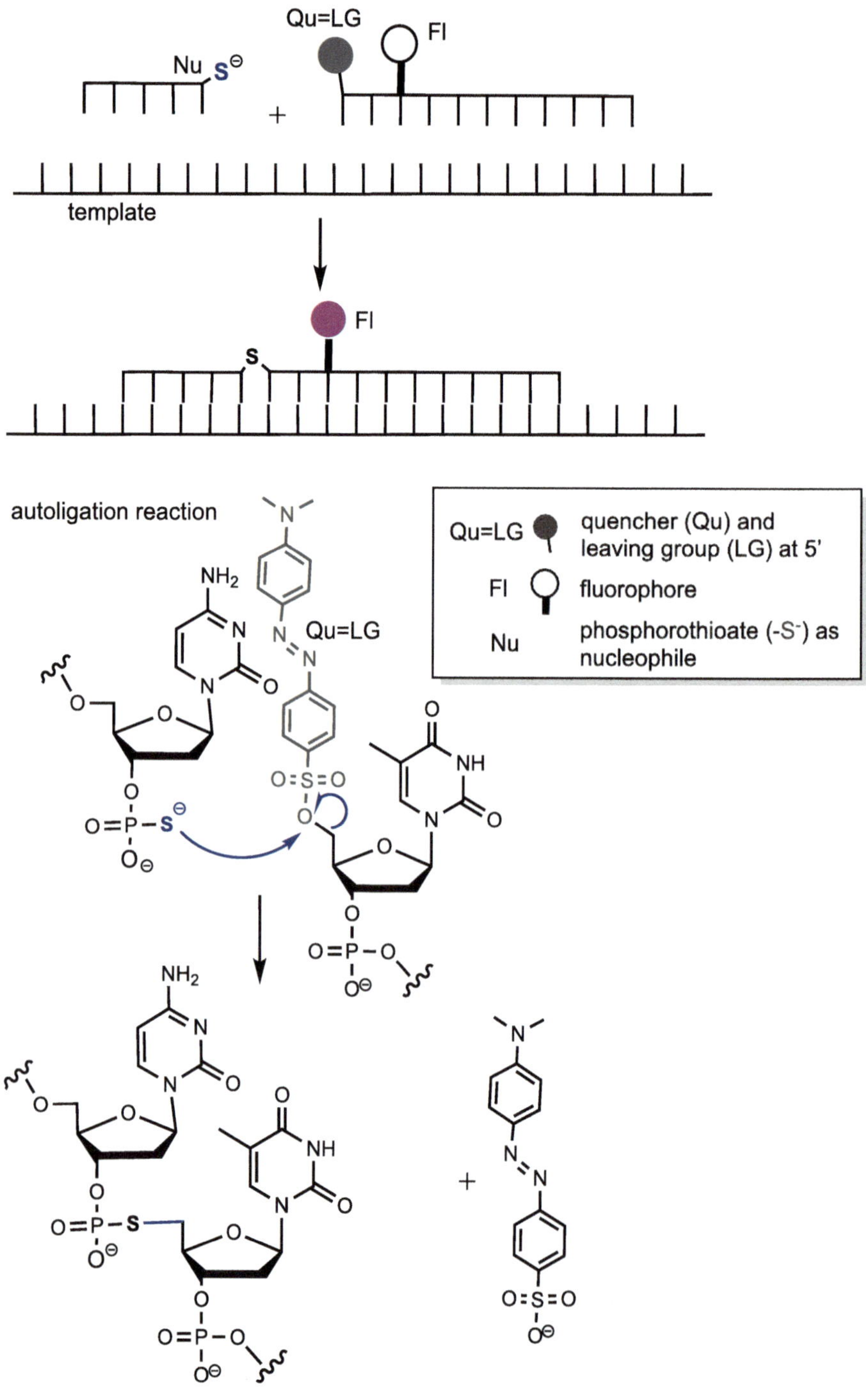

Fig. 11.6 Method for detecting specific RNA sequences by autoligation of two oligonucleotide reagents (QUAL probes) via an S_N2 reaction

A good example of this in the review [8] is taken from Ref. [13]. Here, shorter oligonucleotides carrying different fluorophores on a thymine are used. The aim is to distinguish between four variants of the H-*ras* gene (for mutations in Ras proteins, see Sect. 9.2) in a single experiment using differently colored fluorescence. Thus, a longer oligonucleotide with a 3'-phosphorothioate group was incubated together with four heptanucleotides, each with a dabsyl quencher at the 5' end and different fluorophores on the next-but-one thymine, each correlated with *one* point mutation, with different DNA fragments of the H-*ras* gene. Although the dabsyl chromophore is not considered an optimal quencher for all four fluorophores, the fluorescence was largely suppressed before the reaction (probably also due to *charge-transfer quenching* in addition to RET). In fact, based on the fluorescence wavelength, a clear assignment could be made for all four sequence variants (Ref. [8], Fig. 4). Further experiments showed that the method can also be applied to whole fixed bacterial cells (Ref. [8], Fig. 5) and even to non-fixed bacterial cells in the presence of a detergent.

A review article on the more general topic of "template-guided reactions" was published by Eric T. Kool's group in 2006 in *Chemical Reviews* [14].

The research group then presented another example of template-guided quencher release in 2009 [15]. In this case, it is not an S_N2 reaction, but a variant of the Staudinger reaction (see Sect. 7.5) (Fig. 11.7). The fluorescent component is called "*Q-STAR (quenched Staudinger-triggered azidoether release) probe*."

The Q-STAR component is a short oligonucleotide with a fluorophore on a thymine. At the 5′ end, an azidoether group with a fluorescence-quenching chromophore (again dabsyl in this case) is linked via an amide bond. The other component (TPP-DNA) is a somewhat longer oligonucleotide with a triphenylphosphine residue at the 3′ end. The two molecules only come into sufficient proximity when they are bound to the template DNA (hybridization). The Staudinger reaction releases N_2, and hydrolysis produces, on the one hand, the triphenylphosphine oxide derivative and, on the other, a hemiaminal, which further hydrolyzes to an aldehyde in aqueous solution. This cleaves the bond between the quencher and the oligonucleotide with the fluorophore, resulting in fluorescence.

Experimental Details of the Example from Ref. [15]

At a concentration of the Q-STAR component of c(Q-STAR) = 200 nM, with a threefold excess of TPP-DNA = 600 nM, the fluorescence signal reached full intensity after about 0.5 h in the presence of 200 nM complementary DNA and remained constant for hours. Even with only a small amount of template DNA (c = 10 nM), almost the same signal intensity was reached after about 5 h. This means that one template DNA molecule converted about 20 Q-STAR molecules in the Staudinger reaction. Thus, this method is much faster than the *QUAL probes*; in Ref. [12], the maximum signal is still far from being reached even after 12 h.

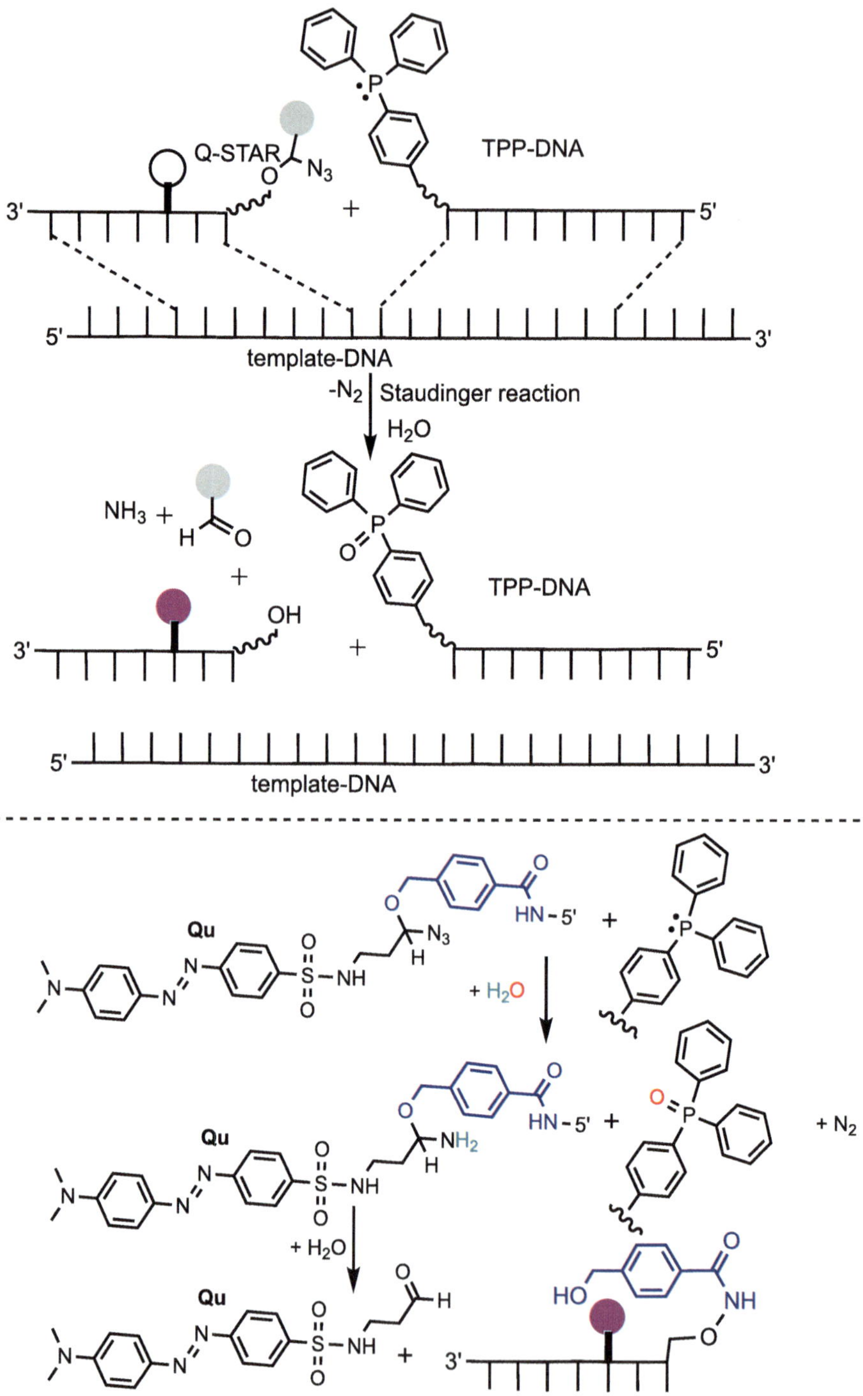

Fig. 11.7 Detection of DNA sequences by release of a quencher after the Staudinger reaction. (After Ref. [15])

At higher concentrations of target DNA, detection is quite rapid; at low amounts of target DNA, however, an amplified signal can be achieved. The same reference also shows that this method works with intact bacterial cells and that two bacterial types differing by a single point mutation in a DNA could be distinguished, using Förster resonance energy transfer (FRET) to excite at a single wavelength but obtain different fluorescence colors (Ref. [15], Fig. 5).

11.2.3 Template-Guided De Novo Fluorophore Synthesis as a Method for DNA Detection [16]

In this work, two oligonucleotides are positioned by the target template DNA such that an aldehyde component and a phosphonium salt are brought close enough together to allow ylide formation and a Wittig reaction (Fig. 11.8). The structures are chosen so that the resulting aryl-substituted alkene forms a fluorophore (stilbene fluorophore, λ_{exc} = 330 nm, λ_{em} = 380 nm). In the presence of α-cyclodextrin, signal amplification occurs because the lipophilic fluorophore formed is incorporated into the hydrophobic interior of the cyclodextrin. As we will also see in Sect. 14.1, this is not an ideal fluorophore for detection with biomolecules, since the excitation and emission ranges are still in the UV region, but in this case, the focus is on the principle. It demonstrates a way to generate fluorophores by template guidance and thus obtain a signal—without the problem of background signals caused by premature cleavage of the oligonucleotides by nucleases, which in the previous examples can lead to removal of the quencher.

Summary and Outlook

In this section, we have learned about several methods that exploit the occurrence of fluorescence as evidence for specific oligonucleotide sequences (RNA or DNA). There are many more examples of this. What they all have in common is that they require oligonucleotide synthesis and conjugation to fluorophores or other special functionalities (quenchers or reactive groups). From an organic synthetic perspective, the methods discussed in Sects. 11.2.2 and 11.2.3 were particularly interesting. However, these are less commonly used by biologists or physicians than those based solely on association, such as *molecular beacons*. One reason for this is probably the lack of understanding of chemical reactions. This is precisely where this book aims to foster a broader mutual understanding.

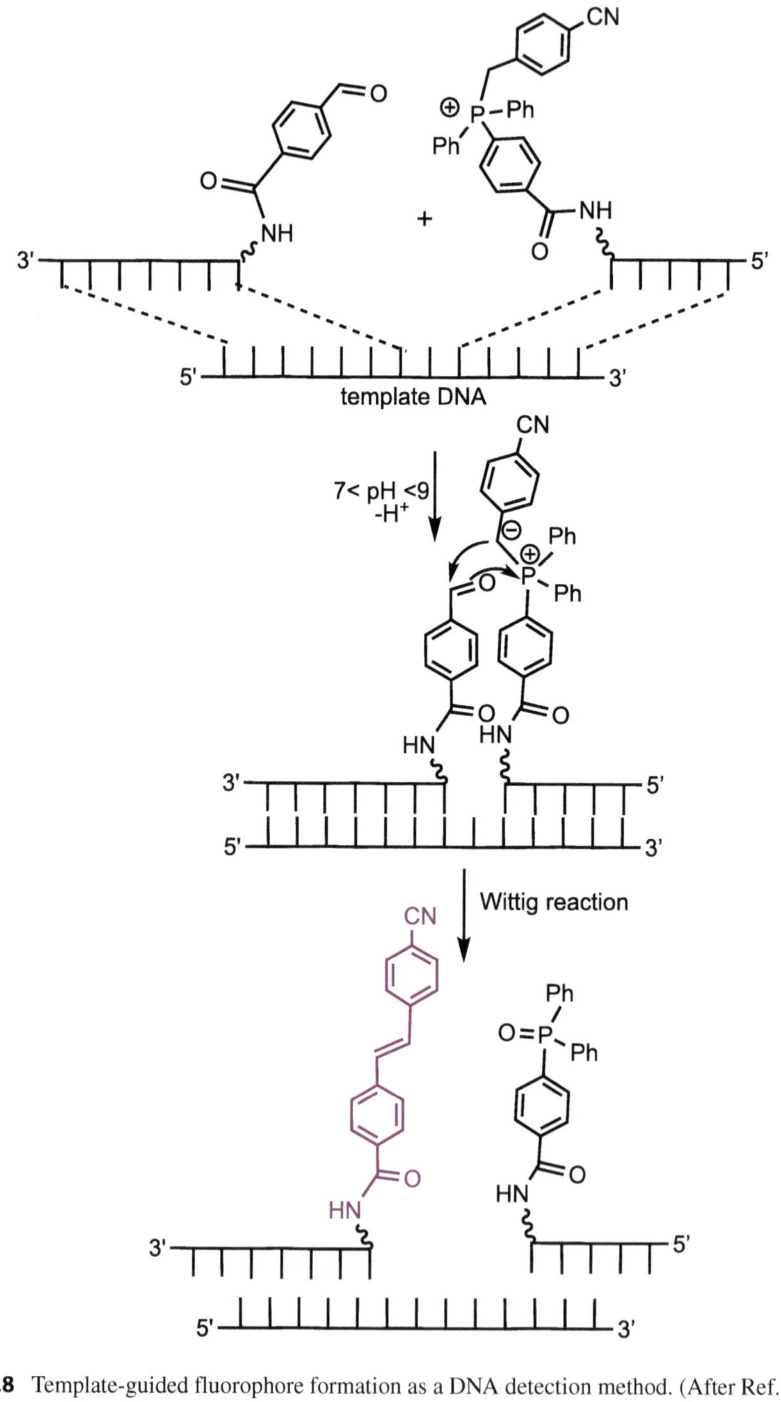

Fig. 11.8 Template-guided fluorophore formation as a DNA detection method. (After Ref. [16])

11.3 Modified Oligonucleotides for Studying Radicals in DNA

We now turn to a completely different application area of modified oligonucleotides, namely the elucidation of processes that damage DNA: photochemical reactions and the effects of radicals. The topic was also of interest because it raised the question of whether DNA behaves similarly to an electrical conductor. This was ruled out in a study from 1999 [17]. The research group of Prof. Bernd Giese in Basel then addressed the question of whether and how radicals and charges can propagate in DNA double helices, not with purely spectroscopic methods, but using chemical approaches [18].

▶ As an example on this topic, we have selected the publication: Excess electron transport through DNA: A single electron repairs more than one UV-induced lesion, Giese, B., Carl, B., Carl, T., Carell, T., Behrens, C., Hennecke, U., Schiemann, O., Feresin, E., *Angew. Chem., Int. Ed.*, **43** (14), 1848–1851, https://doi.org/10.1002/anie.200353264 (2004) and *Angew. Chem.*, **116** (14), 1884–1887, https://doi.org/10.1002/ange.200353264 (2004) [19].

A method is presented that allows the generation of a radical anion at a specific site of a double-stranded model oligonucleotide (charge injection). Modifications at other sites serve as detectors to determine whether the radical anion has been transferred to those positions.

Figure 11.9 first shows the thymine derivative $\mathbf{T_x}$ modified for "charge injection." Excitation of the carbonyl group of the ketone by UV radiation leads to cleavage of the *tert*-butylacylium radical, which then decomposes into carbon monoxide and the relatively "stable" *tert*-butyl radical. At the same time, a radical is formed in the side chain on the thymine, which quickly rearranges to a radical in the heterocycle. Evidence for this was provided by the ESR spectrum (Ref. [19], Fig. 1), obtained after about 10 min and in good agreement with the calculated spectrum for the bold structure in Fig. 11.9. Thus, this system allows targeted generation of a radical (or radical anion, depending on pH) at a thymine. What happens to this radical? To answer this question, a thymine dimer **T=T** was used as a detection system, which is formed photochemically by [2+2] cycloaddition. The two thymines linked via the cyclobutane ring lack the phosphodiester bridge. If the dimer is cleaved by transfer of the radical anion, the strand is also cleaved. The fragments can then be identified by HPLC.

The results of radical transfer correspond to a *hopping* mechanism, in which the electron jumps from one **T** to the next; if it reaches a **T=T** dimer, this can be cleaved. In the series of experiments (Fig. 11.10) with increasing numbers of **AT** pairs between the injection site ($\mathbf{T_x}$) and the detection site (**T=T**), the decrease in cleavage yield is consistent with a multistep mechanism, not direct transfer. When cleavage of the **T=T** dimer is triggered, the electron is not consumed in principle and can continue to migrate. This is illustrated in the experiment with two separate detection sites. Cleavage occurs at both

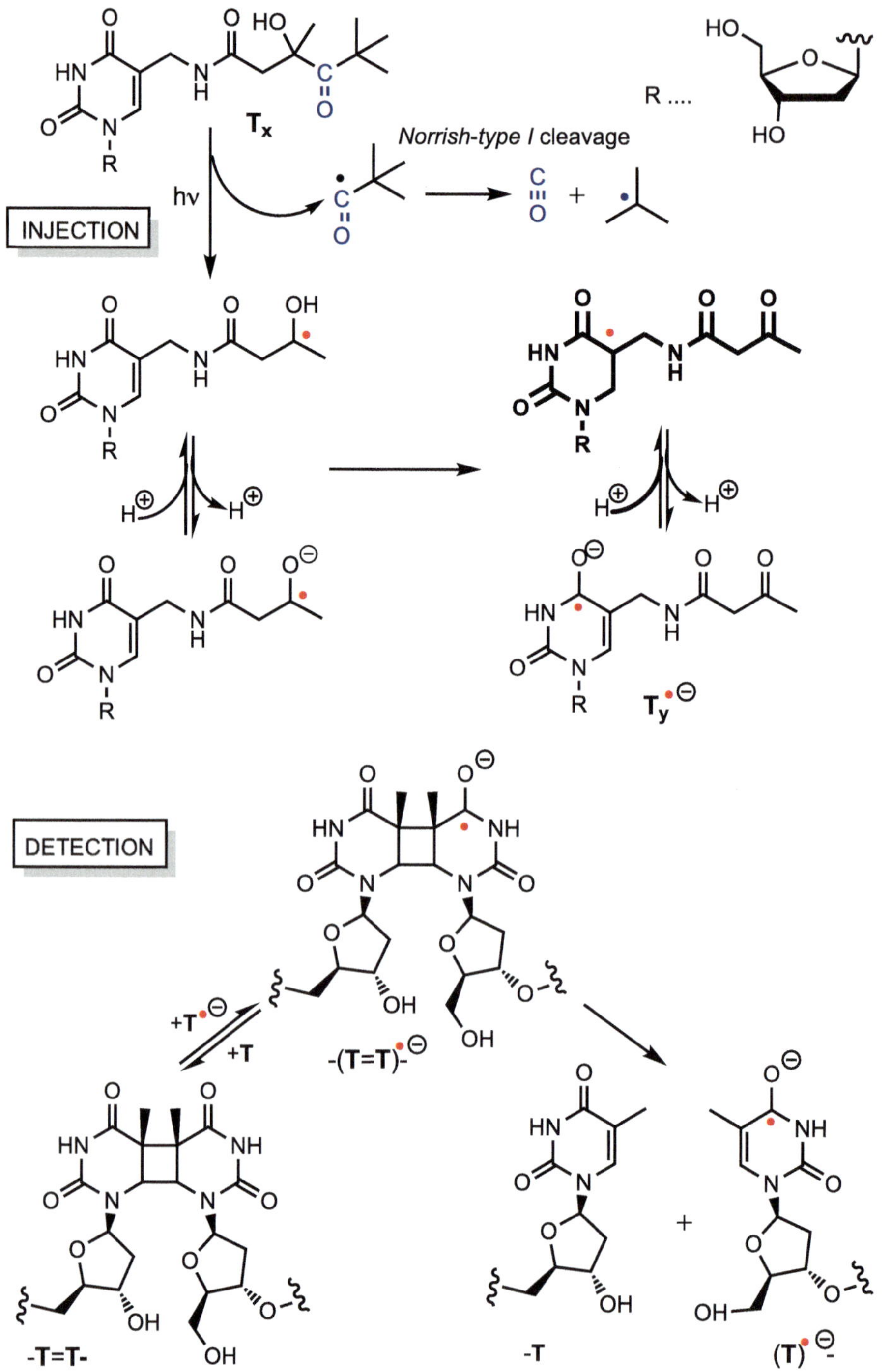

Fig. 11.9 Modified nucleobases for charge injection and detection (**T** … thymine, **T=T** … thymine dimer). (After Ref. [19])

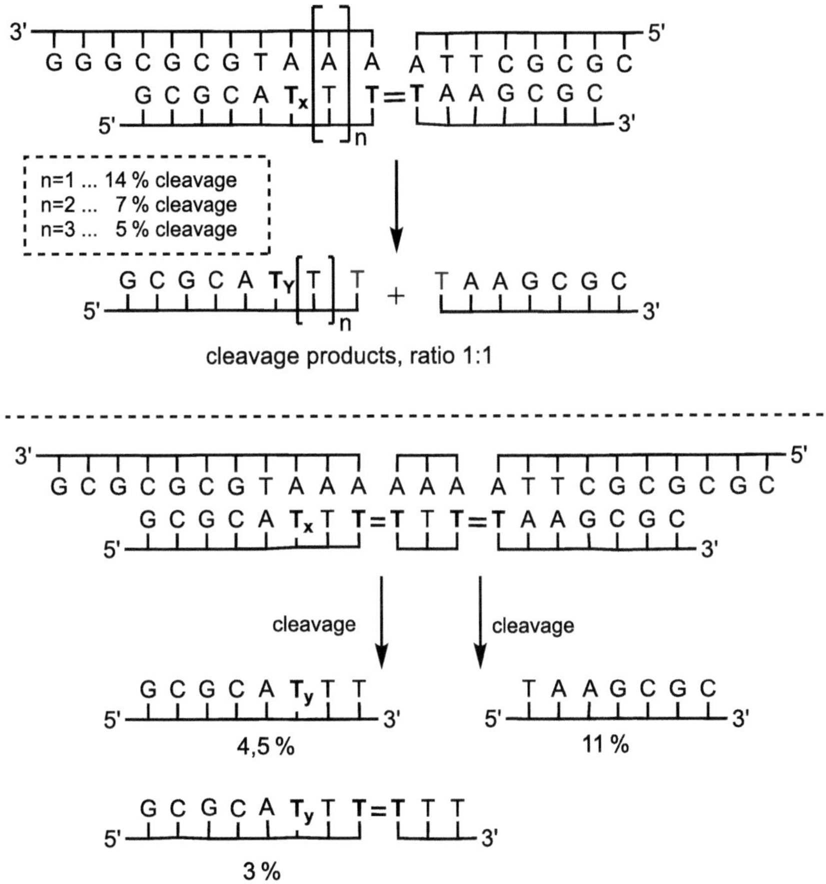

Fig. 11.10 Product distribution by cleavage at the thymine dimer **T=T** after generation of a radical at the modified thymine $\mathbf{T}_x$ ($\mathbf{T}_y$... modified thymine after transfer of the radical anion) (formulas see Fig. 11.9). (After Ref. [19])

sites, but the appearance of the longer 5′ fragment in 3% yield proves that not every **T=T** radical must actually be cleaved if the transfer is fast enough.

A summary of the work by Prof. Giese's group on the topic of radical cation transfer (*holes*) appeared in Ref. [20] in 2004. Ref. [21] from 2010 provides a brief overview of the elucidation of radical reactions in nucleic acids and peptides conducted by this group at the University of Basel. Bernd Giese has also continued to investigate electron transport, especially in relation to proteins. This mainly concerns redox reactions in cells, and it is the aromatic amino acid residues that enable transport over greater distances [22].

Significance of this Publication [19]

This work, like others from this research group, is interesting and inspiring because it demonstrates how targeted "measurements" can be performed on single molecules using, in principle, relatively simple synthetic and analytical methods. This is because every oligonucleotide that possesses such a "charge injector" (**T**$_x$) can accept exactly *one* and only *one* electron.

What does the second part of the title, "… *a single electron repairs more than one UV-induced lesion*," refer to? The **T=T** dimer, which is used as a detection system, is formed by harmful UV radiation in the DNA of skin cells (where, in this case, the phosphodiester backbone of course remains intact) and renders this DNA nonfunctional. As shown in this work, a single electron can cleave several such **T=T** dimers and could thus restore the DNA's function.

11.4 Questions

Let's test our understanding of the content of this chapter!

11.1 Why is the difluorotoluene nucleobase a perfect analogue of thymine for testing the significance of hydrogen bonds?

11.2 How can the stability of a DNA double helix be checked and measured?

11.3 Why is the non-hydrogen-bonding nucleobase Z not as good an analogue for A as F is for T?

11.4 What is an abasic site? Which nucleotide triphosphate is preferentially incorporated opposite such a site?

11.5 What are *molecular beacons* and what is the advantage compared to the FISH method?

11.6 Why should *molecular beacons* preferably be constructed from oligonucleotides with a modified *backbone*?

11.7 What type of reaction is involved in the autoligation method between a phosphothioester and a sulfonate?

11.8 What is the driving force in the Staudinger reaction?

11.9 What advantage do the discussed autoligation methods offer in principle compared to *molecular beacons*?

11.10 What is meant by charge injection in the publication by Bernd Giese, and how is it accomplished?

11.11 Task: Read the three specifically discussed publications (Lit. [1, 8] and [19]), form your own opinion, and discuss it.

References

1. Kool, E.T., Morales, J.C., Guckian, K.M.: Mimicking the structure and function of DNA: insights into DNA stability and replication. Angew. Chem. Int. Ed. **39**(6), 990–1009 (2000). https://doi.org/10.1002/(SICI)1521-3773(20000317)39:6<990::AID-ANIE990>3.0.CO;2-0; Kool, E.T., Morales, J.C., Guckian, K.M.: Einsichten in Stabilität und Replikation der DNA durch Nachahmung ihrer Struktur und Funktion. Angew. Chem. **112**(6), 1046–1068 (2000). https://doi.org/10.1002/(SICI)1521-3757(20000317)112:6<1046::AID-ANGE1046>3.0.CO;2-9
2. Guerra, C. F., Bickelhaupt, F. M., Snijders, J. G., Baerends, E.J.: The nature of the hydrogen bond in DNA base pairs: the role of charge transfer and resonance assistance. Chem. Eur. J. **5**(12), 3581–3594 (1999). https://doi.org/10.1002/(SICI)1521-3765(19991203)5:12<3581::AID-CHEM3581>3.0.CO;2-Y
3. Guckian, K.M., Krugh, T.R., Kool, E.T.: Solution structure of a non-hydrogen-bonded base pair surrogate in DNA. J. Am. Chem. Soc. **122**(29), 6841–6847 (2000). https://doi.org/10.1021/ja994164v
4. Kool, E.T., Sintim, H.O.: The difluorotoluene debate – a decade later. Chem. Commun., 3665–3675 (2006). https://doi.org/10.1039/B605414E
5. Lee, H.R., Helquist, S.A., Kool, E.T., Johnson, K.A.: Importance of hydrogen bonding for efficiency and specificity of the human mitochondrial DNA polymerase. J. Biol. Chem. **283**(21), 14402–14410 (2008). https://doi.org/10.1074/jbc.M705007200; Lee, H.R., Helquist, S.A., Kool, E.T., Johnson, K.A.: Base pair hydrogen bonds are essential for proofreading selectivity by the human mitochondrial DNA polymerase. J. Biol. Chem. **283**(21), 14411–14416 (2008). https://doi.org/10.1074/jbc.M705006200
6. Krueger, A.T., Lu, H., Lee, A.H.F., Kool, E.T.: Synthesis and properties of size-expanded DNAs: toward designed, functional genetic systems. Acc. Chem. Res. **40**(2), 141–150 (2007). https://doi.org/10.1021/ar068200o; Krueger, A.T., Kool, E.T.: Redesigning the architecture of the base pair: toward biochemical and biological function of new genetic sets. Chem. Biol. **16**(3), 242–248 (2009). https://doi.org/10.1016/j.chembiol.2008.12.004
7. Jun, Y.W., Kool, E.T.: Chemical tools for the study of DNA repair. Acc. Chem. Res. **55**(23), 3495–3506 (2022). https://doi.org/10.1021/acs.accounts.2c00608
8. Silverman, A.P., Kool, E.T.: Quenched probes for highly specific detection of cellular RNAs. Trends Biotechnol. **23**(5), 225–230 (2005). https://doi.org/10.1016/j.tibtech.2005.03.007
9. Li, J.J., Fang, X., Schuster, S.M., Tan, W.: Molecular beacons: a novel approach to detect protein – DNA interactions. Angew. Chem. Int. Ed. **39**(6), 1049–1052 (2000). https://doi.org/10.1002/(SICI)1521-3773(20000317)39:6<1049::AID-ANIE1049>3.0.CO;2-2; Li, J.J., Fang, X., Schuster, S.M., Tan, W.: Molecular beacons: a novel approach to detect protein – DNA interactions. Angew. Chem. **112**(6), 1091–1094 (2000). https://doi.org/10.1002/(SICI)1521-3757(20000317)112:6<1091::AID-ANGE1091>3.0.CO;2-T; Tan, W., Fang, X., Li, J., Liu, X.: Molecular beacons: a novel DNA probe for nucleic acid and protein studies. Chem. Eur. J. **6**(7), 1107–1111 (2000). https://doi.org/10.1002/(SICI)1521-3765(20000403)6:7<1107::AID-CHEM1107>3.0.CO;2-9; Brazil, M.: Molecular beacons for DNA binding. Nat. Rev. Drug Discov. **1**(2), 98–99 (2002). https://doi.org/10.1038/nrd737
10. Vilaivan, T.: Fluorogenic PNA probes. Beilstein J. Org. Chem. **14**, 253–281 (2018). https://doi.org/10.3762/bjoc.14.17
11. Zheng, J., Yang, R., Shi, M.,…, Tan, W.: Rationally designed molecular beacons for bioanalytical and biomedical applications. Chem. Soc. Rev. **44**(10), 3036–3055 (2015). https://doi.org/10.1039/C5CS00020C; Huang, J., Wu, J., Li, Z.: Biosensing using hairpin DNA probes.

Rev. Anal. Chem. **34**(1-2), 1–27 (2015). https://doi.org/10.1515/revac-2015-0010; Roth, M., Seitz, O.: A self-immolative molecular beacon for amplified nucleic acid detection. Chem. Eur. J. **27**(57), 14189–14194 (2021). https://doi.org/10.1002/chem.202102600; Yang, G., Song, T., Wang, M.,..., Yang, H.: Recent advancements in nanosystem-based molecular beacons for RNA detection and imaging. ACS Appl. Nano Mater. **5**(3), 3065–3086 (2022). https://doi.org/10.1021/acsanm.1c03966; Zhang, S., Xiong, W., Xu, S., Qian, R.: Recent advances in DNA systems for in situ telomerase activity detection and imaging. Chemosensors **13**(1), 17 (2025). https://doi.org/10.3390/chemosensors13010017

12. Sando, S., Kool, E.T.: Quencher as leaving group: efficient detection of DNA-joining reactions. J. Am. Chem. Soc. **124**(10), 2096–2097 (2002). https://doi.org/10.1021/ja017328s
13. Sando, S., Abe, H., Kool, E.T.: Quenched auto-ligating DNAs: multicolor identification of nucleic acids at single nucleotide resolution. J. Am. Chem. Soc. **126**(4), 1081–1087 (2004). https://doi.org/10.1021/ja038665z
14. Silverman, A.P., Kool, E.T.: Detecting RNA and DNA with templated chemical reactions. Chem. Rev. **106**(9), 3775–3789 (2006). https://doi.org/10.1021/cr050057+
15. Franzini, R.M., Kool, E.T.: Efficient nucleic acid detection by templated reductive quencher release. J. Am. Chem. Soc. **131**(44), 16021–16023 (2009). https://doi.org/10.1021/ja904138v
16. Chen, X.-H., Roloff, A., Seitz, O.: Consecutive signal amplification for DNA detection based on de novo fluorophore synthesis and host–guest chemistry. Angew. Chem. Int. Ed. **51**(18), 4479–4483 (2012). https://doi.org/10.1002/anie.201108845; Chen, X.-H., Roloff, A., Seitz, O.: Konsekutive Signalverstärkung für die DNA-Detektion basierend auf einer De-novo-Fluorophorsynthese und Wirt-Gast-Chemie. Angew. Chem. **124**(18), 4556–4561 (2012). https://doi.org/10.1002/ange.201108845
17. Debije, M.G., Milano, M.T., Bernhard, W.A.: DNA responds to ionizing radiation as an insulator, not as a "molecular wire". Angew. Chem. Int. Ed. **38**(18), 2752–2756 (1999). https://doi.org/10.1002/(SICI)1521-3773(19990917)38:18<2752::AID-ANIE2752>3.0.CO;2-U; Debije, M.G., Milano, M.T., Bernhard, W.A.: DNA verhält sich nach ionisierender Bestrahlung wie ein Isolator und nicht wie ein „moleularer Draht". Angew. Chem. **111**(18), 2926–2930 (1999). http://onlinelibrary.wiley.com/doi/10.1002/(SICI)1521-3757(19990917)111:18%3C2926::AID-ANGE2926%3E3.0.CO;2-7/full. Zugegriffen am 05.01.2026
18. Giese, B.: Long-distance charge transport in DNA: the hopping mechanism. Acc. Chem. Res. **33**(9), 631–636 (2000). https://doi.org/10.1021/ar990040b; Giese, B., Amaudrut, J., Köhler, A.-K.,.., Wessely, S.: Direct observation of hole transfer through DNA by hopping between adenine bases and by tunnelling. Nature **412**(6844), 318320 (2001). https://doi.org/10.1038/35085542; Giese, B.: Radicals and the birth and death of DNA. Chimia **55**(4), 275–280 (2001). https://doi.org/10.2533/chimia.2001.275; Giese, B., Biland, A.: Recent developments of charge injection and charge transfer in DNA. Chem. Commun. (7), 667–672 (2002). https://doi.org/10.1039/b111044f; Kendrick, T., Giese, B.: Charge transfer through DNA triggered by site selective charge injection into adenine. Chem. Commun. (18), 2016–2017 (2002). https://doi.org/10.1039/b205928b
19. Giese, B., Carl, B., Carl, T.,..., Feresin, E.: Excess electron transport through DNA: a single electron repairs more than one UV-induced lesion. Angew. Chem. Int. Ed. **43**(14), 1848–1851 (2004). https://doi.org/10.1002/anie.200353264; Giese, B., Carl, B., Carl, T.,..., Feresin, E.: Excess electron transport through DNA: a single electron repairs more than one UV-induced lesion. Angew. Chem. **116**(14), 1884–1887 (2004). https://doi.org/10.1002/ange.200353264

20. Giese, B.: Hole injection and hole transfer through DNA: the hopping mechanism. Top. Curr. Chem. **236**, 27–44., Schuster, G. B. (Hrsg), Springer-Verlag Berlin-Heidelberg (2004). https://doi.org/10.1007/b94410
21. Giese, B.: Electron transfer in biopolymers. Chimia. **64**(12), 849–850 (2010). https://doi.org/10.2533/chimia.2010.846
22. Giese, B., Karamash, M., Fromm, K.M.: Chances and challenges of long-distance electron transfer for cellular redox reactions. FEBS Lett. **597**(1), 166–173 (2023). https://doi.org/10.1002/1873-3468.14493

Methods for the Modification of Proteins—Introduction of Non-natural Amino Acids

12

Abstract

Methods for introducing unnatural amino acids at defined positions in proteins are discussed. The primary focus is on identifying unused codons in an mRNA that can be assigned to the non-proteinogenic "new" amino acid. A corresponding tRNA must then be conjugated with the new amino acid, either by synthesis or with the help of a specialized aminoacyl-tRNA synthetase. Initially, *in vitro* expression systems were developed, but these were subsequently replaced by *in vivo* expression. Examples of unnatural amino acids include those carrying fluorophores, functional groups for photoaffinity labeling, or *in vivo* cleavable protecting groups. This approach also allows for the direct introduction and testing of amino acids with post-translational modifications. Recently, specialized cell-free expression systems have also experienced a renaissance.

There are several options for the synthesis of modified proteins, some of which we have already encountered or mentioned (Sect. 3.1.3.3, Chap. 8, Sect. 9.2). In this chapter, we will address the now almost routine introduction of individual non-natural amino acids at specific positions in a protein that is produced molecular biologically via an expression system. I would like to approach the topic with a bit of historical perspective, as I have witnessed this development over the years in my lectures.

In the 1990s, there were two approaches for this. For example, if a single amino acid containing heavier atoms was needed for X-ray structure analysis, the protein was expressed in a specific *E. coli* strain that was unable to synthesize the corresponding natural amino acid itself, and an analogous amino acid containing a heavy atom was added [1]. In fact, the ribosomal system often accepts slightly modified amino acids, such as telluromethionine, selenocysteine (Sec), tellurocysteine (each with a Se or Te atom instead of S), or thiaproline (a proline with an S atom instead of $-CH_2-$ at the γ-position

D. Krois, *Organic Chemical Methods*, https://doi.org/10.1007/978-3-662-73573-2_12

in the ring). The disadvantage of this method is that all positions of the suppressed natural amino acid are replaced by the non-natural variant.

Peter G. Schultz was the first to introduce the method of using the rarely utilized stop codon (*nonsense codon*) UAG for the targeted incorporation of a non-natural amino acid (AA_x) at a specific site in a protein, in an article in *Science* [2]. This will be the first example we will examine in more detail here.

12.1 Incorporation of Unnatural Amino Acids via an In Vitro Protein Biosynthesis System

12.1.1 Principle of the Method

The method is described in more detail in a review by Peter G. Schultz from 1995 [3].

Ribosomal Protein Biosynthesis

Although this is probably well known to all chemists, I would like to briefly recall the key steps of protein biosynthesis here. The sequence of a protein to be synthesized in a cell is encoded in an *open reading frame* in the DNA. When this gene is activated for transcription, a *messenger* RNA (mRNA) is transcribed in the nucleus, which then migrates to the ribosome. However, the crucial genetic code is the specific charging of the *transfer* RNA (tRNA) with a particular amino acid. This is carried out by **a**mino**a**cyl-t**R**NA **s**ynthetases (AARS). These enzymes recognize the appropriate tRNA, which carries an anticodon complementary to the codon on the mRNA, and attach the corresponding amino acid to the 3′ terminus of the tRNA. This charged tRNA is then aligned in the ribosome with its anticodon to the codon of the mRNA. The ribosomal enzymes then form the peptide bond to the neighboring amino acid of the previously aligned tRNA, simultaneously releasing the bond between this penultimate amino acid and the tRNA. The now uncharged tRNA then dissociates. This process is schematically illustrated in review [3].

For a given amino acid, there are often several codons and tRNAs. In addition, there are stop codons (TERM) that indicate the end of protein synthesis (at the C-terminus) (Table 12.1). One of these stop codons (UAG) is rarely used, and there exists a so-called *amber suppressor* tRNA, which enables read-through of this codon. This tRNA is now used, following the approach of P. G. Schultz, to introduce unnatural amino acids AS_x.

Figure 12.1 illustrates the method: By targeted mutation in the DNA, the nucleotide sequence TAG is generated at the desired position. This is then transcribed in vitro to yield an mRNA with UAG. On the other hand, *run-off transcription* is used to obtain the *suppressor* $tRNA_{CUA}$ shortened by two nucleotides at the 3′ end. Using organic synthesis, the unnatural amino acid, usually protected with the Nvoc group (Sect. 6.1.3, Fig. 6.6) at the N-terminus, is coupled as a cyanoethyl ester (= activated ester) to the 2′- or 3′-OH group of a pdCpA dinucleotide. The amino groups of the nucleobases do not need to be protected for this. As the penultimate nucleotide, a 2′-*deoxy*cytidine (dC) is used, which does not affect the biological activity of the tRNA but simplifies the synthesis—a 2′-OH group would lead to side products during coupling with the amino acid.

Table 12.1 The genetic code of mRNA

Codon	AA	Codon	AA	Codon	AA	Codon	AA
UUU	Phe	UCU	Ser	UAU	Tyr	UGU	Cys
UUC	Phe	UCC	Ser	UAC	Tyr	UGC	Cys
UUA	Leu	UCA	Ser	UAA	TERM	UGA	TERM
UUG	Leu	UCG	Ser	**UAG**	**TERM**	UGG	Trp
CUU	Leu	CCU	Pro	CAU	His	CGU	Arg
CUC	Leu	CCC	Pro	CAC	His	CGC	Arg
CUA	Leu	CCA	Pro	CAA	Gln	CGA	Arg
CUG	Leu	CCG	Pro	CAG	Gln	CGG	Arg
AUU	Ile	ACU	Thr	AAU	Asn	AGU	Ser
AUC	Ile	ACC	Thr	AAC	Asn	AGC	Ser
AUA	Met	ACA	Thr	AAA	Lys	AGA	Arg
AUG	Met (Start)	ACG	Thr	AAG	Lys	AGG	Arg
GUU	Val	GCU	Ala	GAU	Asp	GGU	Gly
GUC	Val	GCC	Ala	GAC	Asp	GGC	Gly
GUA	Val	GCA	Ala	GAA	Glu	GGA	Gly
GUG	Val	GCG	Ala	GAG	Glu	GGG	Gly

AA ... amino acid**; A** ... adenine; **G** ... guanine; **U** ... uracil **C** ... cytosine; TERM ... *termination*, stop codon; 3-letter AA code see Chap. 2, Fig. 2.1

Enzymatically—using T4 RNA ligase—the dinucleotide loaded with AS_x is attached to the 3′ end of the truncated *suppressor* tRNA, and the Nvoc protecting group is removed photolytically. The mRNA is then added, and in vitro, in the presence of all natural tRNAs, the desired protein containing the unnatural amino acid AS_x is produced by the ribosomal protein synthesis system obtained from cells.

12.1.2 An Early Application Example: Modification of a Ras Protein

▶ A first example of this is described by the group of Peter G. Schultz in Ref. [4], "Probing the Role of Lysine 16 in Ras p21 Protein with Unnatural Amino Acids, Chung, H.-H., Benson, D. R., Schultz, P. G., *J. Am. Chem. Soc.*, **115** (14), 6414–6415, https://doi.org/10.1021/ja00067a068 (1993)".

I have also chosen this publication because it addresses a biological topic we have already discussed, namely the activity of Ras proteins (see Chap. 9). Due to the importance of reduced GTPase activity for the oncogenic effect of Ras protein mutants, this study produced lysine analogs at position 16 of the Ras-p^{21} protein and examined their GTPase activity. This lysine was thought to play a crucial role in GTP binding and

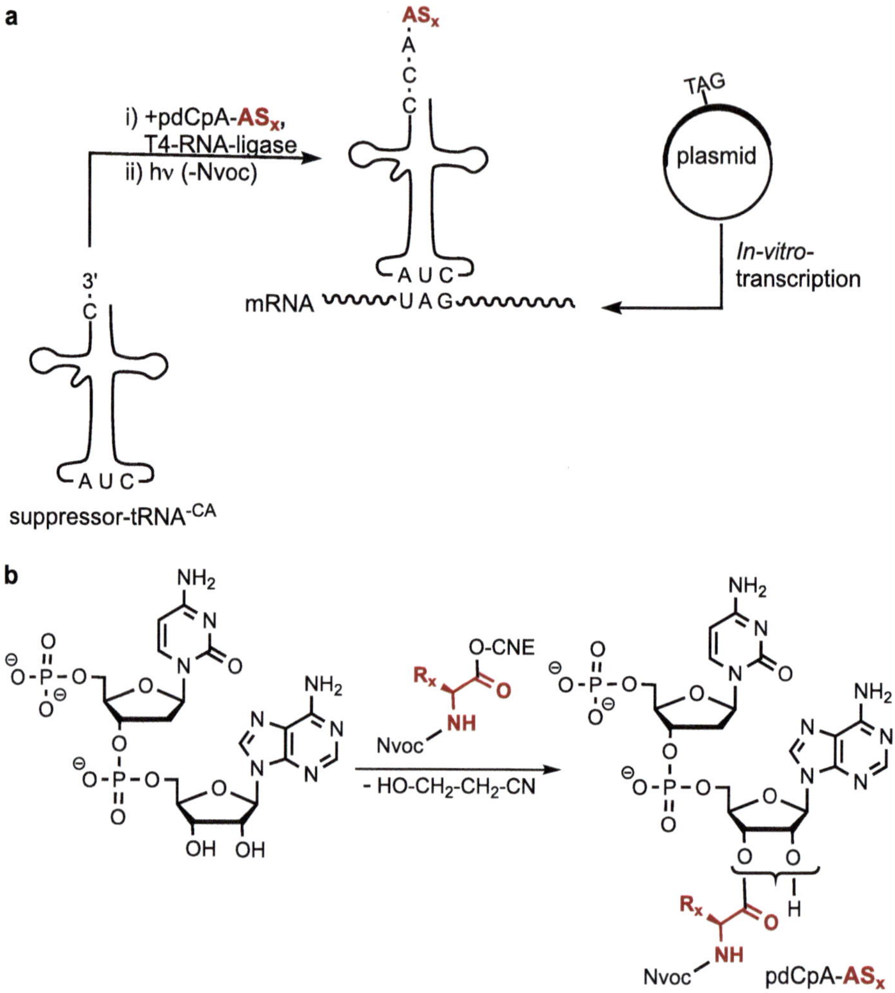

Fig. 12.1 Introduction of an unnatural $\mathbf{AA}_x$ using an in vitro expression system according to Ref. [3]; **a** biological in vitro overall process, **b** chemical synthesis of the dinucleotide (for the 3′ end of the suppressor tRNA) with the N-terminally protected unnatural $\mathbf{AA}_x$

catalysis of phosphate cleavage. Lysine analogs with different side chain lengths were successfully incorporated at position 16 using the described method: aminoethylcysteine (R = $H_2N–CH_2–CH_2–S–CH_2–$), aminoethylhomocysteine (R = $H_2N–CH_2–CH_2–S–CH_2–CH_2–$), and ornithine (R = $H_2N–CH_2–CH_2–CH_2–$, Orn), as well as a variant without an amino group, namely hydroxyethylcysteine (R = $HO–CH_2–CH_2–S–CH_2–$). Although differences in GTPase activity were observed (see Ref. [4], Tab. I), they were not as pronounced as expected. Only the absence of an amino group in the case of hydroxyethylcysteine led to a complete loss of GTPase activity, even in the presence of the activating protein (GAP). Thus, it appears that not the exact length of the side chain, but only its charge, is crucial for activity.

Of methodological interest is the reported protein yield: For the natural protein (WT … *wild-type*), 100 μg of active protein per mL was obtained using the *in vitro* expression system. The yields for the unnatural mutants were 25–50% of this. Thus, the efficiency of the modified suppressor tRNA lies in this range. About three chromatographic steps were required for purification, with an overall yield of about 15% of homogeneous protein.

The overall low yields despite relatively high methodological effort were one reason for developing a methodology based on *in vivo* expression (Sect. 12.2).

12.1.3 Further Examples of In Vitro Expression of Modified Proteins

Further examples of this method can be found in Ref. [5] and [6], with Ref. [5] using an alanine tRNA with altered anticodons. Since the above-mentioned *amber* stop codon UAG only allows the introduction of a single type of unnatural amino acid, and all other codons in mRNA for proteins occur more or less frequently, suppressor tRNAs with four-nucleotide-long anticodons were targeted as a possible solution. This *in vitro frame-shift suppression* was developed by the group of M. Sisido [7]. The fact that this principle allows two different unnatural amino acids to be introduced at two distinct defined positions was demonstrated in a 2006 publication:

- FRET analysis of protein conformational change through position-specific incorporation of fluorescent amino acids. Kajihara, D., Abe, R., Iijima, I., Komiyama, C., Sisido, M., Hohsaka, T., *Nature Meth.*, **3** (11), 923–929, https://doi.org/10.1038/nmeth945 (2006) [8].

Quadruplet codons were used to encode the unnatural AS_x, which can be introduced using a tRNA with a 4-base anticodon, a *frame-shift-suppressor* tRNA. This enables the targeted incorporation of two different AS_x into a protein. In the cited publication, two different fluorophores (Sect. 14.1.2) were activated as *N*-hydroxysuccinimide esters and each coupled to the corresponding tRNA carrying a 4′-aminophenylalanine (AF, instead of natural Phe). The DNA of the Ca^{2+}-dependent protein calmodulin was then modified so that, apart from two tags for easier isolation, one of the amino acids Phe^{19}, Gly^{40}, Gly^{59}, Leu^{69}, Phe^{99}, and Leu^{112} was replaced by the unnatural amino acid AF carrying fluorophore$_1$ (Fl_1) (BODIPYFL, $\lambda_{exc} = 490$ nm/$\lambda_{em} = 515$ nm), and at the N-terminus, the amino acid AF with fluorophore$_2$ (Fl_2) (BODIPY558-AF, $\lambda_{exc} = 545$ nm, but also excitable at ca. 530 nm/$\lambda_{Em} = 575$ nm)[1] was to be incorporated. If the 4-base tRNA is

[1] λ_{exc} and λ_{Em} do not refer to the absolute maxima of excitation and emission spectra, but to the wavelengths used for excitation and detection.

not accepted, a stop codon results at both positions, so that truncated proteins are formed (at least without the C-terminal His-tag) (Fig. 12.2).

His-Tag
If a region –(His–)$_{n>5}$ is appended to a protein sequence to be expressed, it is referred to as a "His-tag," a histidine tag. This offers the advantage of very specific binding to Ni(II) ions. If such ions are immobilized on a column material via a chelator, proteins containing His-tags can be specifically retained on this affinity chromatography column. They can then be eluted, for example, by imidazole in the buffer.

T7-Tag
This is a short peptide sequence Met–Ala–Ser–Met–Thr–Gly–Gly–Gln–Gln–Met–Gly, which is usually incorporated at the N-terminus of expressed proteins to facilitate detection, as there are highly specific antibodies for this sequence.

The proteins with both fluorophores could indeed be obtained in pure form. The yield of pure, doubly labeled calmodulin was about 10 pmol, i.e., 0.2 μg per 50 μL of the cell-free translation system. Fluorescence measurements were performed under various conditions: denaturation with urea and upon addition of a binding peptide with/without Ca^{2+} ions. The conformational changes under these different conditions manifested as markedly different fluorescence spectra, since Förster resonance energy transfer (FRET) (Sect. 14.1.1) from the BODIPY-FL to the BODIPY 558 fluorophore only occurs when they are in close spatial proximity. In Fig. 2 [8], it can be seen that upon denaturation

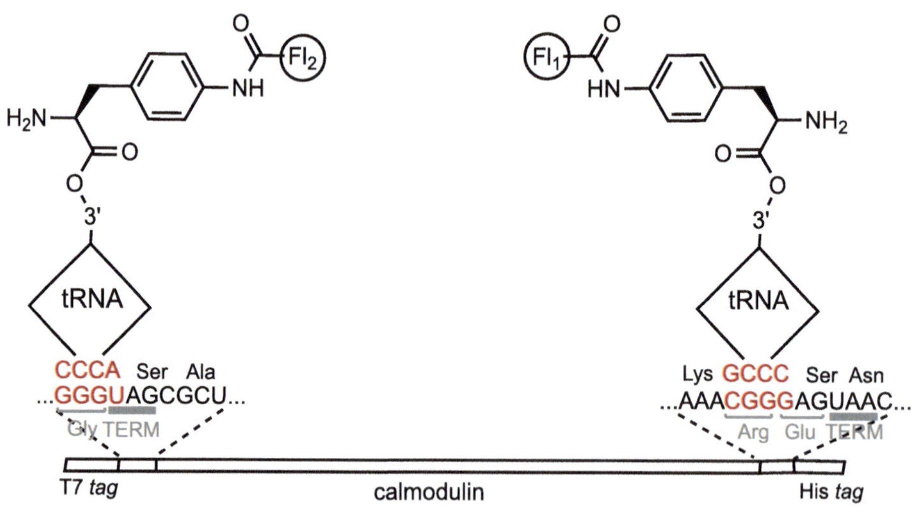

Fig. 12.2 Introduction of two highly fluorescent amino acids into calmodulin using the frameshift suppression method. If the quadruplet codon is read as a triplet, the sequence shown in gray results, which leads to synthesis termination (TERM) at each site—no protein containing a His-tag is produced [$Fl_{1\,...}$ BODIPY-FL, $Fl_{2\,...}$ BODIPY 558]. (According to Ref. [8])

with increasing amounts of urea, the fluorescence at 575 nm decreases. In Fig. 4 [8], the effects of a binding peptide on the conformation of calmodulin proteins with different distances between the fluorophores are shown [N_A99_D, for example, means that the **a**cceptor fluorophore (BODIPY 558) is at the **N**-terminus and the **d**onor fluorophore (BODIPY-FL) is at position **99**.].

12.2 Targeted Incorporation of Non-Natural Amino Acids Into Proteins by In Vivo-Expression

12.2.1 Principle of the Method

In several publications—[9–12]—Peter G. Schultz further developed the method of suppression of the UAG stop codon for the incorporation of non-natural amino acids in living expression systems (*E. coli*, *Saccharomyces cerevisiae*). The key point is: for each non-natural amino acid to be introduced, a matched system of aminoacyl-tRNA synthetase (AARS) and suppressor tRNA must be developed.

▶ When I was confronted with this idea at the beginning of 2000 during my lecture, I thought that it would never catch on and that cell-free expression would be clearly superior. After all, it would be too laborious to develop a modified AARS for each chosen AS_x that exclusively loads the suppressor tRNA with only this AS_x. But I clearly underestimated the methodological possibilities of molecular biology. Such systems are now already routine for many specific AS_x and can be easily purchased as kits.

In Ref. [11], "Expanding the genetic code, Wang, L., Schultz, P. G., *Chem. Commun.*, **2002** (1), 1–11, https://doi.org/10.1039/B108185N (2002)," an overview of various methods discussed in the previous section is first provided. Then, techniques for developing such orthogonal matched pairs of aminoacyl-tRNA synthetase with tRNA are described.

Since there were still no suitable new base pairings for use in living cells, suppression of the UAG stop codon continued to be used for the introduction of the non-natural amino acid. The pair from the archaeabacterium *Methanococcus jannaschii* was chosen as the starting point, because the tyrosyl-tRNA synthetase (*Mj*TyrRS) and the corresponding tRNA are largely orthogonal to the corresponding *E. coli* translation system, i.e., the enzyme does not load any *E. coli* tRNA and the $Mj\text{tRNA}^{Tyr}{}_{CUA}$ is also not loaded by any *E. coli* RS.

But how can the enzyme be made to selectively transfer another, very specific AS_x? The paper presents two possibilities for this. We will discuss one of them here.

The methodology is schematically illustrated in Fig. 12.3. A library of various mutants of *Mj*TyrRS is generated. A TAG stop codon is inserted into an essential protein, usually one that confers resistance to an antibiotic, at a position that is not essential

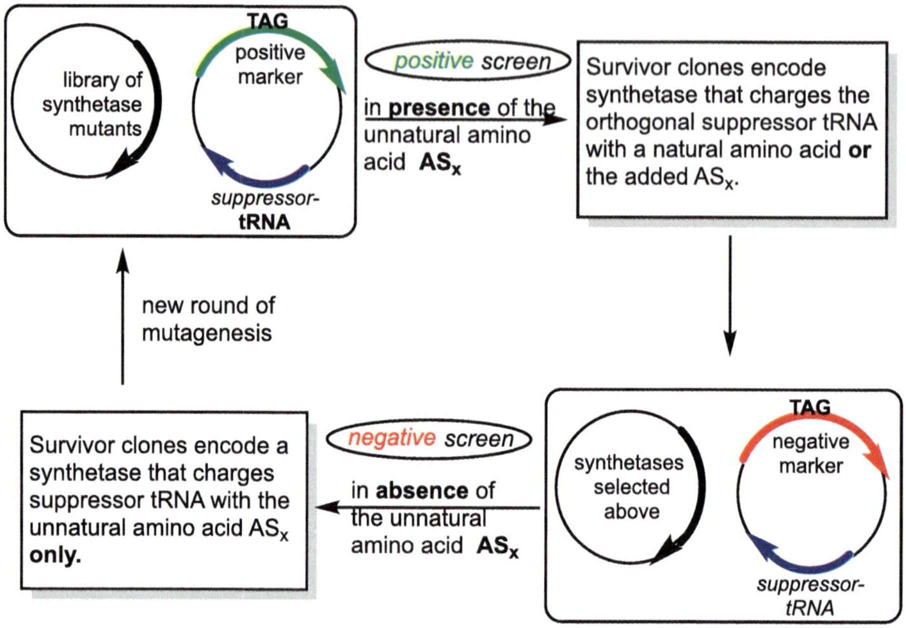

Fig. 12.3 Schematic representation for the optimization and selection of a suitable AARS specific for the chosen AS_x; the AARS is encoded on one plasmid and the suppressor tRNA and marker protein on another plasmid. Positive marker: protein necessary for colony survival; negative marker: protein that is lethal to the organism. (After Ref. [11])

for activity. If the *Mj*tRNA$^{Tyr}{}_{CUA}$—in the presence of the non-natural amino acid—suppresses this stop codon by incorporating **any** amino acid at that position, the colony survives despite the addition of the antibiotic. In a second selection step, the surviving colonies are subjected to negative selection. The DNA for a lethal protein, also with the stop codon, is inserted into the plasmid, but this time the AS_x is **not** added. The lethal protein is therefore only produced if the synthetases can load the suppressor tRNA with natural amino acids. Thus, only those survive that specifically transfer AS_x to the *Mj*tRNA$_{CUA}$. This cycle is repeated until the desired efficiency of the *Mj*AS$_x$RS is achieved. Corresponding plasmids for specific AS_x are now commercially available.

Below are some examples of the application of this methodology.

12.2.2 Incorporation of an Amino Acid for Targeted Photochemical Crosslinking (*Photo-Crosslinking*)

▶ **The Basis of This Section is the Publication** Protein photo-cross-linking in mammalian cells by site-specific incorporation of a photoreactive amino acid, Hino, N., Okazaki, Y., Kobayashi, T., Hayashi, A., Sakamoto, K., Yokoyama, S., *Nature Methods*, **2** (3), 201–206, https://doi.org/10.1038/nmeth739 (2005) [13].

This study [13] is also based on the Ras/MAP kinase signal transduction cascade—thus again indirectly on the Ras protein—(see Chap. 9). It concerns the interactions of the adaptor protein Grb2 with the receptor tyrosine kinase (RTK). The *epidermal growth factor receptor* (EGFR) was chosen as the RTK. After activation by the ligand EGF, a close association with the adaptor protein Grb2 is formed. By targeted incorporation of a photoreactive amino acid, *p*-benzoyl-L-phenylalanine (pBpa) (Fig. 12.4a), which contains the benzophenone chromophore (see Sect. 7.6.1), this is to be investigated.

For this, *in vivo* expression in mammalian cells was necessary, and therefore a different pair, *E. coli* pBpaRS/*Bacillus stearothermophilus* $tRNA_{CUA}^{Tyr}$, had to be used, since the established *Mj*TyrRS (see above) is not orthogonal to the $tRNA^{Tyr}$ of mammalian cells. The amino acid pBpa was incorporated into various positions of the binding pocket of the SH2 domain of Grb2. In addition, the modified Grb2 was provided with a FLAG tag [14]. Figs. 3 and 4 of Ref. [13] show the results of these studies. A protein band with the sum of the molecular masses of EGFR and Grb2 is seen in Fig. 3 only in the case of the modified protein with pBpa at position 111, when the receptor was activated with EGF and irradiated for 15–30 min at $\lambda = 365$ nm. The protein band could be stained with an antibody against EGFR and with an anti-FLAG antibody (M2). Fig. 4 shows that incorporation of pBpa at very specific sites leads to crosslinking of Grb2 with EGFR. The natural amino acid at position 107, histidine, is crucial for the interaction of Grb2 with phosphorylated (activated) EGFR. If pBpa is present at this position, this recognition site is missing, and thus no contact and no crosslinking can occur. The strongest effects are seen when pBpa is incorporated at positions 109 and 111. According to the X-ray structure, the side chain of valine at position 110 of native Grb2 points in a different direction; indeed, with pBpa at this position, no crosslinking (*crosslink*) occurs. These results demonstrate that targeted incorporation of a photochemically activatable amino acid such as pBpa can be used to very finely probe contact zones between proteins.

Flag Tag

Short peptide sequences attached to an expressed protein facilitate identification of the gene product. For this purpose, short peptides are used, for example, against which a specific monoclonal antibody (Ab) is available. The anti-FLAG Abs M1 and M2 recognize very specifically a peptide with the sequence Asp–Tyr–Lys–Asp–Asp–Asp–Asp–Lys [14]

Ref. [15] publishes a detailed protocol for the incorporation of this very amino acid pBpa into proteins for expression in *E. coli*. Such a publication shows that this methodology had indeed become routine—even as early as 2005.

This amino acid was also used to elucidate protein-DNA interactions that are important for gene regulation [16].

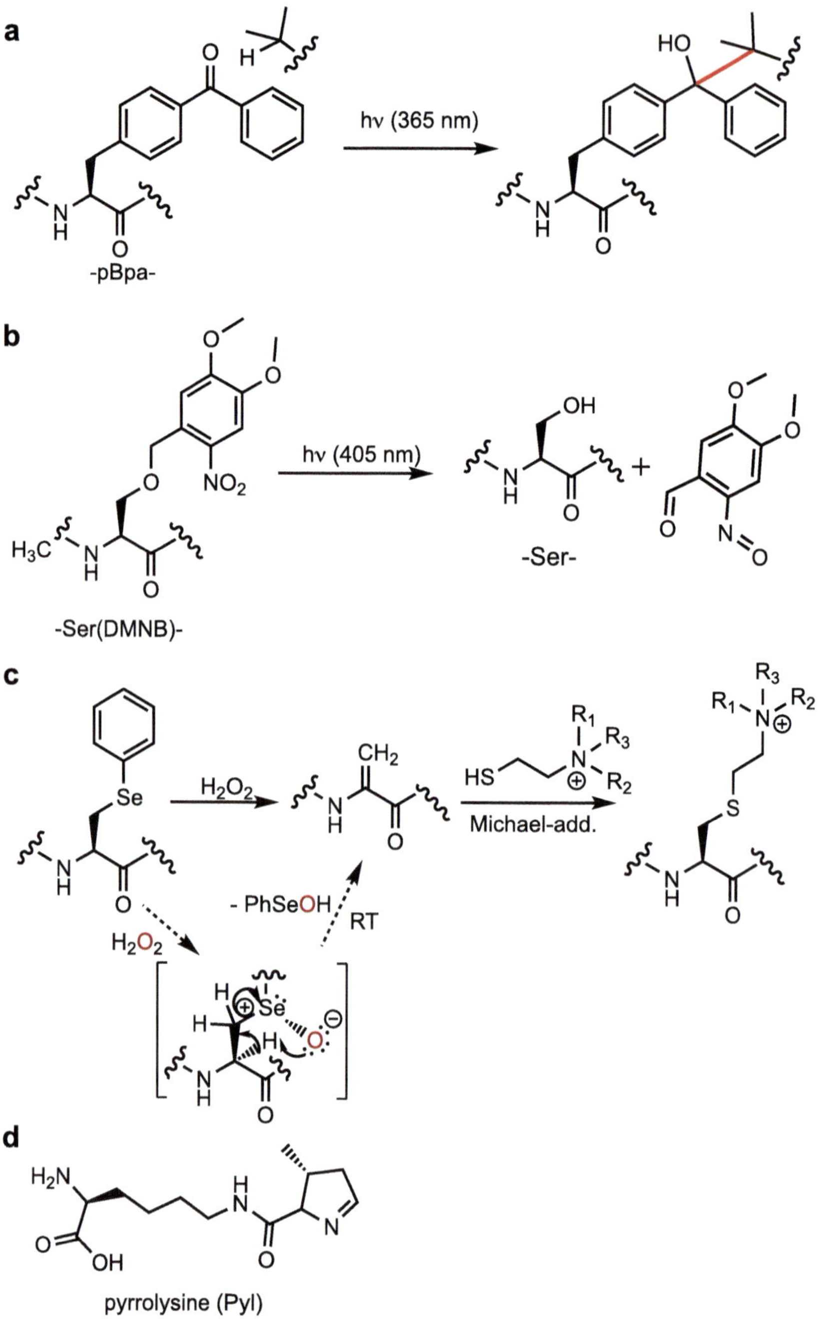

Fig. 12.4 Examples of the incorporation of non-natural amino acids by the method of *in vivo* protein expression: **a** *p*-benzoylphenylalanine (pBpa) and the possibility of photochemical crosslinking with contacting C–H bonds; **b** the photolytic release of a serine from the blocked 4,5-dimethoxy-2-nitrobenzyl ether; **c** posttranslational conversion of phenylselenocysteine into other non-natural amino acids; **d** structure of the amino acid pyrrolysine (Pyl) encoded in archaebacteria

12.2.3 Incorporation of a Serine Analogue With Reversibly Blocked Hydroxyl Function to Control Protein Phosphorylation

▶ **Let Us Consider the Following Publication** Control of protein phosphorylation with a genetically encoded photocaged amino acid, Lemke, E. A., Summerer, D., Geierstanger, B. H., Brittain, S. M., Schultz, P. G., *Nature Chem. Biol.*, **3** (12), 769–772, https://doi.org/10.1038/nchembio.2007.44 (2007) [17].

The introduction of non-natural amino acids into proteins also offers the possibility to temporarily block a side chain function of an amino acid, and then release it *in vivo* under controlled conditions, so that the original native protein is restored. In this work, a method is developed for the first time to introduce such a reversibly blocked (*caged*) serine into proteins. It is essentially that the protecting group should be cleavable under *in vivo* conditions. Photolabile protecting groups are particularly suitable for this purpose. Here, a 4,5-dimethoxy-2-nitrobenzyl protecting group (DMNB) is used for the primary alcohol of serine (Fig. 12.4b). Compared to the more commonly used 2-nitrobenzyl protecting groups, this one has the advantage of being cleavable even with "blue" visible light. Especially for studies in living cells, efforts are made to avoid harmful UV light (the mechanism is the same as for the Nvoc protecting group, see Sect. 6.1.3, Fig. 6.6). The expression of this new AS_x was first tested on human superoxide dismutase (hSOD-33TAG-His_6), with Ser(DMNB) at position 33 and a His_6 tag for isolation.

This AS_x was then incorporated into a protein that is located either in the nucleus or in the cytosol depending on the phosphorylation of various serines. Pho4 is a yeast transcription factor that adapts yeast growth to the available amount of inorganic phosphate (P_i). At low phosphate concentrations, it is located in the nucleus and regulates transcription (mRNA formation) for proteins that serve adaptation to these conditions. If sufficient phosphate is present, Pho4 is phosphorylated by the ***cyclin-cyclin-dependent kinase*** (cyclin-CDK) and thereby inactivated. Phosphorylation of the various serines of Pho4 has different significance. For example, if Ser 114 (S2) and/or Ser128 (S3) are phosphorylated, Pho4 is transported from the nucleus to the cytosol.

Pho4 mutants with Ser(DMNB) at S2 or S3 and with green fluorescent protein (GFP) (see Sect. 14.1.3) at the C-terminus were expressed in *Saccharomyces cerevisiae* using the $tRNA^{Leu}_{CUA}$/LeuRS (from *E. coli*) orthogonal pair optimized for Ser(DMNB). Proteins without the blocked serine also lacked GFP, i.e., synthesis was terminated at the stop codon. Fig. 2 of Ref. [17] shows cells with WT (*wild-type*) Pho4-GFP and mutants under various conditions. (a) The normal protein is distributed in the cytosol in the presence of inorganic phosphate (P_i), but in the nucleus in the absence of phosphate. (b) Mutants with Ala at position 114 (S2) or 128 (S3) are found only in the nucleus even in the presence of phosphate, since they cannot be phosphorylated. (c) The mutant with the TAG codon for amino acid 128 (S3) expresses fluorescent Pho4, which is found only in the nucleus when the non-natural AS_x is added to the medium—in the absence of this amino acid, there is no fluorescent Pho4. When the mutants with Ser(DMNB) at S3 (d) or S2 (f) are irradiated with light of wavelength $\lambda = 405$ nm, fluorescence is no longer found in the nucleus but in the cytosol about 2 min after irradiation—the now free serine was thus phosphorylated in the presence of phosphate. In (e), a scan across two cells is shown. Before irradiation, there is a peak (in the nuclear region), afterwards there is no longer any fluorescence measurable there.

Further experiments allowed the contributions of phosphorylation of S2 and S3 to export from the nucleus to be specified in more detail (Fig. 3 and text). Overall, the incorporation of this temporarily blocked serine is highly interesting for clarifying the effects of serine phosphorylation in various proteins.

12.2.4 Specific Introduction of Posttranslational Modifications at Lysine

In Ref. [18] "Site-specific incorporation of methyl- and acetyl-lysine analogues into recombinant proteins, Guo, J., Wang, J., Lee, J. S., Schultz, P. G., *Angew. Chem. Int. Ed.*, **47** (34), 6399–6401, https://doi.org/10.1002/anie.200802336 (2008)," an interesting detour is used to generate a methylated or acetylated lysine analogue in proteins. These posttranslational modifications at lysine can influence and regulate the biological activities, interactions, and lifetimes of the corresponding proteins. Their targeted investigation is therefore of particular interest. An already established system of *Mj*TyrRS/*Mj*tRNA$^{\mathrm{Tyr}}{}_{\mathrm{CUA}}$ was used for the targeted introduction of phenylselenocysteine [19], to introduce this amino acid instead of lysine at specific sites of a histone protein. This amino acid, even under mild oxidation with H_2O_2 (0.1 M, pH 8.5, room temp.), forms a selenoxide, which at room temperature undergoes a Cope-like rearrangement to eliminate dehydroalanine (Fig. 12.4c). Michael addition with various thiols then yields mono-, di-, and trimethylated lysine analogues or acetyllysine analogues in good yield. According to Ref. [18], the Michael addition to the double bond of dehydroalanine in proteins proceeds diastereoselectively, i.e., the original *R* configuration at the α-C atom is restored, corresponding to an L-lysine residue (see Sect. 2.1: $-CH_2-S-$ has higher priority than $-C{=}O$).

One disadvantage of this method, however, is that methionine residues are partially oxidized to sulfoxides during oxidation with hydrogen peroxide. To avoid these additional heterogeneities, the methionine residues were replaced by leucine, which did not change the biological activity. The protein with the acetylated lysine analogue was deacetylated by the same enzyme as the natural histone and only subsequently phosphorylated.

This example was chosen primarily because, starting from an AS_x (phenylselenocysteine) introduced *in vivo*, a series of different modified proteins could be made accessible by posttranslational organic-chemical transformations. In the same year, another group [20] published a direct approach to acetylated lysine in proteins, underscoring the importance of being able to specifically investigate such posttranslational modifications (PTMs).

In Ref. [21], as additional examples, the direct introduction of further non-natural amino acids via the UAG stop codon is described.

12.3 Further Developments in Expression of Specifically Modified Proteins

The success brought by the method described in Sect. 12.2 naturally sparked the desire for even more possibilities. That's always the way: once a goal is reached, an even higher goal is set—the selective targeted introduction of multiple non-natural amino acids into a protein. But since there are no more free triplet codons, this could only be done with quadruplet codons. This has already been practiced for *in vitro* expression (see Sect. 12.1.3). However, natural ribosomes recognize these codons very poorly. The group of Jason W. Chin developed, through mutation and selection, modified ribosomes (Ribo-Q1, -Q2, and -Q3) that recognize quadruplet codons with high efficiency and thus enable the incorporation of additional non-natural amino acids using corresponding orthogonal tRNAs (with a quadruplet anticodon) [22]. A general review on the development and application of modified ribosomes (Flexizymes) can be found in Ref. [23].

Further possibilities have been found to incorporate multiple non-natural amino acids into a protein, which, however, required several changes to the other natural stop codons, to less frequently used codons for amino acids (e.g., Arg), or modifications to the AARS/tRNA pair for the archaeabacterial amino acid pyrrolysine (Pyl) (Fig. 12.4d) [24, 25]. The often low protein yield is partly attributed to additional factors of the protein synthesis system (*elongation factors* EFs), which for this reason were partially eliminated. In this respect, so-called cell-free (i.e., *in vitro*) protein synthesis has experienced a renaissance. For this, the necessary individual factors of the system are purified from a cell extract and the system is then reconstituted (so-called PURE system)—a laborious and very costly process, but one that enables more efficient synthesis of modified proteins and also allows components that would be lethal to living microorganisms or cells. Here are several recent reviews on this topic: [26, 27] and [28].

The articles [29] introduce so-called *eCells*. The protein synthesis machinery of *E. coli* is not isolated and purified from the cell (PURE system), but the cells are surrounded with an artificial shell of polyelectrolytes in a layer-by-layer (LbL) assembly, and then the cell wall underneath is dissolved. In this way, compartmentalization is maintained, but with a semipermeable shell. This approach yields outputs that are somewhat lower than with the PURE system, but significantly higher than when using a whole cell extract. The costs, however, are much lower than with the PURE system.

Finally, I would like to refer to the most recent review by P. G. Schultz [30], in which he summarizes his work in this article from *Account of Chemical Research*.

Protein Modification—Summary

In this chapter, we have seen examples of the targeted modification of proteins with non-natural amino acids. One might object that these methods had less and less to do with organic chemistry. However, it should have become clear that a great deal of organic chemistry is involved in the development and selection of the new AS_x, quite apart from the fact that they must, of course, also be specially synthesized and introduced into the *in vivo* system in suitable form.

12.4 Questions

Let's Test Our Understanding of the Content of This Chapter!

12.1 What is the *amber* stop codon and what is the triplet codon called at the DNA and mRNA level?

12.2 Explain the principal differences between early *in vitro* and *in-vivo* expression of modified proteins and briefly describe the procedure!

12.3 What was investigated in the first study discussed and why?

12.4 How is it ensured in the second example with calmodulin that only proteins with both fluorescent amino acids are synthesized in full length, even though the *frame-shift-suppressor* tRNAs are less well recognized?

12.5 How does the relative proximity of the N- and C-terminus of calmodulin affect the fluorescence spectra in the modified protein with the two different fluorophores?

12.6 What is the role of the protein Grb2? (see Chap. 9)

12.7 What was intended to be demonstrated in the example from Sect. 12.2.2?

12.8 What is the crucial residue of the non-natural amino acid in the example with EGFR, and what reaction does it mediate?

12.9 What is the general name for the technique used in the example from Sect. 12.2.2 and what other functionalities would also be suitable for it?

12.10 Which protecting group is used for the non-natural serine derivative and for what purpose?

12.11 How is the protein Pho4 visualized?

12.12 From which non-natural amino acid can dehydroalanine be easily generated?

12.13 What is meant by the PURE system?

12.14 Explain the significance of aminoacyl-tRNA synthetases (AARS)!

References

1. Budisa, N., Minks, C., Medrano, F.J., …, Moroder, L.: Residue-specific bioincorporation of non-natural, biologically active amino acids into proteins as possible drug carriers: structure and stability of the per-thiaproline mutant of annexin V. Proc. Nat. Acad. Sci. (USA) **95**(2),

455–459 (1998). https://doi.org/10.1073/pnas.95.2.455; Budisa, N., Karnbrock, W., Steinbacher, S., …, Huber, R.: Bioincorporation of teluromethionine into proteins: a promising new approach for x-ray structure analysis of proteins. J. Mol. Biol. **270**(4), 616–623 (1997). https://doi.org/10.1006/jmbi.1997.1132

2. Noren, C.J., Anthony-Cahill, S.J., Griffith, M.C., Schultz, P.G.: A general method for site-specific incorporation of unnatural amino acids in proteins. Science. **244**(4901), 182–188 (1989). https://doi.org/10.1126/science.2649980
3. Cornish, V.W., Mendel, D., Schultz, P.G.: Probing protein structure and function with an expanded genetic code. Angew. Chem. Int. Ed. **34**(6), 621–633 (1995). https://doi.org/10.1002/anie.199506211; Cornish, V.W., Mendel, D., Schultz, P.G.: Untersuchungen von Struktur und Funktion von Proteinen mit einem erweiterten genetischen Code. Angew. Chem. **107**, 677–690 (1995). https://doi.org/10.1002/ange.19951070604
4. Chung, H.-H., Benson, D.R., Schultz, P.G.: Probing the role of lysine 16 in Ras p21 protein with unnatural amino acids. J. Am. Chem. Soc. **115**(14), 6414–6415 (1993). https://doi.org/10.1021/ja00067a068
5. Ma, C., Kudlicki, W., Odom, O.W., …, Hardesty, B.: In vitro protein engineering using synthetic tRNAAla with different anticodons. Biochemistry. **32**(31), 7939–7945 (1993). https://doi.org/10.1021/bi00082a015
6. Steward, L.E., Collins, C.S., Gilmore, M.A., …, Chamberlin, A.R.: In vitro site-specific incorporation of fluorescent probes into β-galactosidase. J. Am. Chem. Soc. **119**(1), 6–11 (1997). https://doi.org/10.1021/ja963023f
7. Hohsaka, T., Ashizuka, Y., Murakami, H., Sisido, M.: Incorporation of nonnatural amino acids into streptavidin through in vitro frame-shift suppression. J. Am. Chem. Soc. **118**(40), 9778–9779 (1996). https://doi.org/10.1021/ja9614225
8. Kajihara, D., Abe, R., Iijima, I., …, Hohsaka, T.: FRET analysis of protein conformational change through position-specific incorporation of fluorescent amino acids. Nat. Meth. **3**(11), 923–929 (2006). https://doi.org/10.1038/nmeth945
9. Martin, A.B., Schultz, P.G.: Opportunities at the interface of chemistry and biology. Trends in Biochemical Sciences. **24**(12), M24–M28 (1999). https://doi.org/10.1016/S0968-0004(99)01498-X
10. Wang, L., Schultz, P.G.: A general approach for the generation of orthogonal tRNAs. Chemistry & Biology. **8**(9), 883–890 (2001). https://doi.org/10.1016/S1074-5521(01)00063-1
11. Wang, L., Schultz, P.G.: Expanding the genetic code. Chem. Commun. **1**, 1–11 (2002). https://doi.org/10.1039/B108185N
12. Chin, J.W., Cropp, T.A., Chu, S., …, Schultz, P.G.: Progress toward an expanded eukaryotic genetic code. Chemistry & Biology. **10**(6), 511–519 (2003). https://doi.org/10.1016/S1074-5521(03)00123-6
13. Hino, N., Okazaki, Y., Kobayashi, T., …, Yokoyama, S.: Protein photo-cross-linking in mammalian cells by site-specific incorporation of a photoreactive amino acid. Nat. Meth. **2**(3), 201–206 (2005). https://doi.org/10.1038/nmeth739
14. Hopp, T.P., Prickett, K.S., Price, V.L., …, Conlon, P.J.: A short polypeptide marker sequence useful for recombinant protein identification and purification. Nat. Biotechnol. **6**(10), 1204–1210 (1988). https://doi.org/10.1038/nbt1088-1204. https://en.wikipedia.org/wiki/FLAG-tag. Zugegriffen am 05.01.2026
15. Farrell, I.S., Toroney, R., Hazen, J.L., …, Chin, J.W.: Photo-cross-linking interacting proteins with a genetically encoded benzophenone. Nat. Methods. **2**(5), 377–384 (2005). https://doi.org/10.1038/nmeth0505-377
16. Lee, H.S., Dimla, R.D., Schultz, P.G.: Protein-DNA photo-crosslinking with a genetically encoded benzophenone-containing amino acid. Bioorg. Med. Chem. Lett. **19**(17), 5222–5224 (2009). https://doi.org/10.1016/j.bmcl.2009.07.011

17. Lemke, E.A., Summerer, D., Geierstanger, B.H., Brittain, S.M., Schultz, P.G.: Control of protein phosphorylation with a genetically encoded photocaged amino acid. Nat. Chem. Biol. **3**(12), 769–772 (2007). https://doi.org/10.1038/nchembio.2007.44
18. Guo, J., Wang, J., Lee, J.S., Schultz, P.G.: Site-specific incorporation of methyl- and acetyl-lysine analogues into recombinant proteins. Angew. Chem. Int. Ed. **47**(34), 6399–6401 (2008). https://doi.org/10.1002/anie.200802336; Guo, J., Wang, J., Lee, J.S., Schultz, P.G.: Site-specific incorporation of methyl- and acetyl-lysine analogues into recombinant proteins. Angew. Chem. **120**(34), 6499–6501 (2008). https://doi.org/10.1002/ange.200802336
19. Wang, J., Schiller, S.M., Schultz, P.G.: A biosynthetic route to dehydroalanine-containing proteins. Angew. Chem. Int. Ed. **46**(36), 6849–6851 (2007). https://doi.org/10.1002/anie.200702305; Wang, J., Schiller, S.M., Schultz, P.G.: A biosynthetic route to dehydroalanine-containing proteins. Angew. Chem. **119**(36), 6973–6975 (2007). https://doi.org/10.1002/ange.200702305
20. Neumann, H., Peak-Chew, S.Y., Chin, J.W.: Genetically encoding N^{ε}-acetyllysine in recombinant proteins. Nat. Chem. Biol. **4**(4), 232–234 (2008). https://doi.org/10.1038/nchembio.73
21. Tippmann, E.M., Liu, W., Summerer, D., …, Schultz, P.G.: A genetically encoded diazirine photo-crosslinker in Escherichia coli. ChemBioChem **8**(18), 2210–2214 (2007). https://doi.org/10.1002/cbic.200700460; Xie, J., Supekova, L., Schultz, P.G.: A genetically encoded metabolically stable analogue of phosphotyrosine in Escherichia coli. ACS Chem. Biol. **2**(7), 474–478 (2007). https://doi.org/10.1021/cb700083w
22. Neumann, H., Wang, K., Davis, L., …, Chin, J.W.: Encoding multiple unnatural amino acids via evolution of a quadruplet-decoding ribosome. Nature. **464**(7287), 441–444 (2010). https://doi.org/10.1038/nature08817
23. Hecht, S.M.: Expansion of the genetic code through the use of modified bacterial ribosomes. J. Mol. Biol. **434**(8), 167211 (2022). https://doi.org/10.1016/j.jmb.2021.167211
24. Dunkelmann, D.L., Chin, J.W.: Engineering pyrrolysine systems for genetic code expansion and reprogramming. Chem. Rev. **124**(19), 11008–11062 (2024). https://doi.org/10.1021/acs.chemrev.4c00243
25. Dakhnevich, A., Kazakova, A., Iliushin, D., Ivanov, R.A.: Pyrrolysine aminoacyl-tRNA synthetase as a tool for expanding the genetic code. Int. J. Mol. Sci. **26**(2), 539 (2025). https://doi.org/10.3390/ijms26020539
26. Wu, Y., Wang, Z., Qiao, X., …, Qi, H.: Emerging methods for efficient and extensive incorporation of non-canonical amino acids using cell-free systems. Front. Bioeng. Biotechnol. **8**, 863 (2020). https://doi.org/10.3389/fbioe.2020.00863; Wu, Y., Tang, M., Wang, Z., …, Qi, H.: Efficient in vitro full-sense-codons protein synthesis. Adv. Biol. **6**(10), 2200023 (2022). https://doi.org/10.1002/adbi.202200023
27. Liao, C., Li, H., Zheng, H., …, Ni, J.: Optimization and application of the purified cell-free system. ChemBioChem. **26**(11), e202500176 (2025). https://doi.org/10.1002/cbic.202500176
28. Zhang, L., Liu, W.-Q., Li, J.: Establishing a eukaryotic Pichia pastoris cell-free protein synthesis system. Front. Bioeng. Biotechnol. **8**, 536 (2020). https://doi.org/10.3389/fbioe.2020.00536
29. Van Raad, D., Huber, T.: In vitro protein synthesis in semipermeable artificial cells. ACS Syn. Biol. **10**(5), 1237–1244 (2021). https://doi.org/10.1021/acssynbio.1c00044; Van Raad, D., Huber, T.: eCell technology for cell-free protein synthesis, biosensing, and remediation. In: Lu, Y., Jewett, M.C. (Hrsg) Cell-Free Macromolecular Synthesis. Advances in Biochemical Engineering/Biotechnology, vol 185,, Kap: 6, pp. 129–146, Springer Nature Switzerland. https://doi.org/10.1007/10_2023_225 (2023)
30. Schultz, P.G.: Synthesis at the interface of chemistry and biology. Acc. Chem. Res. **57**(18), 2631–2642 (2024). https://doi.org/10.1021/acs.accounts.4c00320

Optimization of Structure-activity Relationships (SAR): Development of Inhibitors and Agonists

13

Abstract

The specificity of enzyme-substrate and receptor-ligand interactions is governed by so-called structure-activity relationships (SAR). The first part of this chapter deals with the optimization of an inhibitor of the specific protease renin, which was the first and so far only antihypertensive agent of its kind to be approved. In addition to the methods of optimization, a synthesis of this peptidomimetic is also described. Medical and economic considerations in this context are discussed. In the second part, we will first follow the somewhat confusing and obscure search for sweeteners, which until the year 2000 proceeded without any clue as to a receptor. But even now, with the receptor identified, the model for the sweet effect of various molecules is still largely based on indirect studies of the mechanism of action; there are still no experimental structural determinations of receptor-ligand complexes. The development of low-calorie sweeteners is of great economic interest, especially in view of the increasing prevalence of overweight in the global population, although it is not clear whether these will actually provide a remedy.

▶ The interactions between enzyme and substrate or receptor and ligand are of great importance for biological processes. Their specificity is governed by so-called structure-activity relationships (SAR). Enzyme substrates and receptor ligands are usually rather low-molecular-weight compounds. When developing inhibitors for enzymes and agonists or antagonists for receptors, the aim is to specifically enhance these interactions while simultaneously reducing the structure to the necessary minimum. In the following, we will examine this in more detail using two completely different examples. We will follow the

D. Krois, *Organic Chemical Methods*, https://doi.org/10.1007/978-3-662-73573-2_13

introduction of the first renin inhibitor as a drug against high blood pressure and, in the second part, as a "sweet" conclusion to the second part of this textbook, discuss the mode of action of sweeteners on the corresponding receptor.

13.1 From Peptide to Peptidomimetic: Aliskiren, the First Approved Renin Inhibitor

13.1.1 Antihypertensive Agents

High blood pressure is considered responsible for many cardiovascular diseases, and as more and more people tend to develop it—probably also due to a lifestyle that requires less and less physical activity—drugs that can lower blood pressure have become particularly important. They are therefore also of great economic interest to pharmaceutical companies. Biologically, the renin-angiotensin system (RAS) or renin-angiotensin-aldosterone system (RAAS) is of great importance for the regulation of blood pressure (Fig. 13.1). The biologically active octapeptide angiotensin II, which leads to an increase in blood pressure by binding to the corresponding receptor (AT-II receptor), is formed from a protein, angiotensinogen, by two cleavages with the help of specific proteases: renin and ACE (*angiotensin converting enzyme*).

Until a few years ago, only ACE inhibitors (ACEI) and AT-II antagonists (acting at the AT-II receptor, *angiotensin II receptor blockers* ARB) were approved as antihypertensive agents. However, ACEIs have the disadvantage that ACE is also involved in other processes, such as the deactivation of kinins like bradykinin. In contrast, the Asp protease renin has only a single biological substrate, angiotensinogen, and its inhibition should result in fewer side effects. In this section, we will closely follow the development of the first renin inhibitor, which began about 50 years ago [1] and led to the only approval as a drug to date in 2007.

13.1.2 Development of an Inhibitor

The development of enzyme inhibitors [2] was initially mostly based on the *trial-and-error* principle. Sometimes, the starting point is a randomly identified natural product or some peptide. In the case of renin, development began with RIP (*renin inhibiting peptide*), which is very similar to the N-terminal region of the natural substrate, angiotensinogen, but instead of the cleavable sequence Leu–Val, it has the sequence Phe–Phe (at P1–P1′), which cannot be cleaved by the enzyme (Fig. 13.2). Therefore, the inhibitor remains bound; only the cleavage products, which are not formed, could easily diffuse away.

A significant advance toward the rational development of inhibitors was brought about by the concept of mimicking the transition state of the enzyme-catalyzed reaction.

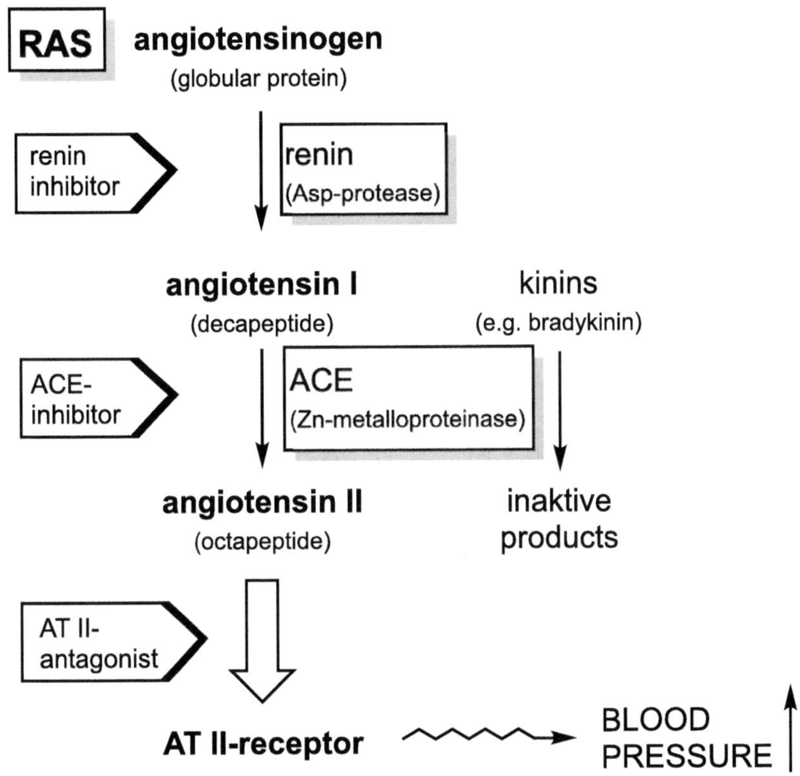

Fig. 13.1 The renin-angiotensin system (RAS): The proteases renin and ACE (*angiotensin converting enzyme*) are responsible for the formation of the active blood pressure-increasing octapeptide angiotensin II (AT II)

▶ **Transition State Analogs** Enzymes are catalysts of chemical reactions. Their effect is therefore based on lowering the activation energy, the highest "peak" of the reaction profile. This stabilization of the transition state can only be achieved through particularly favorable interactions of the enzyme with the substrate(s) at this point in the reaction sequence. Accordingly, compounds that mimic the structure of the transition state but are stable must have particularly high affinity for the enzyme. These are called "transition state analogs."

The cleavage of a peptide bond proceeds via the mechanism of nucleophilic substitution of acyl compounds and is usually catalyzed by acids. This is certainly the case for proteases that have aspartic acid in the active site. If we consider this addition-elimination mechanism, as shown in Fig. 13.3, we see that there is a transition from an sp^2-hybridized carbonyl C atom to an sp^3-hybridized intermediate (with a hydroxyl group, formulas highlighted) before the C=O group is restored

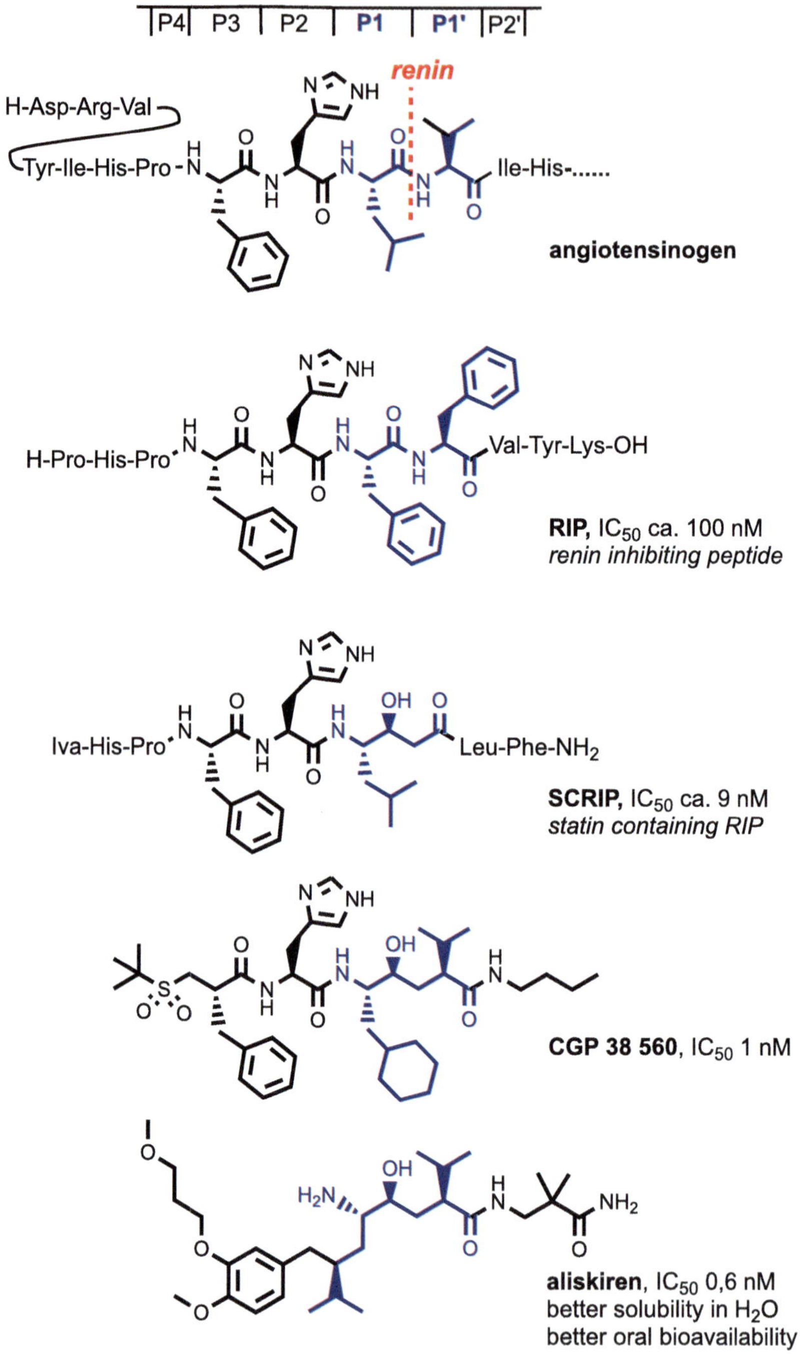

Fig. 13.2 Development of the first renin inhibitor approved as a drug, aliskiren. The designations P1, 2, 3, 4 and P1′, 2′, etc. refer to the numbering of amino acid residues from the cleavage site toward the N- and C-terminus, respectively; IC_{50} values are the inhibition constants

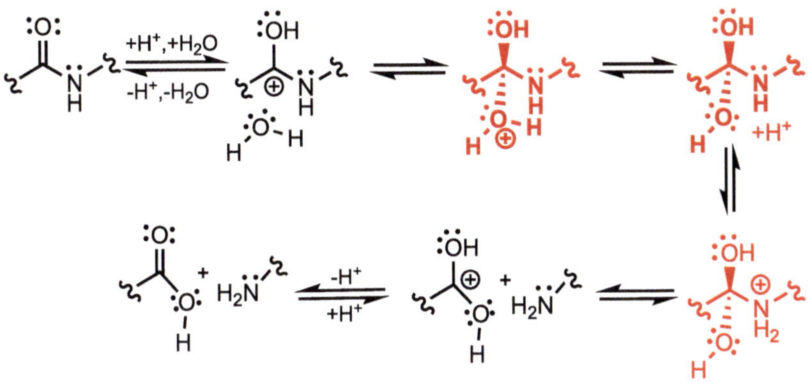

Fig. 13.3 Addition-elimination mechanism of the cleavage of carboxylic acid amides under acid catalysis; structures in red … sp^3-hybridized former acyl C atom with hydroxyl groups

by elimination. Therefore, corresponding stable compounds with a HO group best mimic the transition state of the rate-determining step.

Efficacy of Inhibitors—IC_{50} Values The efficacy of inhibitors is always initially determined using *in vitro* assays. In these, enzyme activity is measured as a function of increasing inhibitor concentrations. This generally results in sigmoidal curves (see also Sect. 10.5 titer determination by ELISA or Sect. 11.1 DNA melting curves), and the characteristic value used is the inhibitor concentration that leads to 50% enzyme activity → IC_{50} value (Sect. 9.3).

For this purpose, the non-natural γ-amino acid statin was first incorporated at P1–P1′, and the N- and C-termini were further varied, resulting in SCRIP (*statin containing renin inhibiting peptide*). Finally, to restore the correct length of the dipeptide segment (Leu–Val), a δ-amino acid was used at this position. In the inhibitor CGP 38560, developed at Ciba-Geigy at the time, only the amino acids Phe–His (P3–P2) before the cleavage site remained. The inhibition constant IC_{50} is already very good at 1.0 nM. However, the oral bioavailability of CGP 38560, as well as many other developed renin inhibitors, is low. The presence of peptide bonds also leads to faster degradation in the organism; this naturally reduces efficacy and requires high dosing. But since the synthesis costs of such compounds are also relatively high, neither CGP 38560 nor other inhibitors of this type were developed to the stage of a drug [3]. Ciba-Geigy was taken over by Novartis, and the optimization of inhibitor properties continued [3]. However, synthesis was too expensive and laborious for the necessary clinical tests. Novartis was already planning to abandon the project in 1999. It was then spun off into an independent company, Speedel, where another attempt at optimization was made, which was so successful that Novartis took over the "Aliskiren" project again in 2002 and finally achieved market approval in the USA and Europe in 2007. To date (2025), this peptidomimetic remains the only renin inhibitor approved for therapy.

13.1.3 Structure-Based Design of the First Renin Inhibitor

▶ In the publication: "Structure-based drug design: the discovery of novel non-peptide orally active inhibitors of human renin, Rahuel, J., Rasetti, V., Maibaum, J., Rueger, H., Göschke, R., Cohen, N. C., Stutz, S., Cumin, F., Fuhrer, W., Wood, J. M., Grütter, M. G., *Chemistry & Biology*, **7** (7), 493–504, https://doi.org/10.1016/S1074-5521(00)00134-4 (2000)" [4], this new approach to optimizing inhibitor properties is described in detail.

Starting from the inhibitor CGP 38560 (Fig. 13.2), repeated X-ray structural analysis of a recombinant glycosylated human renin with various inhibitors in the active site was used in an evolutionary process to achieve optimal adaptation to the spatial features of the binding pocket, while at the same time finally eliminating the peptide portion. Table 1 of the paper lists a number of such inhibitors and their IC_{50} values for human renin and porcine renin. Inhibition of human renin was further improved, but above all, specificity was increased: porcine renin is practically no longer inhibited by compound **10**, aliskiren. Fig. 6a shows the structure of CGP 38560 (**1**) and aliskiren (**10**), embedded in renin. One can see the good fit of the cyclohexyl group and isopropyl group into the hydrophobic S1 pocket and the phenyl groups into S3. The linear group CH_3–O–$(CH_2)_3$–O– of **10** fits optimally into the $S3^{sp}$ extension of the S3 binding pocket newly identified by X-ray studies. Compared to earlier renin inhibitors with a still significant peptide portion, aliskiren has significantly improved physical and pharmacokinetic properties. It is highly water-soluble and has a long half-life *in vivo*, as it cannot be degraded by proteases and peptidases. Moreover, it had already shown oral bioavailability in primates (marmosets) before clinical trials (see also Ref. [5]).

The clinical trials initially conducted by Speedel and, in the third phase, again by Novartis, then confirmed the good tolerability and antihypertensive effect (comparable to ACE inhibitors and AT-II receptor antagonists). This led to approval in the USA (FDA … *Food and Drug Administration*) and Europe (EMEA … *European Medicines Agency*) in 2007 [3].

13.1.4 Clinical Experience with the Renin Inhibitor

The use of aliskiren as monotherapy for lowering mild to moderate blood pressure has generally proven effective and has usually shown fewer side effects than other antihypertensive agents. The advantage of renin inhibitors compared to ACEIs (ACE inhibitors) and ARBs (AT-II receptor blockers) in therapy is that plasma renin activity (PRA) and the concentration of angiotensin I (AT I) are low. However, as with any blockade of the RAS system, more renin is produced in response, and the concentration of inactive prorenin in the serum and that of inhibited renin increase. This could lead to increased activation of the prorenin/renin receptor [(P)RR], which triggers signals in the MAP kinase cascade; ultimately, this can lead to organ damage. However, there is evidence that aliskiren reduces the synthesis of (P)RR and thus counteracts this possibility [6].

After approval, a number of studies were initiated to investigate improved efficacy and tolerability for particularly problematic clinical cases involving a combination of hypertension, diabetes, and kidney damage. Improved results were expected from combination therapy of aliskiren with ACEIs or ARBs. However, these studies were partially discontinued early, and in 2013, such combination therapy in these cases was discouraged [6], [7]. Further clinical studies also failed to provide a convincing picture. As monotherapy, aliskiren is effective, but in combination with ACEIs or ARBs, no further advantages were found in difficult clinical cases (hypertension + diabetes and/or kidney problems). However, it is noted that the composition of the studies was too heterogeneous; it is possible that aliskiren does indeed offer advantages for some groups, but not for others. However, the diagnostic tools to make such distinctions are still lacking [8–11]. The historical and evolution-based perspective in Ref. [11] is also interesting, where the question is raised as to whether the RAAS (which arose about 400 million years ago) still offers any advantages for humans today.

Meanwhile, another renin inhibitor is in development (clinical trials phase II), namely imarikiren from the pharmaceutical company Takeda [12].

13.1.5 On the Synthesis of Aliskiren

However, after this more medical digression, let us now return to the significance of organic chemistry in this example. A major obstacle in the development of renin inhibitors, including Aliskiren, was the high cost of synthesis. Therefore, in this section, we will at least partially address the synthesis; it is rare enough that syntheses of pharmaceutical products are published in detail in journals. The cost-effective final synthesis (see below), on the other hand, has only been published as a patent.

Figure 13.4 shows the retrosynthetic analysis for Aliskiren according to Ref. [13]. The greatest challenge was the stereospecific introduction of the correct absolute configurations into the carbon framework. Figure 13.5 first shows the simple conversion of commercially available isovanillin (**2**) to the bromomethyl derivative **6**. The stereospecific introduction of the isopropyl group was carried out using the Evans method, with the aid of a chiral auxiliary [14], which, after specific hydrolysis of the exocyclic amide bond with LiOOH (LiOH + H_2O_2) [15], yielded the carboxylic acid (*R*)-**9a** with virtually 100% e.e. (*enantiomeric excess* ≙ enantiomeric excess) in high yield. The chiral auxiliary could be recovered almost quantitatively in larger scale reactions. Reduction of the carboxylic acid with sodium borohydride/I_2 according to Ref. [16], followed by bromi-

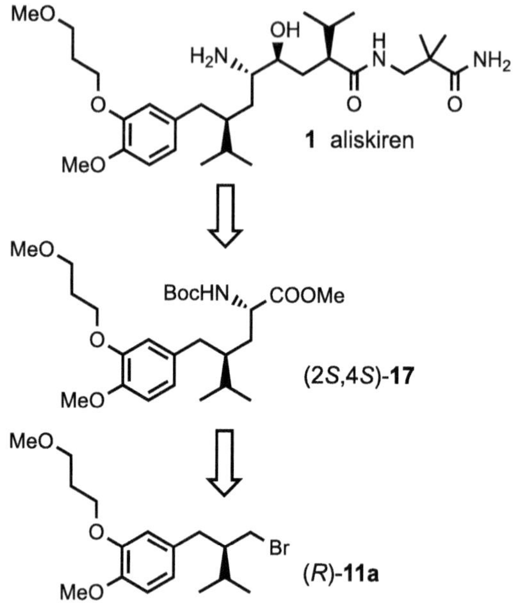

Fig. 13.4 Retrosynthesis for Aliskiren (1)

nation of the alcohol with *N*-bromosuccinimide, finally yielded the first synthon of the retrosynthesis (cf. Fig. 13.4), (*R*)-**11a**.

The second center of asymmetry was to be introduced according to the Schöllkopf method via a chiral bislactim ether [17] (Fig. 13.6). Using the reagent of (5*R)*-configuration (**12a**), the desired diastereomer was obtained almost quantitatively, but with the enantiomer (**12b**), a ca. 1:1 mixture resulted. This outcome suggested that the 2′-chiral center in **11a** was already exerting a significant influence. Accordingly, the reaction was alternatively carried out with the achiral bislactim ether **21**. In fact, 95% of the desired diastereomer could also be obtained this way. After hydrolysis, the amino group was protected with Boc, and finally the carboxylic acid ester was reduced to the aldehyde **2** with diisobutylaluminum hydride (iBu_2AlH, DIBAH) [18].

In the final part of the synthesis (Fig. 13.7), (2*S*)-2-bromomethyl-3-methylbutyl benzyl ether (also prepared by the Evans method [13, 14]) was reacted as a Grignard reagent with the aldehyde **2** [18]. However, no desired diastereoselectivity was achieved. Cyclization to the aminal facilitated the separation of the desired product **4a**. After hydrogenolysis of the benzyl ether, the alcohol was oxidized to the carboxylic acid in two steps. This was then reacted with various amines to synthesize different inhibitors with variations at the P2′ substrate binding site (see also Ref. [19]). The coupling for the synthesis of Aliskiren was carried out with the coupling reagent diethyl cyanophosphonate (diethylphosphoryl cyanide, DEPC) [20] or HBTU (analogous to HATU, Sect. 3.1.2.1, pyridine ring replaced by benzene ring). After cleavage of the aminal with *p*-toluenesulfonic acid (*p*TsOH) and removal of the Boc protecting group with HCl in dioxane, the hydrochloride was obtained, which was converted to the free base with sodium hydroxide solution. Addition of fumaric acid then yielded Aliskiren as the semifumarate.

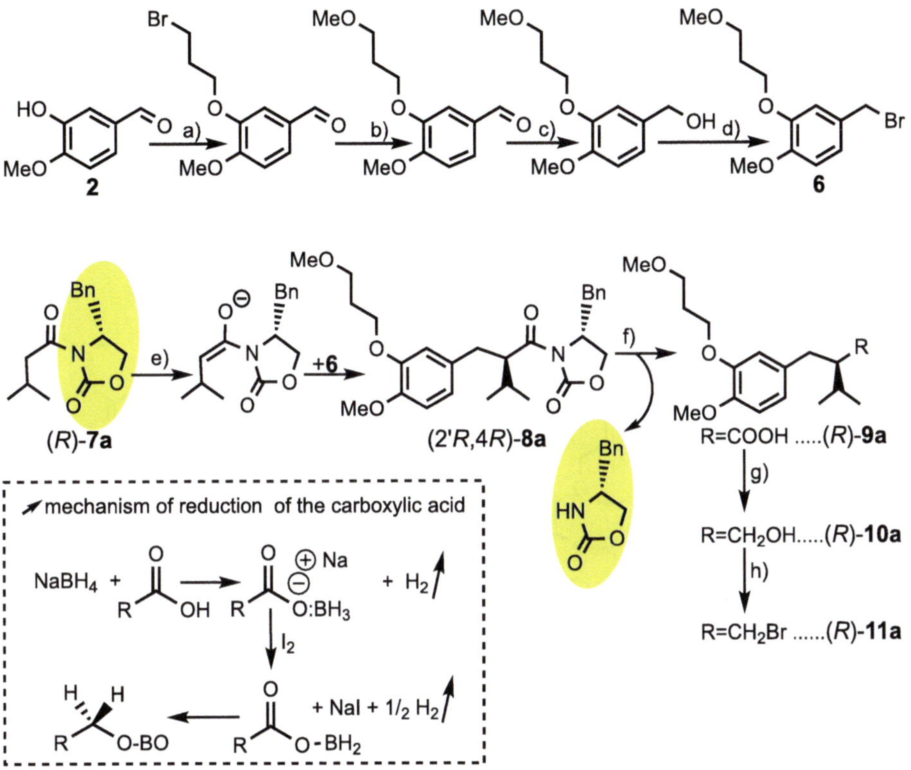

Fig. 13.5 Aliskiren synthesis part 1. Reagents: a) 1,3-dibromopropane, K_2CO_3, MeCN, 93%; b) NaOMe, MeOH, 72%; c) $NaBH_4$, MeOH, 96%; d) Me_3SiBr, $CHCl_3$, 100%; e) LiHMDS (lithium hexamethyldisilazide, strong base), −70 °C, after addition of **6** up to RT, 86%; f) LiOH, 30% H_2O_2, 95%; g) $NaBH_4$, I_2, 90%; h) *N*-bromosuccinimide (NBS); Bn … benzyl. The box shows the mechanism for the reduction of carboxylic acids with $NaBH_4/I_2$ according to Ref. [16]. The chiral auxiliary (highlighted in yellow) is prepared from phenylalaninol (in this case D-configured) and can be recovered. (According to Ref. [13])

However, this published synthesis would have been too laborious for large-scale production. Today, the drug is manufactured by a different route. Ref. [3] outlines the new synthetic route in broad terms. The three synthons are shown in Fig. 13.8. For synthon A, the configuration at the asymmetric carbon atom is introduced by an enantioselective hydrogenation [21]. The enantiomers of synthon B are separated by an enzymatic kinetic resolution of an ester. The coupling of synthons A and B is achieved by a Ni-catalyzed *cross-coupling* reaction. A diastereoselective lactonization establishes the two missing chiral centers, and opening of the lactone by synthon C ultimately leads to the target compound. These reactions have only been published as patents. However, it is evidently a very effective, cost-saving synthesis that made it possible to bring the drug to market in the first place.

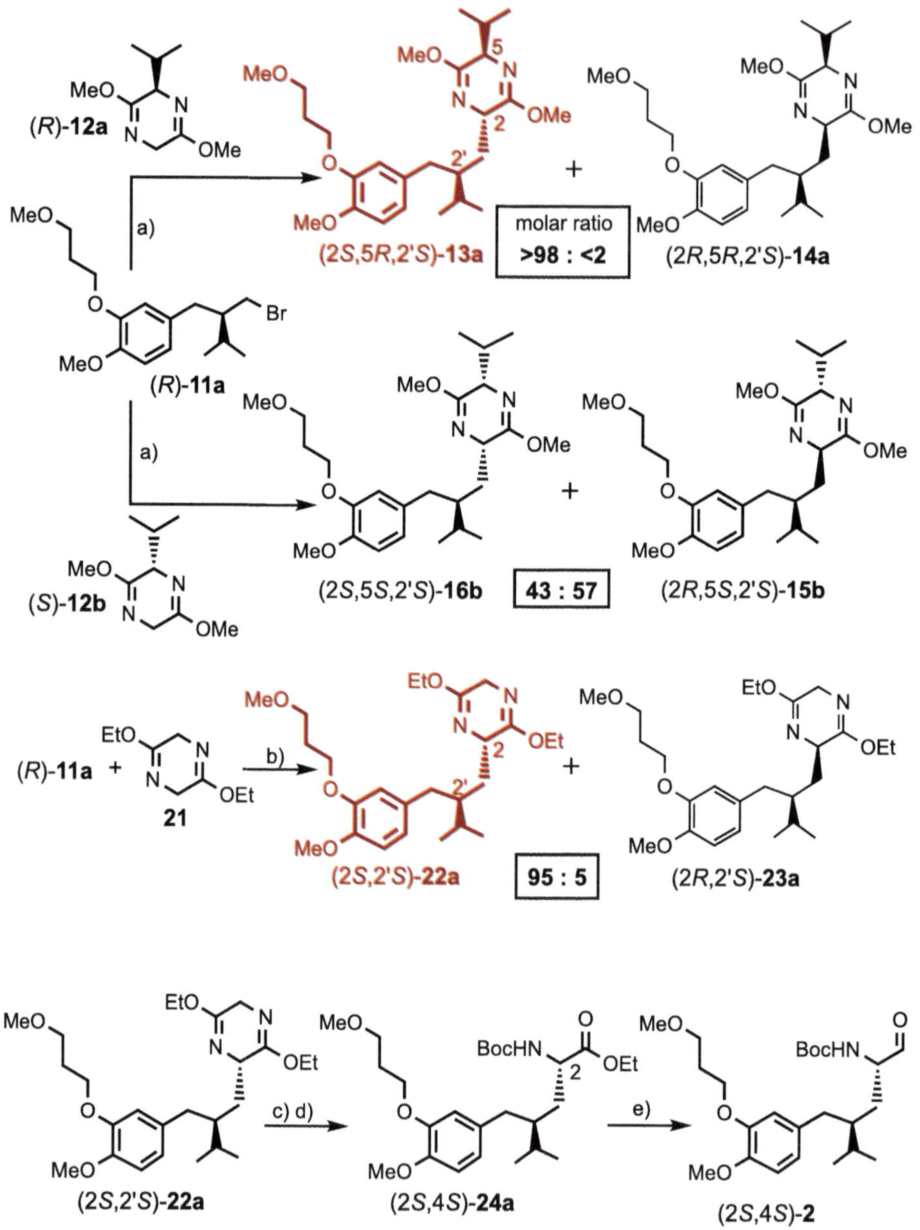

Fig. 13.6 Aliskiren synthesis part 2. Reagents: a) **12a** or **12b** + BuLi (n-butyllithium), −75 °C, 75 min, then +(*R*)-**11a**; b) **21** + BuLi, −75 °C, 75 min, then +(*R*)-**11a**; c) 1 M HCl, MeCN; d) Boc_2O, Et_3N, CH_2Cl_2, 82% based on (*R*)-**11a**; e) DIBAH, toluene, −75 °C, 97% [18]. (According to Ref. [13])

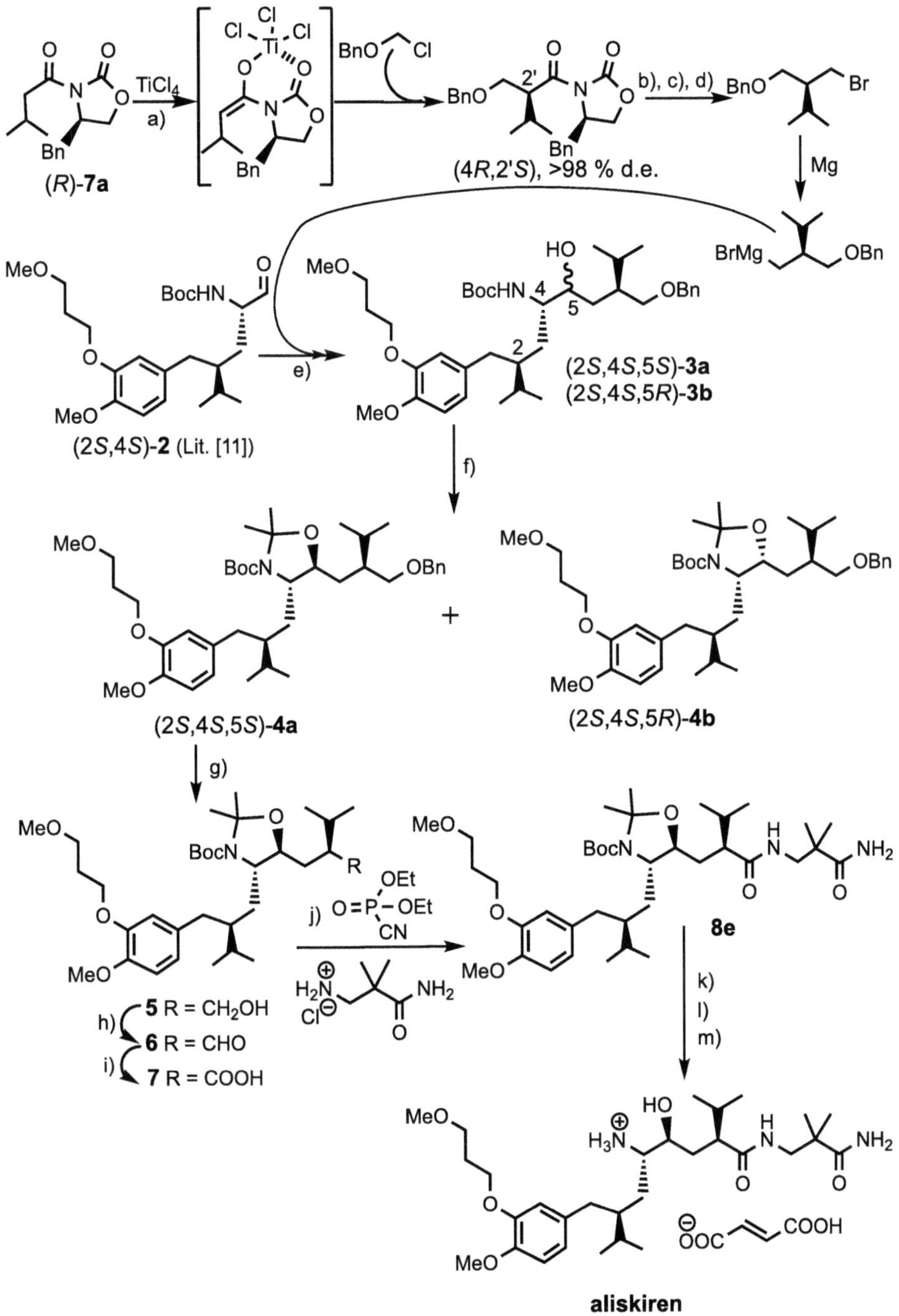

Fig. 13.7 Aliskiren synthesis part 3: Reagents: a) $(iPr)_2EtN$, 50%; b) LiOH, 30% H_2O_2, 95%; c) $NaBH_4$, I_2, 90%; d) NBS; e) + Grignard, THF, 53%, **3a**/**3b** = 4/6; f) 2,2-dimethoxypropane, pTsOH, 50% **4a**, 36% **4b**; g) H_2, Pd/C, 100%; h) *N*-methylmorpholine-*N*-oxide, tetrapropylammonium perruthenate (TPAP, catalytic), CH_2Cl_2; i) $KMnO_4$, $Bu_4N^+Br^-$, toluene/H_2O, 58% (2 steps); j) (DEPC, see formula), Et_3N, THF; k) pTsOH, MeOH, 79% (2 steps); l) 4 M HCl/dioxane; m) NaOH → free amine, then + fumaric acid, 2-propanol. (According to Ref. [13] and [18])

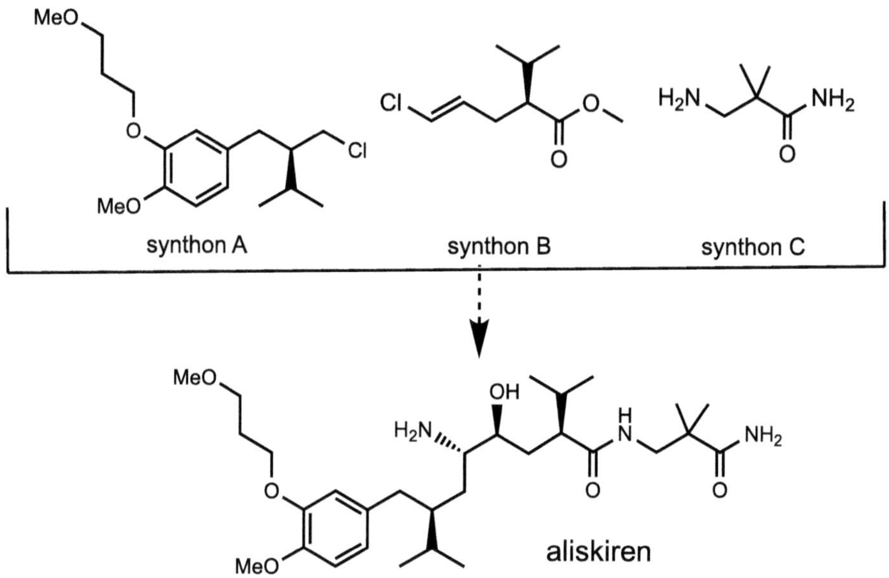

Fig. 13.8 New Aliskiren synthesis from fragments

Summary

We have followed the development of an enzyme inhibitor. At the same time, we have become aware of how difficult it is to develop a drug for clinical use. Due to the complexity of biological processes and the various complications in patients, the applicability of drugs for specific cases is not easy to determine. The real progress achieved by the first renin inhibitor apparently could not be clearly demonstrated even seventeen years after its introduction.

The synthesis section was intended to clearly demonstrate that there can be no research and development in this field without organic synthesis methods. Had it not been possible to efficiently design the synthesis of an inhibitor that had proven promising, Aliskiren would not even have been available for larger clinical trials.

13.2 Sweet Taste and Sweeteners—The Receptor and Its Agonists

13.2.1 Taste—One of the Five Senses in Humans and Animals

Our food intake is determined by the senses of taste and smell. There is consensus that taste sensations have evolved primarily to promote the intake of food useful to the organism and to avoid harmful substances [22]. Humans recognize five basic taste types: salty, sour, sweet, and bitter, as well as the perhaps less well-known "umami" taste (Japanese:

savory taste), which is characteristic of the sodium salt of glutamic acid (monosodium glutamate). Until almost the year 2000, none of the receptors for the taste modalities "sweet," "bitter," or "umami" were known, and, based on the structural differences of various taste substances, theories were developed regarding the nature of the receptors and the mechanism of sweet taste recognition [23–27]. Although, as a result of the complete sequencing of the human genome, most receptors are now known (Sect. 13.2.5), as of 2009 there was still much speculation regarding the correlation of structure and activity (SAR)—that is, taste [28].

For taste perception, the receptor-ligand interaction is crucial on the one hand, but so is the way the signal is transmitted to the brain. It is now believed that taste receptor cells (TRC) or *taste bud cells* (TBC) generally possess only one type of taste receptor, but the signal transmission is quite complex and not yet fully understood [29–32]. In this process, there are also interactions with signals from other TRCs. This could explain the frequently observed switch from "sweet" to "bitter" or vice versa, even with minor changes in the stereochemistry of the taste substance (Sect. 13.2.4) [33].

13.2.2 Structures of Sweet-Tasting Substances—A Bewildering Diversity

Since no receptors for "sweet" or "bitter" were known until the year 2000, the development and use of sweeteners initially relied on serendipitous discoveries, from which models of the receptor binding site were sometimes devised, enabling a more rational further development. Let us first look at a selection of different structures of sweet-tasting substances and their relative sweetness (Fig. 13.9). The search for common structural features seems difficult ….

▶ **Relative Sweetness** "Relative sweetness" refers to a value scale based on the sweetness of sucrose (cane or beet sugar) as a reference point (rel. sweetness: 1). If a solution of a sweetener at a dilution of 1/x (relative to 10% m/v) is perceived by test subjects as equally sweet as a sucrose solution (10% m/v), this sweetener is assigned a relative sweetness of x.

Example: A solution of 0.05 g aspartame (H–Asp–Phe–OMe) in 100 mL volume is perceived as equally sweet as a solution of 10 g sucrose in 100 mL. This corresponds to a dilution of 1/200. Therefore, aspartame has a relative sweetness of 200.

The classification is made by a panel of test subjects, who, in addition to sweetness (*sweetness*), also evaluate taste qualities such as *acid*, *bitter*, *burnt sugar*, *caramel*, *body*, *drying*, *smoothness*, *astringent aftertaste*, *bitter aftertaste*, and *sweet persistence*. Even today, it has not been fully possible to replace these human results with electronic measuring devices (*electronic tongue*).

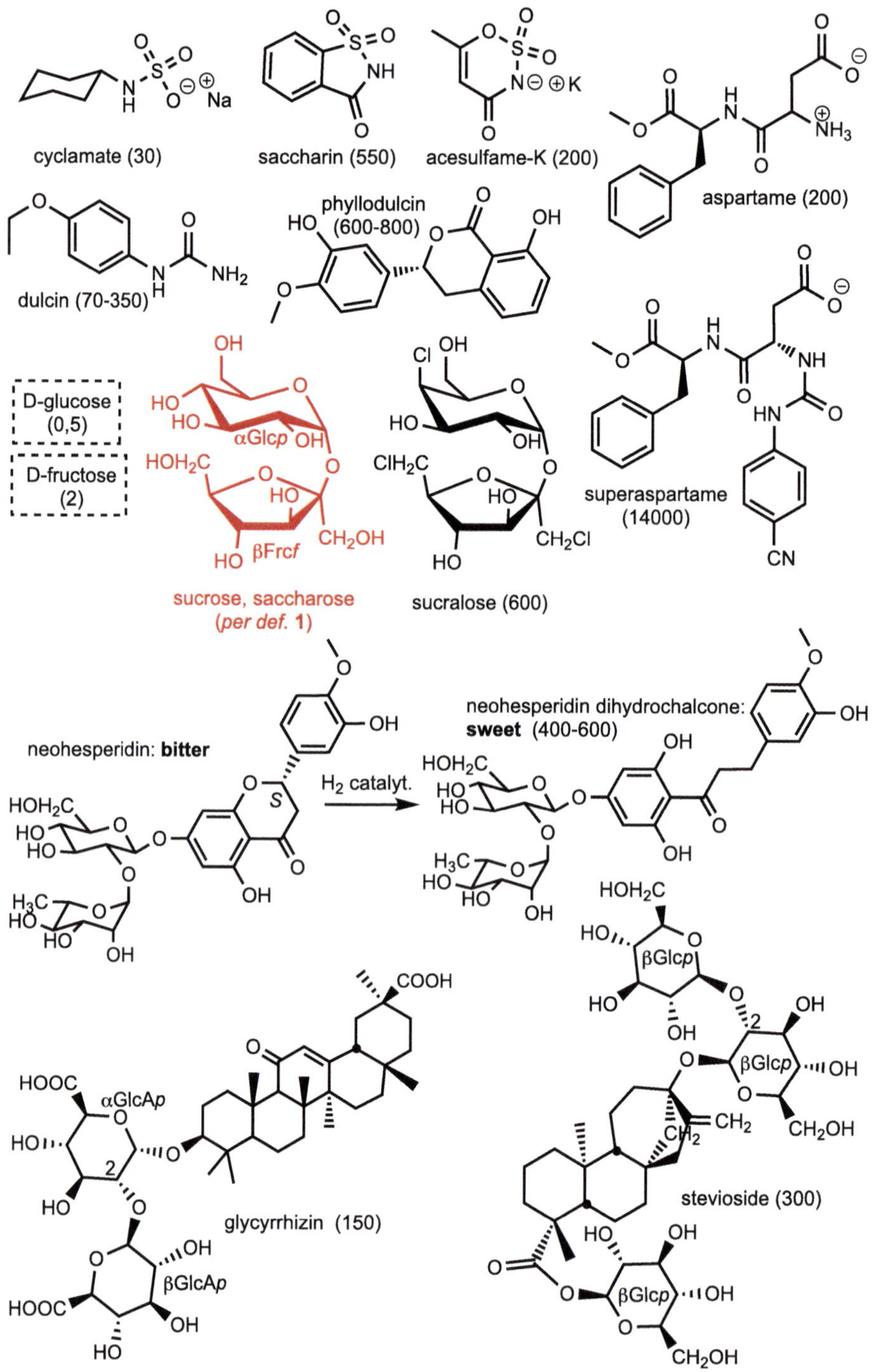

Fig. 13.9 Structures and (relative sweetness) of some natural or synthetic sweeteners

Sodium cyclamate, saccharin, acesulfame K, dulcin, aspartame, and superaspartame are synthetic sweeteners. Sucralose, a sweetener widely used today, is derived from sucrose, but its use has recently come into question (see Ref. [34]). Neohesperidin dihydrochalcone (NHDC) is obtained from the bitter compound in orange peels by catalytic hydrogenation. Phyllodulcin, another natural product with an isovanillin structural element, was first isolated in 1916 from a Japanese plant (*Hydrangea opuloides* Steud. var. Thunbergii) [35]. Glycyrrhizin is found in the root of the licorice plant and, in addition to its sweet taste, also imparts the typical licorice flavor. Due to its similarity to steroid structures, it also has other biological effects, which preclude excessive consumption [36]. Stevioside is the sweet principle from the plant *Stevia rebaudiana* Bertoni, now also approved in the EU—interestingly, steviol, the aglycone without glucose substituents, is tasteless.

There is also a range of proteins, such as brazzein (54 amino acids), monellin (94 amino acids), thaumatin (207 amino acids), and mabinlin (105 amino acids), which are found in exotic fruits and exhibit a relative sweetness of about 700–3000 (based on %-m/v solutions) or even 10,000–100,000 (based on molar concentrations). The proteins miraculin (191 amino acids) and neoculin (227 amino acids), on the other hand, are not sweet themselves but elicit a strong sweet taste at acidic pH and are therefore referred to as taste-modifying proteins. However, neither the sequence nor the structure of these proteins reveals any obvious similarities. And according to current knowledge, all these substances apparently act on the same receptor (Sect. 13.2.5), the T1R2–T1R3 heterodimer [37]. According to recent models, these proteins apparently bind in the cysteine-rich extracellular domain (CRD) of the T1R2–T1R3 receptor [32, 38].

A Side Note on the Sweetness of Chemical Compounds

From a time when, as a chemistry student, one still had to pipette by mouth, I remember the intensely sweet taste of chloroform ($CHCl_3$) and dichloromethane (CH_2Cl_2) when the vapor of these solvents entered the oral cavity—certainly not an experiment to be imitated. However, this sweetness is probably not related to the receptor itself. Rather, it is assumed that these lipophilic molecules penetrate the membrane of taste cells with sweet taste receptors and directly activate the G-proteins.

13.2.3 Structural Theories for Sweeteners

Until 2002, attempts were made to derive theories from the structure of sweeteners in order to develop improved sweeteners. Based on this, models of the receptor binding site were also sometimes proposed. The first such theory regarding the structural requirements for substances with high sweetness was published by Shallenberger and Acree in *Nature* in 1967 and further refined in 1996 [23]. At the core of this model is a hydrogen bond donor (AH) and hydrogen bond acceptor (B) pair in the sweetener at a distance of 2.5–4.0 Å, corresponding to a complementary pair in the receptor. In 1972, L. B. Kier expanded the model by adding a third lipophilic binding site [24]. Tinti and Nofre postulated additional important binding sites to explain the effect of simple sugars; based on this, a model of the receptor was also derived [25]. Murray Goodman and his group worked on sweeteners derived from aspartame. Using the preferred conformations of these dipeptides and analogs, determined by NMR, an "L-shaped" model was developed, in which, apart from the AH/B pair, the orientation of a hydrophobic group is also important and determines whether the taste is bitter or sweet [26, 27].

Figure 13.10 shows some examples of aspartame-like compounds and their taste. The possible orientations of the hydrophobic group relative to the AH/B pair, determined by NMR studies and classified into five classes according to the model, are in good agreement with the observed taste behavior [39]. Especially the so-called retro-inverso-modified dipeptide analogs [40] exhibit not only high sweetness but also greater metabolic stability. This seemed particularly interesting for applications in fruit juices and similar products, since normal dipeptide derivatives like aspartame are degraded relatively quickly by peptidases.

A Personal Experience with Retro-Inverso Dipeptides

During my postdoctoral stay at the Max Planck Institute for Biochemistry in Martinsried in 1986/87, Prof. Murray Goodman was also a visiting professor in Prof. Dr. Erich Wünsch's department. During one of the coffee breaks, he handed out small white pills of such a retro-inverso dipeptide, and we were able to experience its sweetness for ourselves.

The conformational analysis of neotame (RS 7000), an N-substituted aspartame derivative, showed the orientation of an additional hydrophobic group (*tert*-butyl group) in the space between the +y- and +x-axis in Goodman's "L-model," referred to as the D-region. The idea of placing another aromatic group there proved to be very successful. In Ref. [27], compounds **7–10** were synthesized accordingly (Fig. 13.11).

The extreme sweetness of **7** (and also **8**) is explained by the particularly high stability of an L-shaped conformation (according to NMR and crystal structure analysis) due to the interaction of the aromatic groups. Notably, **7** and **8** also display an isovanillin-like substitution pattern (see Ref. [35]).

13.2.4 Bitter or Sweet—Often Just a Matter of (S)- or (R)-Configuration

Since the early days of research on sweet-tasting substances, the abrupt change from "sweet" to "bitter" with only minor modifications to the ligand has been a striking feature. This has led to speculation that either the same receptor or at least two very similar related receptors must be involved. Examples include the diastereomeric pairs in rows 1, 2, and 4 in Fig. 13.10. Goodman's "L-model" also takes this fact into account.

In the development of new isovanillyl sweeteners, a strong dependence of taste on the configuration at asymmetric carbon atoms was also observed, as shown by the enantiomeric pair in Fig. 13.12 [35]. This result is quite to be expected, since, by nature, the

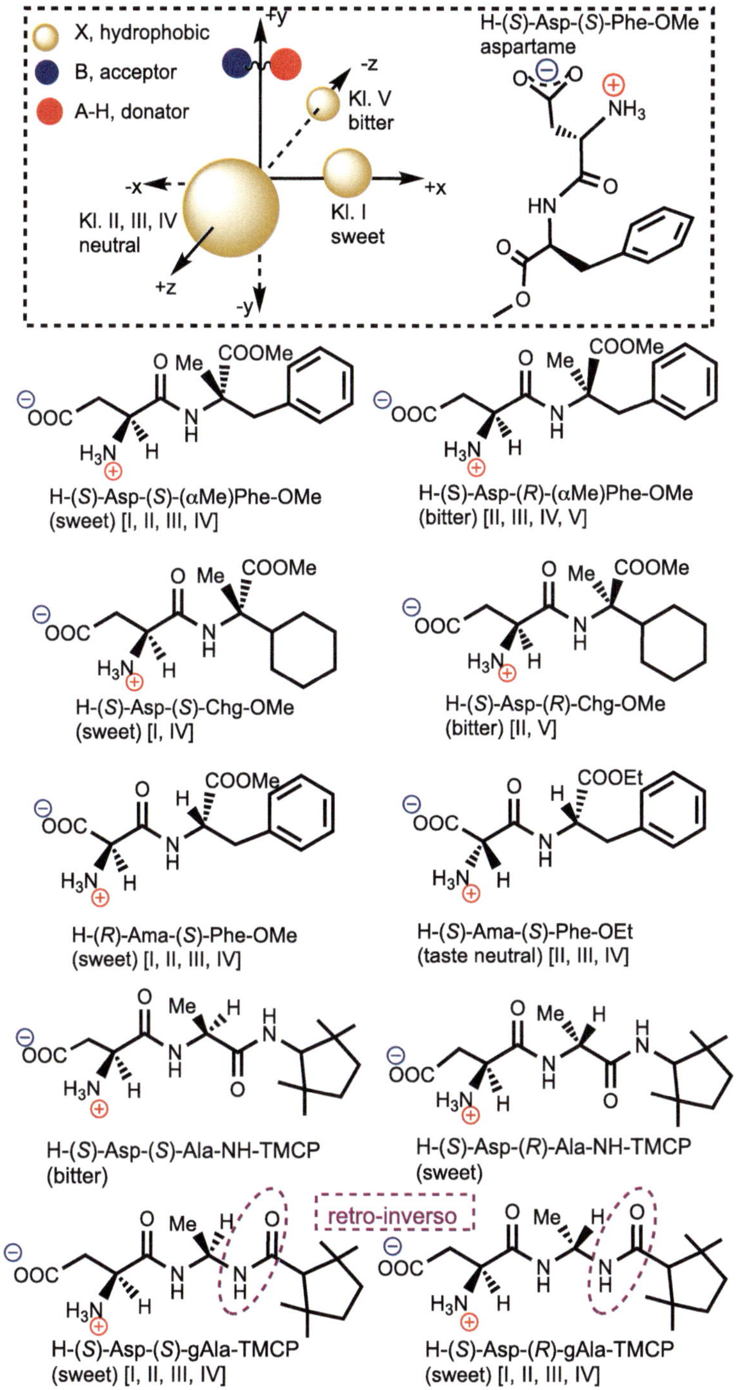

Fig. 13.10 Explanation of sweet or bitter taste properties of dipeptides and analogs according to the L-model of M. Goodman. The classes noted in [……] refer to the preferred conformational space determined by NMR. (According to Ref. [26, 27])

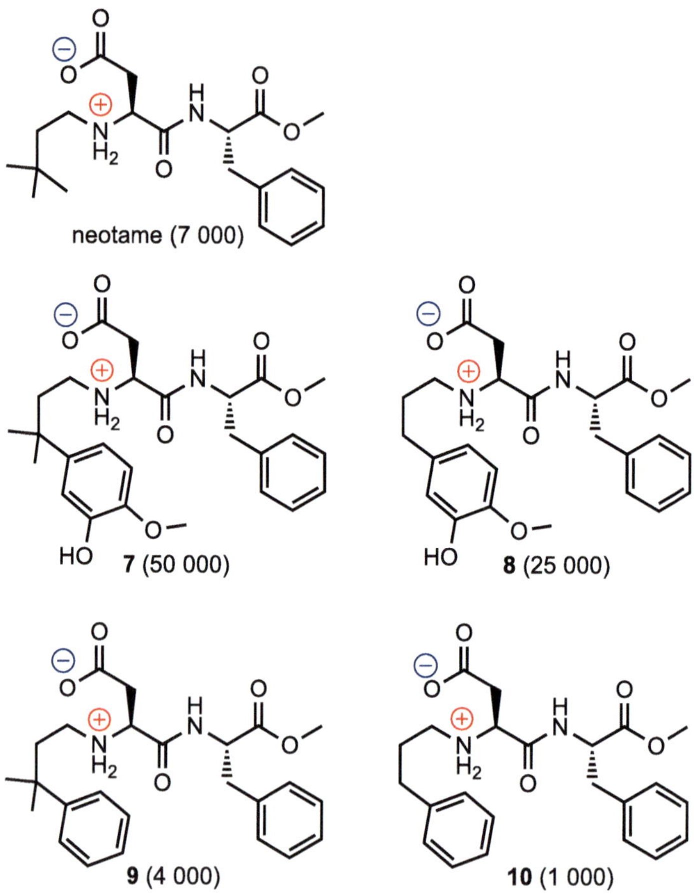

Fig. 13.11 Dipeptide sweeteners with a second aromatic group, derived from neotame (value in parentheses = relative sweetness). (According to Ref. [27])

receptor is *enantiomerically pure*. What is striking, however, is the "flip" in taste quality for the enantiomers of tryptophan or the diastereomeric aspartames (Fig. 13.12). The identification of the responsible receptors for sweet and bitter taste (Sect. 13.2.5) cannot easily explain these phenomena, since they are structurally completely different receptors.

13.2.5 Discovery and Function of the Receptors

With the sequencing of the human genome, the taste receptors for "sweet," "umami," and "bitter" could finally be identified. Two related heterodimeric receptors, T1R2–T1R3 and T1R1–T1R3, are responsible for sweet and umami taste, respectively, while a number of other receptors (T2Rs) were found for bitter taste [41] and [42]. The abrupt switch from "sweet → bitter" is now thought to be explained more by signal transduction [33].

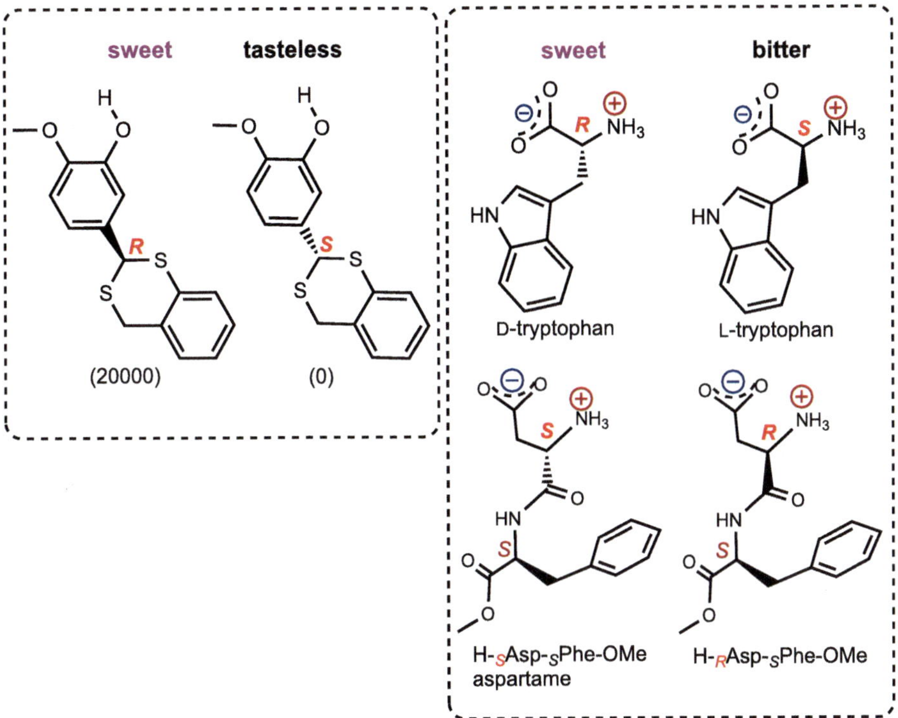

Fig. 13.12 Dependence of taste quality on configuration—enantiomeric and diastereomeric pairs

Through knockout experiments and reconstitution of chimeric receptors from rat and human, binding sites for sweeteners could be identified (see, e.g., Ref. [43, 44]). Many of these results are summarized in the review [32] and in the schematic representation in Fig. 13.13.

As we can see, most sweeteners act on the T1R2 part, while cyclamate and the sweet taste inhibitor lactisol act on the transmembrane domain (TMD) of T1R3, which is identical to the umami receptor T1R1–T1R3 [45]. Accordingly, lactisol also inhibits umami taste, and cyclamate acts as a modulating enhancer there [43].

Phenomenon of Sweet Aftertaste When Drinking Water

An interesting aspect is the sweet aftertaste of fresh water (SWA … *sweet-water aftertaste*). What does this mean? For example, after eating artichokes, tap water tastes distinctly sweet. The same effect has also been observed with various substances that inhibit the T1R2–T1R3 receptor, such as lactisol. In Ref. [32], this is explained as follows: The receptor is basically in a partially active state and thus responds immediately to sweet stimuli. An inhibitor, on the other hand, puts the receptor into a completely inactivated state. If the mouth is then rinsed with tap water, the inhibition is removed and the corresponding partially active state is restored, i.e., a basic sweet taste will recur until one becomes accustomed to it.

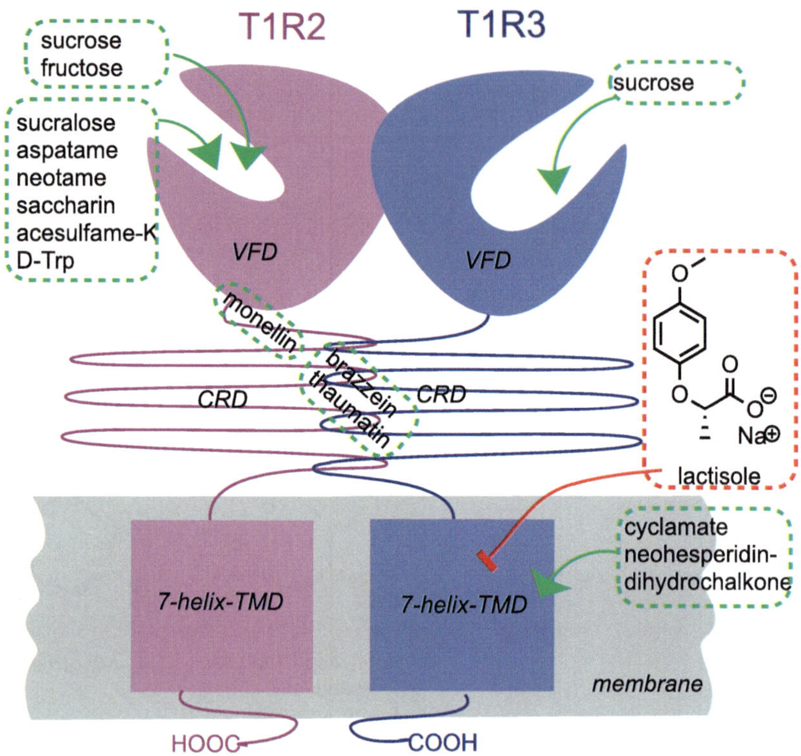

Fig. 13.13 Schematic representation of the human T1R2–T1R3 receptor with the identified binding sites for **sweeteners (in red)** or **taste inhibitors (in green)** (VFD ... *Venus flytrap-like domain*, N-terminal; CRD ... *cysteine-rich domain*; TMD ... *trans-membrane domain*). (According to Ref. [32])

Let us return to the main topic of Chap. 13: receptor-ligand interactions. Structure-based design based on structure-activity relationships (SAR) of agonists or antagonists, as we saw in the example in Sect. 13.1.3, is still not possible for sweeteners, since there is no experimental structure of the human receptor neither alone nor with ligands. So far, only an X-ray structure of the heterodimeric ligand-binding domain from a fish has been obtained [46]. However, today, protein structures can be predicted much better using AI (Sect. 3.2.5). Using such computational methods, attempts have been made to create refined models for the molecular basis of sweetener action [47]. Preliminary work for a more chemical-synthetic approach aims to equip sweeteners with a *photoaffinity label*, a diazirine (Sect. 7.6.1), to identify the binding sites [48].

13.2.6 Sweeteners as a Remedy Against a Looming Increase in Obesity?

The development and use of low-calorie sweeteners as an alternative to sugar has become important for medical reasons as well, given the global increase in overweight (*obesity*) and type II diabetes [49]. However, recent studies have cast some doubt on the success of such a strategy. Sweet taste receptors have been found outside the oral cavity: in the digestive tract, but also in the pancreas [50]. Apparently, a positive signal from such a receptor is also used to regulate metabolic processes. Thus, it is possible that even a calorie-free sweetener could lead to obesity or other metabolic problems [51].

Ultimately, the best option is certainly to simply use less sweetener. If our sense of taste is less strongly stimulated, we become accustomed to the lower level and still achieve the same enjoyment effect.

Summary and Outlook

Whereas in the case of the development of a renin inhibitor it was ultimately possible to design ideal molecules for enzyme inhibition based on structural information, such data are unfortunately still lacking for taste receptors. To date, there appear to be no structures for ligand-activated receptors. Therefore, we are still forced to take the reverse approach, inferring receptor models from the activity of ligands with different structures. But we can be confident that a breakthrough will occur here in the near future. Then our understanding of the mechanisms of taste ligands will certainly improve significantly. However, the mode of action of low-calorie sweeteners in the body is more complex than previously assumed, and so they may not be the ideal solution in the fight against overweight (*obesity*).

13.3 Questions

Let's Test Our Understanding of the Content of This Chapter!

13.1 Structure-activity relationships (SAR) refer, among other things, to what?

13.2 How is blood pressure regulated via the RAS (also RAAS)?

13.3 How and why do the statin-containing peptide SCRIP, the renin inhibitor CPG 38560 developed by Ciba-Geigy, and aliskiren inhibit the cleavage between Leu–Val of angiotensinogen?

13.4 Why, in the development of renin inhibitors, is there an attempt to avoid true peptide bonds and only mimic the side chains of amino acids?

13.5 Since no natural amino acids were used in the synthesis anymore, what challenges arose?

13.6 Approximately when was the human receptor for sweet taste discovered?

13.7 What do the sweeteners saccharin, acesulfame K, and aspartame have in common?
13.8 Monosodium glutamate is the ligand of which receptor?
13.9 How is "relative sweetness" defined, and how is it determined?
13.10 Explain the structure of the sweet taste receptor and indicate the presumed binding sites of some sweet-tasting substances!
13.11 What is lactisol and how does it act on taste receptors?
13.12 By what method, other than crystal structure analysis, could one identify the contact sites between sweetener and receptor?
13.13 Why might low-calorie sweeteners still lead to weight gain?

References

1. Burton, J., Poulsen, K., Haber, E.: Competitive inhibitors of renin. Inhibitors effective at physiological pH. Biochemistry **14**(17), 3892–3898 (1975). https://doi.org/10.1021/bi00688a024; Burton, J., Cody, R.J., Herd, J.A., Haber, E.: Specific inhibition of renin by an angiotensinogen analog: studies in sodium depletion and renin-dependent hypertension. Proc. Natl. Acad. Sci. U.S.A. **77**(9), 5476–5479 (1980). https://doi.org/10.1073/pnas.77.9.5476
2. Gante, J.: Peptidomimetics – tailored enzyme inhibitors. Angew. Chem. Int. Ed. **33**(17), 1699–1720 (1994). https://doi.org/10.1002/anie.199416991; Gante, J.: Peptidmimetica – maßgeschneiderte Enzyminhibitoren. Angew. Chem. **106**(17), 1780–1802 1994). https://doi.org/10.1002/ange.19941061705
3. Jensen, C., Herold, P., Brunner, H.R.: Aliskiren: the first renin inhibitor for clinical treatment. Nat. Rev. Drug Discov. **7**(5), 399–410 (2008). https://doi.org/10.1038/nrd2550
4. Rahuel, J., Rasetti, V., Maibaum, J., …, Grütter, M.G.: Structure-based drug design: the discovery of novel nonpeptide orally active inhibitors of human renin. Chem. Biol. **7**(7), 493–504 (2000). https://doi.org/10.1016/S1074-5521(00)00134-4
5. Wood, J.M., Maibaum, J., Rahuel, J., …, Bedigian, M.P.: Structure-based design of aliskiren, a novel orally effective renin inhibitor. Biochem. Biophys. Res. Commun. **308**(4), 698–705 (2003). https://doi.org/10.1016/S0006-291X(03)01451-7
6. Lizakowski, S., Tylicki, L., Rutkowski, B.: Direct renin inhibition – a promising strategy for renal protection? Med. Sci. Monit. **19**, 451–457 (2013). https://doi.org/10.12659/MSM.883949
7. Friedrich, S., Schmieder, R.E.: Review of direct renin inhibition by aliskiren. J. Renin Angiotensin Aldosterone Syst. **14**(3), 193–196 (2013). https://doi.org/10.1177/1470320313497328; Riccioni, G.: The role of direct renin inhibitors in the treatment of the hypertensive diabetic patient. Ther. Adv. Endocrinol. Metab. **4**(5), 139–145 (2013). https://doi.org/10.1177/2042018813490779
8. Pantzaris, N.-D., Karanikolas, E., Tsiotsios, K., Velissaris, D.: Renin inhibition with aliskiren: a decade of clinical experience. J. Clin. Med. **6**(6) 61 (2017). https://doi.org/10.3390/jcm6060061
9. Azharruddin, M., Akthar, S.M.: A detailed updated review on antihypertensive drugs and their classification. Eur. J. Biomed. Pharmaceut. Sci. **7**(10), 489–492 (2020)
10. Akbariromani, H., Haseeb, R., Nazly, S., et al.: Efficacy of direct renin inhibitors in slowing down the progression of diabetic kidney disease: a meta-analysis. Cureus. **14**(8), e28608 (2022). https://doi.org/10.7759/cureus.28608

11. Ksiazek, S.H., Hu, L., Andò, S., …, Provenzano, M.: Renin–angiotensin–aldosterone system: from history to practice of a secular topic. Int. J. Mol. Sci. **25**(7), 4035 (2024). https://doi.org/10.3390/ijms25074035
12. Imaeda, Y., Tokuhara, H., Fukase, Y., …, Kuroita, T.: Discovery of TAK-272: a novel, potent, and orally active renin inhibitor. ACS Med. Chem. Lett. **7**(10), 933–938 (2016). https://doi.org/10.1021/acsmedchemlett.6b00251; Ito, S., Kagawa, T., Saiki, T., …, Umeda, Y.: Efficacy and safety of imarikiren in patients with type 2 diabetes and microalbuminuria: a randomized, controlled trial. Clin. J. Amer. Soc. Nephrol. **14**(3), 354–363 (2019). https://doi.org/10.2215/cjn.07720618
13. Göschke, R., Stutz, S., Heinzelmann, W., Maibaum, J.: The nonchiral bislactim diethoxy ether as highly stereo-inducing synthon for sterically hindered g-branched a-amino acids: a practical, large scale route to an intermediate of the novel renin inhibitor aliskiren. Helv. Chim. Acta. **86**(8), 2848–2870 (2003). https://doi.org/10.1002/hlca.200390235
14. Evans, D.A., Ennis, M.D., Mathre, D.J.: Asymmetric alkylation reactions of chiral imide enolates. A practical approach to the enantioselective synthesis of.alpha.-substituted carboxylic acid derivatives. J. Am. Chem. Soc. **104**(6), 1737–1739 (1982). https://doi.org/10.1021/ja00370a050; Evans, D.A., Britton, T.C., Dorow, R.L., Dellaria, J.F.: Stereoselective amination of chiral enolates. A new approach to the asymmetric synthesis of.alpha.-hydrazino and. alpha.-amino acid derivatives. J. Am. Chem. Soc. **108**(20), 6395–6397 (1986). https://doi.org/10.1021/ja00280a050
15. Evans, D.A., Britton, T.C., Ellman, J.A.: Contrasteric carboximide hydrolysis with lithium hydroperoxide. Tetrahedron Lett. **28**(49), 6141–6144 (1987). https://doi.org/10.1016/S0040-4039(00)61830-0
16. Kanth, J.V.B., Periasamy, M.: Selective reduction of carboxylic acids into alcohols using sodium borohydride and iodine. J. Org. Chem. **56**(20), 5964–5965 (1991). https://doi.org/10.1021/jo00020a052; McKennon, M.J., Meyers, A.I., Drauz, K., Schwarm, M.: A convenient reduction of amino acids and their derivatives. J. Org. Chem. **58**(13), 3568–3571 (1993). https://doi.org/10.1021/jo00065a020
17. Schöllkopf, U., Groth, U., Deng, C.: Enantioselective syntheses of (R)-amino acids using L-valine as chiral agent. Angew. Chem. Int. Ed. **20**(9), 798–799 (1981), https://doi.org/10.1002/anie.198107981; Schöllkopf, U., Groth, U., Deng, C.: Enantioselektive Synthese von (R)-Aminosäuren unter Verwendung von L-Valin als chiralem Hilfsstoff. Angew. Chem. **93**(9), 793–795 (1981). https://doi.org/10.1002/ange.19810930915; Schöllkopf, U.: Asymmetric syntheses of amino acids via metalated bis-lactim ethers of 2,5-diketopiperazines. Pure Appl. Chem. **55**(11), 1799-1806 (1983). https://doi.org/10.1351/pac198355111799
18. Maibaum, J., Stutz, S., Göschke, R., …, Wood, J.M.: Structural modification of the P2′ position of 2,7-dialkyl-substituted 5(S)-amino-4(S)-hydroxy-8-phenyl-octanecarboxamides: the discovery of aliskiren, a potent nonpeptide human renin inhibitor active after once daily dosing in marmosets. J. Med. Chem. **50**(20), 4832–4844 (2007). https://doi.org/10.1021/jm070316i
19. Göschke, R., Stutz, S., Rasetti, V., …, Maibaum, J.: Novel 2,7-dialkyl-substituted 5(S)-amino-4(S)-hydroxy-8-phenyl-octanecarboxamide transition state peptidomimetics are potent and orally active inhibitors of human renin. J. Med. Chem. **50**(20), 4818–4831 (2007). https://doi.org/10.1021/jm070314y
20. Yamada, S.-i., Kasai, Y., Shioiri, T.: Diethylphosphoryl cyanide. A new reagent for the synthesis of amides. Tetrahedron Lett. **14**(18), 1595–1598 (1973). https://doi.org/10.1016/S0040-4039(01)96005-8
21. Boogers, J.A.F., Felfer, U., Kotthaus, M., …, de Vries, J.G.: A mixed-ligand approach enables the asymmetric hydrogenation of an alpha-isopropylcinnamic acid en route to the renin

inhibitor aliskiren. Org. Process Res. Dev. **11**(3), 585–591 (2007). https://doi.org/10.1021/op0602369

22. Beauchamp, G.K.: Why do we like sweet taste: a bitter tale? Physiol. Behav. **164**, 432–437 (2016). https://doi.org/10.1016/j.physbeh.2016.05.007
23. Shallenberger, R.S., Acree, T.E.: Molecular theory of sweet taste. Nature **216**(5114), 480–482 (1967). https://doi.org/10.1038/216480a0; Shallenberger, R.S.: The AH,B glycophore and general taste chemistry. Food Chem. **56**(3), 209–214 (1996). https://doi.org/10.1016/0308-8146(96)00016-7
24. Kier, L.B.: A molecular theory of sweet taste. J. Pharm. Sci. **61**(9), 1394–1397 (1972). https://doi.org/10.1002/jps.2600610910
25. Nofre, C., Tinti, J.-M.: Sweetness reception in man: the multipoint attachment theory. Food Chem. **56**(3), 263–274 (1996). https://doi.org/10.1016/0308-8146(96)00023-4
26. Yamazaki, T., Benedetti, E., Kent, D., Goodman, M.: Conformational requirements for sweet-tasting peptides and peptidomimetics. Angew. Chem. Int. Ed. **33**(14), 1437–1451 (1994). https://doi.org/10.1002/anie.199414371; Yamazaki, T., Benedetti, E., Kent, D., Goodman, M.: Konformationsvoraussetzungen für den süßen Geschmack von Dipeptiden und Dipeptidmimetica. Angew. Chem. **106**(14), 1502–1517 (1994). https://doi.org/10.1002/ange.19941061405
27. Goodman, M., Del Valle, J.R., Del Valle, J.R., Benedetti, E.: Molecular basis of sweet taste dipeptide taste ligands. Pure Appl. Chem. **74**(7), 1109–1116 (2002). https://doi.org/10.1351/pac200274071109
28. Temussi, P.A.: Sweet, bitter and umami receptors: a complex relationship. Trends Biochem. Sci. **34**(6), 296–302 (2009). https://doi.org/10.1016/j.tibs.2009.02.005
29. Margolskee, R.F.: Molecular mechanisms of taste transduction. Pure Appl. Chem. **74**(7), 1125-1133 (2002). https://doi.org/10.1351/pac200274071125
30. Damak, S., Rong, M., Yasumatsu, K., …, Margolskee, R.F.: Detection of sweet and umami taste in the absence of taste receptor T1r3. Science. **301**(5634), 850–853 (2003). https://doi.org/10.1126/science.1087155
31. Delay, E.R., Hernandez, N.P., Bromley, K., Margolskee, R.F.: Sucrose and monosodium glutamate taste thresholds and discrimination ability of T1R3 knockout mice. Chem. Senses. **31**(4), 351–357 (2006). https://doi.org/10.1093/chemse/bjj039
32. DuBois, G.E.: Molecular mechanism of sweetness sensation. Physiol. Behav. **164**, 453–463 (2016). https://doi.org/10.1016/j.physbeh.2016.03.015
33. Amrein, H., Bray, S.: Bitter-sweet solution in taste transduction. Cell. **112**(3), 283–284 (2003). https://doi.org/10.1016/S0092-8674(03)00077-1; Talavera, K., Yasumatsu, K., Yoshida, R.,… Nilius, B.: The taste transduction channel TRPM5 is a locus for bitter-sweet taste interactions. FASEB J. **22**(5), 1343–1355 (2008). https://doi.org/10.1096/fj.07-9591com
34. Aguayo-Guerrero, J.A., Méndez-García, L.A., Solleiro-Villavicencio, H., …, Escobedo, G.: Sucralose: from sweet success to metabolic controversies – unraveling the global health implications of a pervasive non-caloric artificial sweetener. Life. **14**(3), 323 (2024). https://doi.org/10.3390/life14030323
35. Bassoli, A., Merlini, L., Morini, G.: Isovanillyl sweeteners. From molecules to receptors. Pure Appl. Chem. **74**(7), 1181–1187 (2002). https://doi.org/10.1351/pac200274071181
36. Kitagawa, I.: Licorice root. A natural sweetener and an important ingredient in Chinese medicine. Pure Appl. Chem. **74**(7), 1189–1198 (2002). https://doi.org/10.1351/pac200274071189
37. Temussi, P.A.: Why are sweet proteins sweet? Interaction of brazzein, monellin and thaumatin with the T1R2-T1R3 receptor. FEBS Lett. **526**(1–3), 1–4 (2002). https://doi.org/10.1016/S0014-5793(02)03155-1; Picone, D., Temussi, P.A.: Dissimilar sweet proteins from plants: oddities or normal components? Plant Sci. **195**(Oct.), 135–142 (2012). https://doi.org/10.1016/j.plantsci.2012.07.001

38. Kim, T.-Y., Woo, E.-j., Yoon, T.-S.: Binding mode of brazzein to the taste receptor based on crystal structure and docking simulation. Biochem. Biophys. Res. Commun. **592**, 119–124 (2022). https://doi.org/10.1016/j.bbrc.2022.01.004
39. Mattern, R.-H., Amino, Y., Benedetti, E., Goodman, M.: Conformational analysis of potent sweet taste ligands by NMR and computer simulations. Biopolymers. **49**(6), 525–539 (1999). https://doi.org/10.1002/(SICI)1097-0282(199905)49:6<525::AID-BIP10>3.0.CO;2-3
40. Goodman, M., Coddington, J., Mierke, D.F., Fuller, W.D.: A model for the sweet taste of stereoisomeric retro-inverso and dipeptide amides. J. Am. Chem. Soc. **109**(15), 4712–4714 (1987). https://doi.org/10.1021/ja00249a040
41. Nelson, G., Hoon, M.A., Chandrashekar, J., …, Zuker, C.S.: Mammalian sweet taste receptors. Cell. **106**(3), 381–390 (2001). https://doi.org/10.1016/S0092-8674(01)00451-2
42. Li, X., Staszewski, L., Xu, H., …, Adler, E.: Human receptors for sweet and umami taste. Proc. Natl. Acad. Sci. **99**(7), 4692–4696 (2002). https://doi.org/10.1073/pnas.072090199
43. Xu, H., Staszewski, L., Tang, H., …, Li, X.: From the cover: different functional roles of T1R subunits in the heteromeric taste receptors. Proc. Natl. Acad. Sci. USA. **101**(39), 14258–14263 (2004). https://doi.org/10.1073/pnas.040438410
44. Winnig, M., Bufe, B., Kratochwil, N.A., Slack, J.P., Meyerhof, W.: The binding site for neohesperidin dihydrochalcone at the human sweet taste receptor. BMC Struct. Biol. **7**(1), 66 (2007). https://doi.org/10.1186/1472-6807-7-66
45. Nakagita, T., Akiko, I., Matsuya, T., …, Misaka, T.: Structural insights into the differences among lactisole derivatives in inhibitory mechanisms against the human sweet taste receptor. PLoS One. **14**(3) e0213552 (2019). https://doi.org/10.1371/journal.pone.0213552
46. Nuemket, N., Yasui, N., Kusakabe, Y., …, Yamashita, A.: Structural basis for perception of diverse chemical substances by T1r taste receptors. Nat. Commun. **8**(1), 15530 (2017). https://doi.org/10.1038/ncomms15530
47. Zheng, H., Xu, X., Fang, Y., Sun, R., Liu, B.: The molecular theory of sweet taste: revisit, update, and beyond. J. Med. Chem. **67**(5), 3232–3243 (2024). https://doi.org/10.1021/acs.jmedchem.3c02055
48. Hashimoto, M., Nakagita, T., Misaka, T.: Recent progress in the use of diaziridine-based sweetener derivatives to elucidate the chemoreception mechanism of the sweet taste receptor. RSC Adv. **11**(51), 32236–32247 (2021). https://doi.org/10.1039/D1RA04831G
49. Sylvetsky, A.C., Rother, K.I.: Trends in the consumption of low-calorie sweeteners. Physiol. Behav. **164**, 446–450 (2016). https://doi.org/10.1016/j.physbeh.2016.03.030
50. Laffitte, A., Neiers, F., Briand, L.: Functional roles of the sweet taste receptor in oral and extraoral tissues. Curr. Opin. Clin. Nutr. Metab. Care. **17**(4), 379–385 (2014). https://doi.org/10.1097/MCO.0000000000000058; Kojima, I., Nakagawa, Y., Ohtsu, Y., …, Nagasawa, M.: Return of the glucoreceptor: glucose activates the glucose-sensing receptor T1R3 and facilitates metabolism in pancreatic β-cells. J. Diabetes Investig. **6**(3), 256–263 (2015). https://doi.org/10.1111/jdi.12304
51. Burke, M.V., Small, D.M.: Physiological mechanisms by which non-nutritive sweeteners may impact body weight and metabolism. Physiol. Behav. **152**(Part_B), 381–388 (2015). https://doi.org/10.1016/j.physbeh.2015.05.036; Pepino, M.Y.: Metabolic effects of non-nutritive sweeteners. Physiol. Behav. **152**(Part_B), 450–455 (2015). https://doi.org/10.1016/j.physbeh.2015.06.024; Fowler, S.P.G.: Low-calorie sweetener use and energy balance: results from experimental studies in animals, and large-scale prospective studies in humans. Physiol. Behav. **164**(B), 517–523 (2016). https://doi.org/10.1016/j.physbeh.2016.04.047

Fluorescent Compounds 14

Abstract

Chapters 14 and 15 are a kind of Appendix. In chapter 14, we aim to address and briefly discuss a phenomenon that has been mentioned and utilized in many previous chapters, in order to facilitate understanding. Fluorescent compounds have become indispensable in analytics, biology, and medicine, and therefore deserve their own dedicated treatment in this book. In Chap. 15 the solutions and explanations for the exercises posed in the individual chapters are listed. These not only serve to check one's own answers, but can also contribute to a deeper understanding of the chapter texts.

14.1 Fluorophores and Fluorescence

▶ In the preceding chapters, we have repeatedly encountered the use of fluorescent groups. They are increasingly being used to visualize biomolecules *in vitro* as well as *in vivo* (fluorescence microscopy, gel electrophoresis, …), to quantify biomolecules (ELISA, …), to monitor kinetics, or to document conformational changes via FRET or fluorophore-quencher relationships. Now we want to take a look at the most important fluorophores and discuss some of their properties. What is the role of organic chemistry in this context?—On the one hand, efforts are made to improve the properties of fluorophores through chemical-synthetic modifications, but chemical reactions are also used *in vivo* to, for example, visualize and quantify the presence of reactive molecules or ions in the cell.

D. Krois, *Organic Chemical Methods*, https://doi.org/10.1007/978-3-662-73573-2_14

14.1.1 Fluorescence—Some Important Terms

Why is fluorescence so well suited as a detection or imaging method? The emission of electromagnetic radiation has a much better signal-to-background noise ratio (*signal/noise ratio*) than absorption spectroscopic methods. In principle, every single emitted photon can be "seen." Nevertheless, absorption in the UV-Vis range is also important in this context—because only then can fluorescence occur. Therefore, we should consider both effects together.

Relationship UV-Vis—Fluorescence—Phosphorescence - Photochemistry
To recall some spectroscopic terms, let us take a closer look at the excitation of electrons by radiation in the UV and visible light range (Fig. 14.1). When electromagnetic radiation with wavelengths λ between 180 nm (ν approx. 1700 THz) and 1000 nm (ν approx. 300 THz) is absorbed by molecules, electrons from the outer molecular orbitals (MO, usually π-bonds) or non-bonding electron pairs (n) are excited to higher antibonding orbitals (π*). In this way, an electronically excited overall state ($\boldsymbol{S_1}$, $\boldsymbol{S_2}$, …) is reached. During this excitation, the spin state of the electron does not change, so it remains in the singlet state (only in radicals are not all electron spins paired—so no singlet state). Subsequently, a so-called *intersystem crossing* (ISC) to an excited triplet state ($\boldsymbol{T_1}$) can occur, which is actually a forbidden process.

Through internal conversion (i.e., by releasing thermal radiation) or fluorescence, the molecule generally returns to the ground state $\boldsymbol{S_0}$ within nanoseconds (ns). Only the triplet state has a longer lifetime, and phosphorescence occurs with a delay after excitation ("afterglow").

Excitation and emission always occur from the vibrational ground state (vibrational relaxation takes place even faster), but can end in different excited vibrational states. Depending on the geometric change ($\boldsymbol{S_0}$, $\boldsymbol{S_1}$), the 0 → 0 transitions are not necessarily favored (Fig. 14.1b)—this leads to a Stokes shift.

Here, we will focus on organic fluorophores—transition metals can also sometimes have very good fluorescence properties (Sect. 7.9.3). Whether a molecule is suitable as a fluorophore depends on a number of factors. First, a so-called chromophore[1] must be present, i.e., a conjugated part of the molecule whose S_0 and S_n states have a small energy difference, so that excitation by UV radiation (above 300 nm) or visible light is possible—these are usually $\pi \rightarrow \pi^*$ transitions. However, there are many absorbing organic compounds that do not exhibit any fluorescence at all. Non-radiative transitions via internal conversion predominate. The proportion of fluorescence, phosphorescence, or photochemical processes is characterized by quantum yields (*quantum yield* … QY or Φ), which is the number of photons leading to fluorescence, phosphorescence, or photochemical transformations, relative to the number of absorbed photons, i.e., a number between 0.0 and 1.0. In the following, we are interested in the fluorescence quantum yield (QY or photoluminescence quantum yield … PLQY). For fluorescence labeling,

[1] From Ancient Greek χρωμα ("chroma")… color, and φορεω ("phoreo") = φερω … to carry; chromophore = color-bearing.

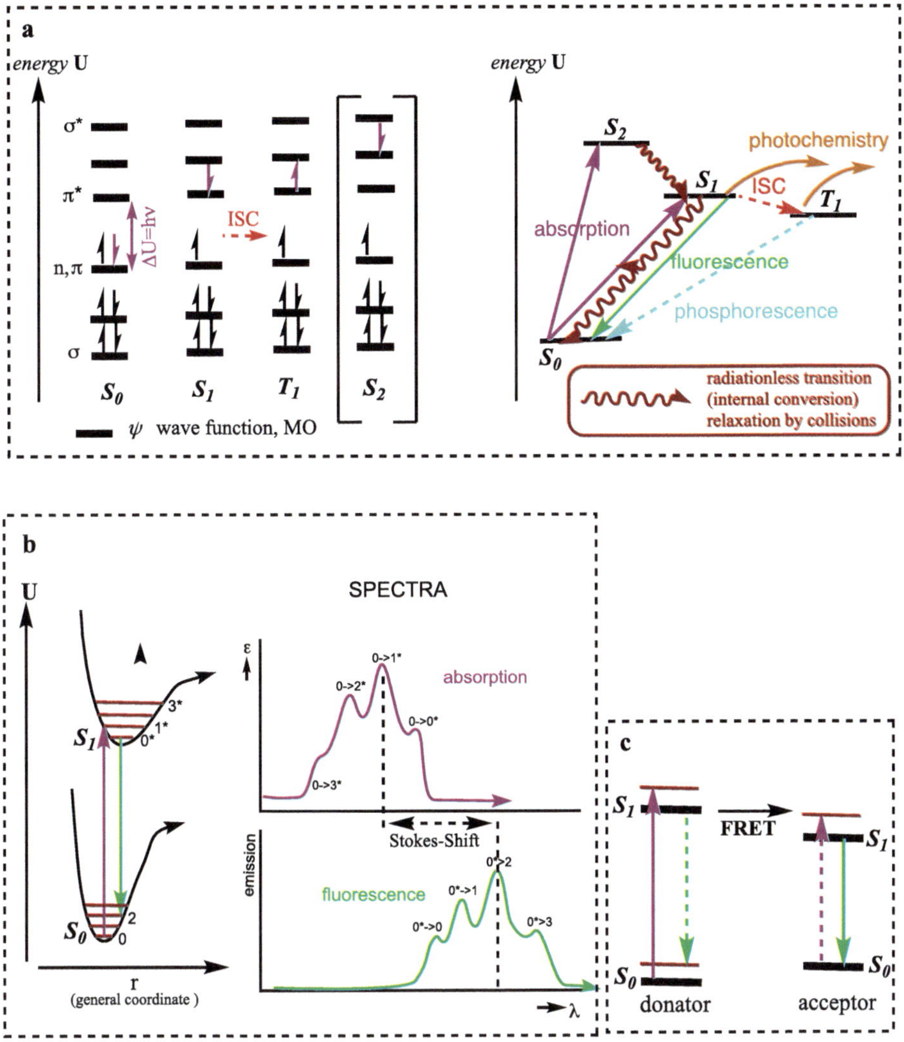

Fig. 14.1 **a** Electron excitation by UV-Vis radiation and the possible processes originating from the electronically excited states (simplified schematic representation); **b** Explanation of the Stokes shift for an excitation $S_0 \rightarrow S_1$ due to vibrational levels (——); **c** Explanation of Förster resonance energy transfer (FRET)

this should be particularly high. However, the "absorption yield," described by the molar extinction coefficient ε ($M^{-1}cm^{-1}$), is also important. Therefore, the term brilliance of fluorescence is defined as the product of QY × ε.

For the use of fluorophores, the so-called "Stokes shift" is also a measure of fluorescence quality; this is the difference between the excitation wavelength (*excitation wavelength*) (i.e., UV-Vis absorption) λ_{exc}, and the emission wavelength λ_{em}. Depending on the structural differences between the excited state and the ground state (S_0) (Fig. 14.1), this difference can range from a few nanometers up to 200 nm. For measurement purposes, a large Stokes shift is advantageous.

Can optimal fluorophores be designed with respect to these criteria?—Only to a certain extent. The extinction coefficient ε is relatively easy to estimate; it depends on the extent of the conjugated system and on the "allowedness" of the transition according to quantum theory. The fluorescence quantum yield and the Stokes shift, on the other hand, are difficult to predict in general. However, for a given chromophore system, modifications for improvements can be planned based on existing experimental data.

What is FRET (Förster resonance energy transfer)? The process is schematically illustrated in Fig. 14.1c. It is the non-radiative energy transfer from a donor to an acceptor, which then emits light. The prerequisites are a small spatial distance (maximum 10 nm) and the overlap of the emission range of the donor and the excitation range of the acceptor—see the dashed arrows of equal length in the energy diagram in Fig. 14.1c. (For examples, see Sect. 12.1.2 and 11.2.2)

14.1.2 Fluorophore Types: Photophysical Properties, Synthesis Notes, and Applications

A brief selection of reviews on various types of fluorophores, both widely used and still under development, is provided in [1–8]. Some examples are illustrated in Fig. 14.2, and their photophysical characteristics are listed in Table 14.1.

In some cases, it is not easy to find all the characteristic data in the literature, so appropriate examples have been selected and the references provided.

Virtually all of these organic molecules have been synthesized. In most cases, higher molecular weight (larger) molecules have also been synthesized to increase brilliance. On the other hand, there are efforts to find the smallest possible fluorophores, which cause minimal changes when introduced into biomolecules. The different types sometimes show very different behavior in response to changes in solvent or the polarity of their immediate environment. Some can be used directly as probes for lipophilic regions, as they only fluoresce in apolar environments—others are completely insensitive to such changes. Since the non-radiative relaxation mechanism (internal conversion) is often related to the mobility of molecular segments, some fluorophores show a strong dependence of quantum yield (QY) on viscosity.

Here are further reviews with information on the synthesis, properties, and application of fluorophores: [19–26].

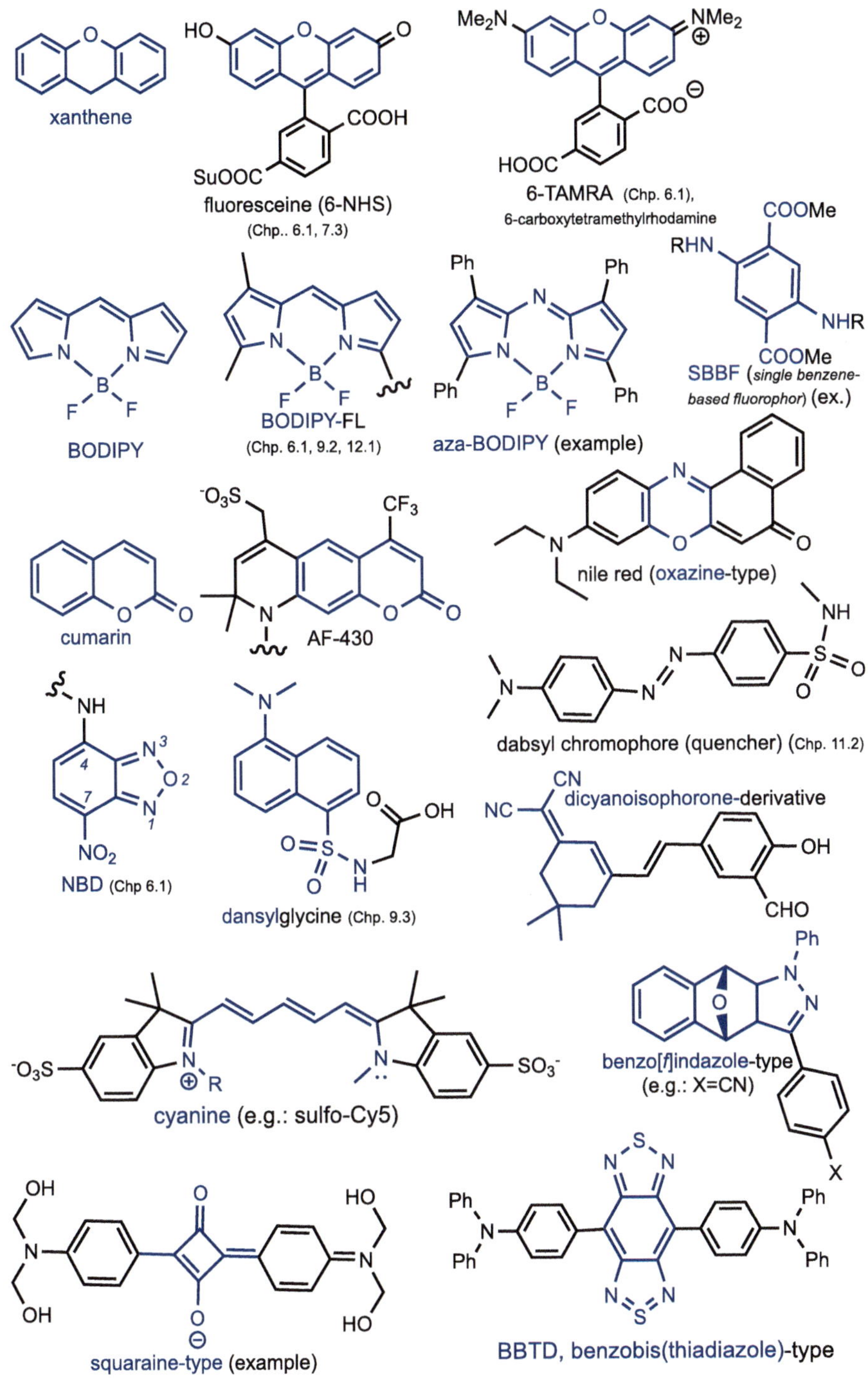

Fig. 14.2 Representatives of selected types of important fluorophores (and quenchers)

Table 14.1 Photophysical data of representatives of several fluorophore types (for structures, see Fig. 14.2)

Fluorophore	Solvent	$\lambda_{abs=exc}$/nm	ε/$M^{-1}cm^{-1}$	λ_{em}/nm	QY	Ref.
Fluorescein-6-NHS (xanthen type)	H_2O	492	74,000	517	0.93	[9]
6-TAMRA (rhodamine type)	EtOH?	541	84,000	567	0.1	[9]
BODIPY-FL	EtOH	503	92,000	509	0.97	[9]
Aza-BODIPY (ex.)	$CHCl_3$	640	79,000	688	0.34	[7]
Coumarin	EtOH	275(311)	11,000	--	--	[10]
AF-430 (coumarin type)	H_2O	430	16,000	542	0.23	[9]
Nile red (oxazine type)	Heptane	500	40,000	538/568	0.48	[11]
NBD (4-amino-7-nitrobenzoxadiazole derivative)	Acetone	461	19,000	529	0.58	[12]
Dansylglycine (dansyl type)	Dioxane	339	4,300	495	0.66	[10]
Dabsylamide	EtOH	435	30,000	--	--	[13]
Sulfo-Cy5-NHS (cyanine type)	H_2O	646	270,000	662	0.28	[9]
Squaraine type (ex.)	EtOH	639	140,000	663	0.15	[14]
Benzo[*f*]indazole (ex.)	MeOH	400	25,000	505	0.84	[15]
BBTD type (example)	Toluene	709	14,000	975	0.07	[16]
Dicyanoisophorone derivative (ex.)	H_2O	505	45,000	664	0.38	[17]
SBBF (ex. R= CH_2–CH_2–NH–Boc)	CH_2Cl_2	434	5,600	535	0.26	[18]

Especially in recent times, many methods have been developed to use fluorescence changes such as "*off-on*" effects or so-called ratiometric methods[2] to detect and quantify specific compounds or ions *in vivo*. Just a few examples:

- Detection of sulfide ions or biothiols (such as Cys, homocysteine Hcy, glutathione GSH), usually exploiting the high nucleophilicity of thiols or SH^- in S_N2, S_NAr, or nucleophilic addition reactions [17, 18, 27–31]. In Fig. 14.3 we show three examples of –SH fluorescence probes: Fig. 14.3a The 3,5-dinitrophenyl ether acts as a quencher of isophorone fluorescence and is specifically cleaved by H_2S. Fig. 14.3b Here, the BODIPY fluorophore emits virtually no fluorescence due to the quaternary (positively charged) pyridine N atom. H_2S reduces the azide group to an amino group, and the *p*-aminobenzyl residue then spontaneously cleaves off (*self-immolative strategy*)—the

[2] Ratiometric method: If the emission spectrum changes due to reaction or association with the target molecule or ion, the ratio of emissions at two specifically selected wavelengths can be used to determine the concentration very precisely.

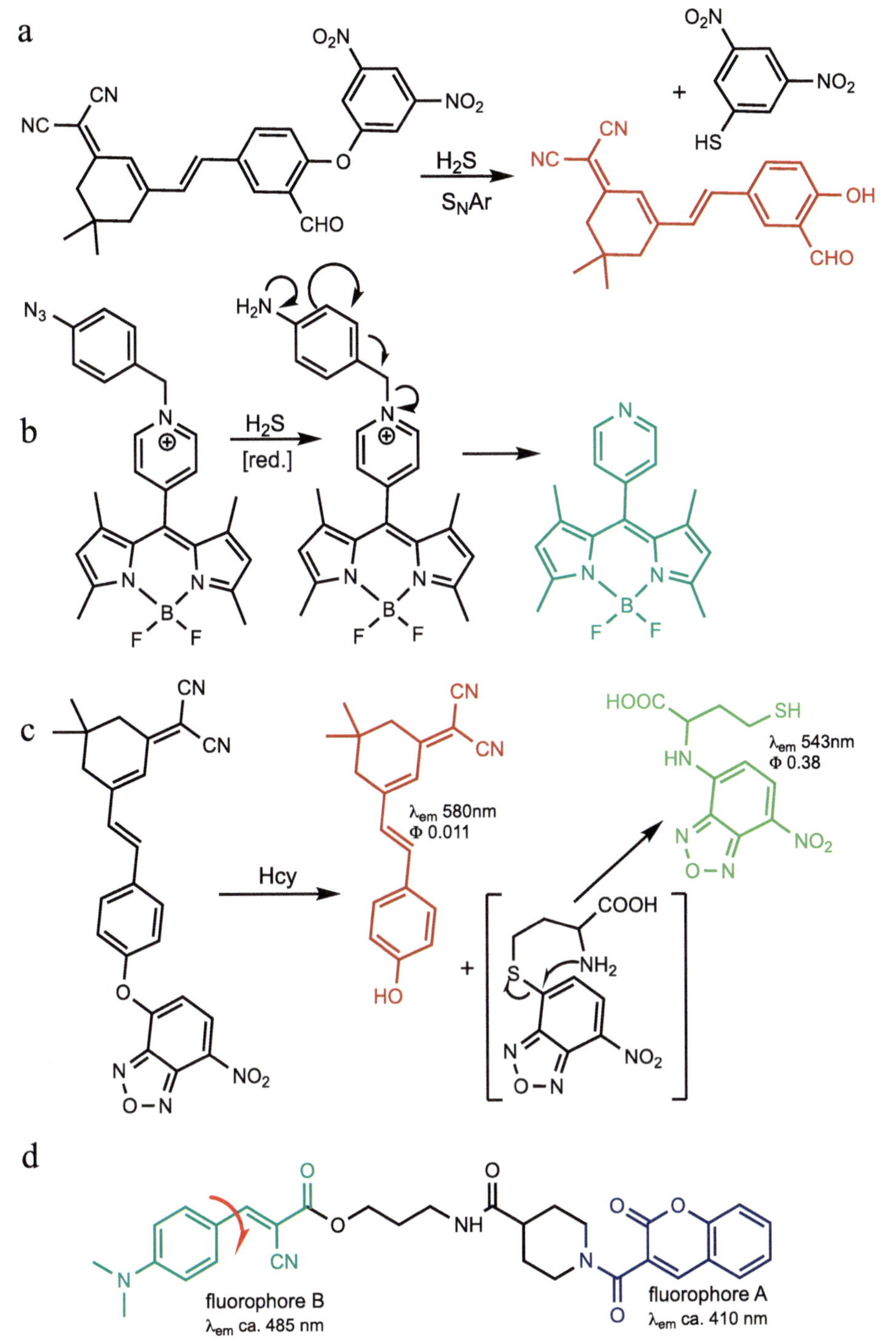

Fig. 14.3 Some examples of specific fluorescence probes: **a** specific detection of H_2S. (According to Ref. [17]) **b** detection of H_2S by reduction of azide and fragmentation (according to Ref. [27]). **c** specific detection of homocysteine Hcy (according to Ref. [31]). **d** ratiometric fluorescent molecular rotor (FMR) (according to Ref. [32])

strongly fluorescent pyridine-BODIPY fluorophore is formed. Fig. 14.3c We encounter two fluorophores, an isophorone type and an NBD type, which do not emit when linked (*quenching*). In the presence of homocysteine Hcy (but also other thiols), the NBD chromophore is cleaved off by S_NAr. However, these NBD derivatives also do not fluoresce. Only the long-wavelength weak fluorescence of the isophorone fluorophore becomes visible. In the case of homocysteine, rearrangement to the *N*-substituted NBD is strongly favored, and this product now fluoresces strongly at 543 nm.

- Detection of carbon monoxide (in cells) via reduction of aromatic nitro groups [32].
- Detection of hypochlorite (HOCl): HOCl very specifically cleaves a thiocarbamate group (Me_2N–CS–O–); this leads to a fragmentation reaction of a linker (*self-immolative spacer*) and release of the strongly fluorescent phenoxazine dye [33]. In Ref. [2], by contrast, an example is given in which HOCl, specifically in lysosomes (a morpholine residue acts as an organelle-specific tag), induces strong fluorescence via cyclization.
- Detection of enzyme activity *in vivo* using specific ratiometric fluorescence probes [34].
- Fluorescence probes for determining viscosity: [24] and [32]. An example in Ref. [35] uses a ratiometric method for viscosity measurement, see Fig. 14.3d: While for the coumarin fluorophore A, rotations play no role in fluorescence, rotation about the marked single bond in the "*push-pull*"[3] fluorophore B greatly reduces the quantum yield. As viscosity increases, rotation is hindered, and fluorescence at 465 nm increases significantly. The ratio of emissions at 465/410 nm is a measure of the local viscosity in the environment of the FMR (*fluorescent molecular rotor*).
- Detection of metal ions, cyanide ions, etc.; see Ref. [2]
- Photoactivatable fluorophores, often also referred to as *photocaged fluorophores*, are those that have a photolytically cleavable "protecting group" (which prevents fluorescence) or a cleavable quencher. A problem for *in vivo* application is that the photochemical reaction usually requires UV or short-wavelength visible light, which is often too damaging for living cells (in Sect. 12.2.3 a protecting group is cleaved *in vivo* with blue visible light, but the fluorescence is constantly "ON", GFP). Further developments will be needed for improved applications [36, 37] and [38].
- Fluorescence labeling of specific cell organelles [39].
- For some examples of conjugation of fluorophores to biomolecules, see Ref. [40–44].

14.1.3 Natural Fluorescent Proteins

It was in 1962 that the first intrinsically fluorescent protein, the *green fluorescent protein* (GFP), was discovered and isolated from a jellyfish (*Aequorea victoria*) [45], but it was only after 1990 that its development for use as an intrinsic fluorescence marker began

[3] "*Push-pull*" means: electron donation at one end of the conjugated system and electron withdrawal at the other end.

[46]. Subsequently, additional fluorescent proteins with different emission maxima were isolated from other marine organisms. What they all have in common is that the emitting chromophore is formed posttranslationally through reactions at the peptide backbone, possibly involving amino acid side chains. This reaction is also of great interest from a chemist's perspective, and we will examine its details below. The advantages of these proteins, which are relatively short—about 250 amino acids—as fluorophores lie in the fact that their DNA can be fused with the DNA of a protein one wishes to visualize, so that translation yields a chimeric fluorescent protein [47]. Thus, this is a "labeling" method that can be performed by molecular biologists without synthetic chemistry and is also quite cost-effective. A disadvantage, which was mentioned in Sect. 9.2, is of course the considerable increase in mass of the construct, which could alter its biological function. However, this risk is, in principle, present with any modification of a biomolecule.

We are now interested in the structure of the fluorophore and its formation from the peptide chain. In Fig. 14.4a, the mechanism of fluorophore formation in GFP is shown [48]. The reaction occurs in the sequence –Ser^{65}–Tyr^{66}–Gly^{67}–, with nearby side chains of other amino acids, such as Arg^{96} and Glu^{222}, playing an important catalytic role. The nucleophilic attack of the amide N atom of glycine on the carbonyl C atom of amino acid 65 (in this case, serine) can only be understood in the context of the folded protein. After all, we know that an amide N atom is generally positively polarized. Cyclization occurs, followed by elimination of water to form an imidazolinone. Oxidation by hydride abstraction enables conjugation with the phenyl ring of tyrosine. Such oxidations of benzylic CH_2 groups are quite common reactions—even outside the protein environment. Formation of the phenolate anion finally generates the active fluorophore. The isolated, synthetically produced fluorophore itself has a negligible quantum yield; apparently, conformational restriction by the protein is essential. By appropriate modifications that restrict mobility, the quantum yield could be increased [49, 50].

Starting around 2000, many naturally occurring variants of GFP were isolated and characterized—as well as some artificially produced variants—so that by now a considerable spectral range of fluorescence is covered (Table 14.2) [47, 51–54].

Mechanistically interesting is the formation of the red-emitting fluorophore of the Kaede protein (Fig. 14.4b). The amino acid triad in this case is –His^{62}–Tyr^{63}–Gly^{64}–. Here, too, a chromophore virtually identical to that of GFP is spontaneously formed posttranslationally. Upon UV irradiation, the peptide chain is cleaved between the –NH and α-C– of His^{62}—a highly unusual process that appears to be enabled by the fully conjugated system and the imidazole ring of histidine. The product, as a phenolate, has an emission shifted almost 100 nm further into the red (Table 14.2). This chromophore, including some derivatives, has also been synthesized and its absorption and emission properties investigated [55]. The fluorescence quantum yield is also very low in this case (QY approx. 0.005), but somewhat higher than for the less extended GFP chromophore (QY <0.0001, each in H_2O).

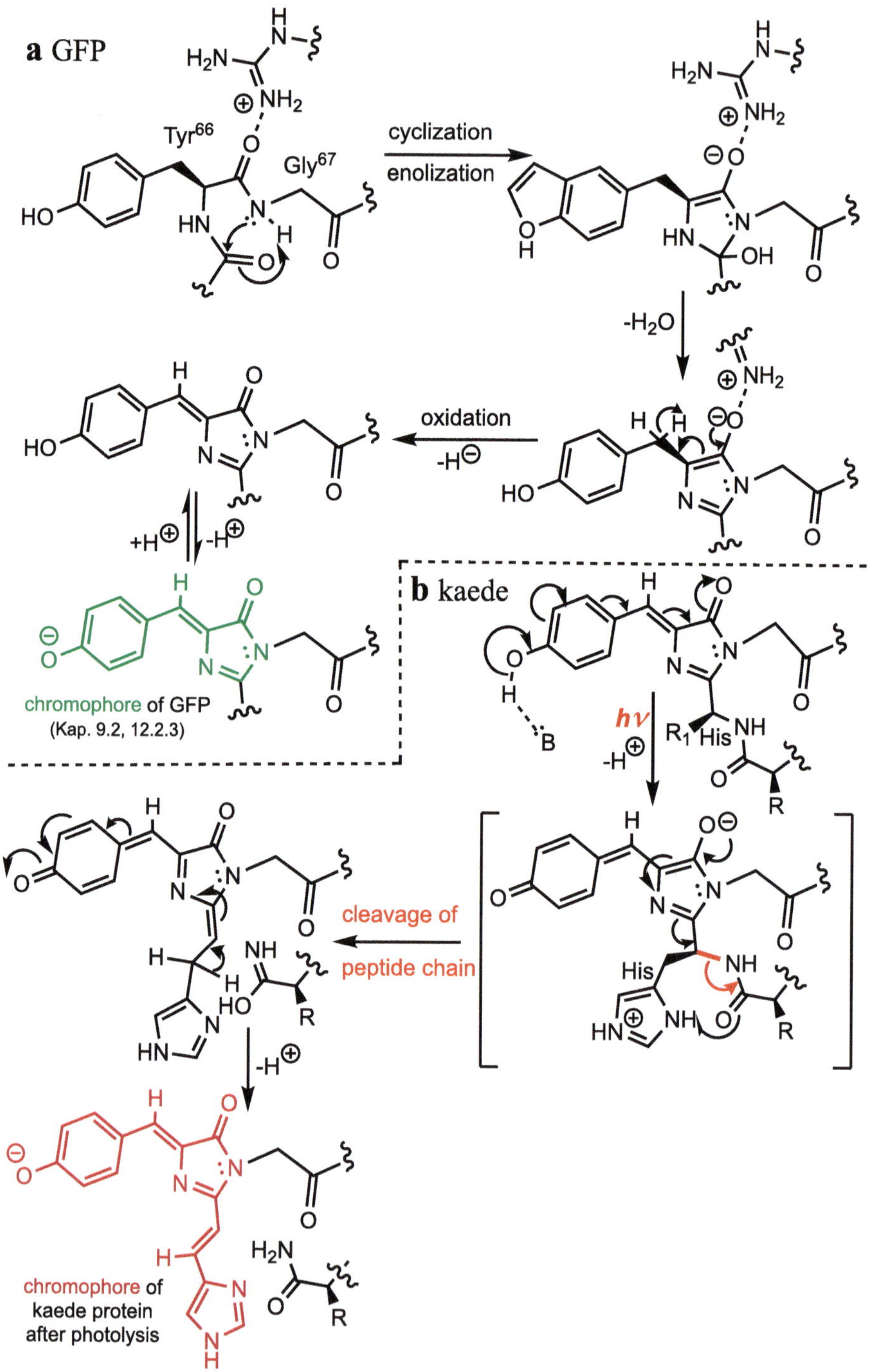

Fig. 14.4 **a** Formation of the chromophores of the fluorescent proteins GFP and **b** the conversion to the red fluorescent chromophore of the Kaede protein. (After Ref. [48])

Table 14.2 Photophysical data of some naturally occurring fluorescent proteins. (After Ref. [47])

Protein	Isolation	$\lambda_{abs=exc}$/nm	ε/$M^{-1}cm^{-1}$	λ_{em}/nm	QY
MiCy	Karasawa et al. 2004 [51]	472	27,300	495	0.90
GFP	Shimomura et al. 1962 [45]	475	25,000	508	0.79
mKO	Karasawa et al. 2004 [51]	548	52,000	561	0.60
Kaede (red)	Ando et al. 2002 [52]	572	60,000	582	0.33
Katushka	Shcherbo et al 2007 [53]	588	65,000	635	0.34

Summary and Outlook

Fluorescence has now, in the broadest sense, become the most important detection method in biology and has long since replaced radioactive labeling. This is due to its high sensitivity and the ability to optimally adapt the corresponding reagents to the requirements. We have seen the wealth of established fluorophores available and how fluorescence properties can be optimized and specific "probes" developed through modifications. Only a few examples were shown in this appendix; the reviews cited will give an impression of the full range. And yet, there is still much room for improvement. I have cited many references in this section—still only a tiny fraction of what is currently being published in this field. In my impression, there is a strong predominance of publications from Chinese research institutes—China is also a leader in the development of LED light sources, a technology closely linked to the phenomenon of fluorescence.

References

1. Samanta, S., Lai, K., Wu, F., …, Yang, Z.: Xanthene, cyanine, oxazine and BODIPY: the four pillars of the fluorophore empire for super-resolution bioimaging. Chem. Soc. Rev. **52**(20), 7197–7261 (2023). https://doi.org/10.1039/D2CS00905F
2. Sarkar, S., Chatterjee, A., Biswas, K.: A recent update on rhodamine dye based sensor molecules: a review. Crit. Rev. Anal. Chem. **54**(7), 2351–2377 (2024). https://doi.org/10.1080/10408347.2023.2169598
3. Dai, L., Zhang, Q., Ma, Q., Lin, W.: Emerging near infrared fluorophore: dicyanoisophorone-based small-molecule fluorescent probes with large stokes shifts for bioimaging. Coord. Chem. Rev. **489**, 215193 (2023). https://doi.org/10.1016/j.ccr.2023.215193
4. Archipowa, N., Wittmann, L., Köckenberger, J., …, Kutta, R.J.: Characterization of fluorescent dyes frequently used for bioimaging: photophysics and photocatalytical reactions with proteins. J. Phys. Chem. B. **127**(44), 9532–9542 (2023). https://doi.org/10.1021/acs.jpcb.3c04484
5. Liu, D., He, Z., Gao, W., …, Shi, W.: Near-infrared II cyanine fluorophores with large stokes shift engineered by regulating respective absorption and emission. Nat. Commun. **16**(1), 4911 (2025). https://doi.org/10.1038/s41467-025-60241-2

6. Liu, T., Kompa, J., Ling, J., …, Chen, Z.: Gentle rhodamines for live-cell fluorescence microscopy. ACS Cent. Sci. **10**(10), 1933–1944 (2024). https://doi.org/10.1021/acscentsci.4c00616
7. Shi, Z., Han, X., Hu, W., …, Huang, W.: Bioapplications of small molecule Aza-BODIPY: from rational structural design to in vivo investigations. Chem. Soc. Rev. **49**(21), 7533–7567 (2020). https://doi.org/10.1039/D0CS00234H
8. Loudet, A., Burgess, K.: BODIPY dyes and their derivatives: syntheses and spectroscopic properties. Chem. Rev. **107**(11), 4891–4932 (2007). https://doi.org/10.1021/cr078381n
9. Lumiprobe Katalog online. https://de.lumiprobe.com/catalog/reactive-fluorophores-quenchers. Zugegriffen am 07.01.2026
10. PhotochemCAD™: https://www.photochemcad.com/databases/common-compounds/miscellaneous-dyes/dansylglycine; https://www.photochemcad.com/databases/common-compounds/coumarins/coumarin. Zugegriffen am 07.01.2026
11. Datta, A., Mandal, D., Pal, S.K., Bhattacharyya, K.: Intramolecular charge transfer processes in confined systems. nile red in reverse micelles†. J. Phys. Chem. B. **101**(49), 10221–10225 (1997). https://doi.org/10.1021/jp971576m
12. Fery-Forgues, S., Fayet, J.-P., Lopez, A.: Drastic changes in the fluorescence properties of NBD probes with the polarity of the medium: involvement of a TICT state? J. Photochem. Photobiol. A. **70**(3), 229–243 (1993). https://doi.org/10.1016/1010-6030(93)85048-D
13. Lin, J.-K., Lai, C.-C.: High performance liquid chromatographic determination of naturally occurring primary and secondary amines with dabsyl chloride. Anal. Chem. **52**(4), 630–635 (1980). https://doi.org/10.1021/ac50054a008
14. Arun, K.T., Ramaiah, D.: Near-infrared fluorescent probes: synthesis and spectroscopic investigations of a few amphiphilic squaraine dyes. J. Phys. Chem. A. **109**(25), 5571–5578 (2005). https://doi.org/10.1021/jp051338o
15. Avellanal-Zaballa, E., Prieto, L., Casado, N., …, Bañuelos, J.: Benzo[f]indazoles: a new class of glow dyes for the generation of versatile fluorophores. Dyes and Pigments. **235**, 112624 (2025). https://doi.org/10.1016/j.dyepig.2024.112624
16. Qian, G., Dai, B., Luo, M., …, Wang, Z.Y.: Band gap tunable, donor–acceptor–donor charge-transfer heteroquinoid-based chromophores: near infrared photoluminescence and electroluminescence. Chem. Mater. **20**(19), 6208–6216 (2008). https://doi.org/10.1021/cm801911n; s. auch: Ono, K., Tanaka, S., Yamashita, Y.: Benzobis(thiadiazole)s containing hypervalent sulfur atoms: novel heterocycles with high electron affinity and short intermolecular contacts between heteroatoms. Angew. Chem. Int. Ed. **33**(19), 1977–1979 (1994). https://doi.org/10.1002/anie.199419771; Zhang, L.e., Liu, C., Zhou, S., …, Jiang, X.: Improving quantum yield of a NIR-II dye by phenylazo group. Adv. Healthcare Mater. **9**(4), 1901470 (2020). https://doi.org/10.1002/adhm.201901470
17. Qian, M., Zhang, L., Pu, Z., …, Peng, X.: A NIR fluorescent probe for the detection and visualization of hydrogen sulfide using the aldehyde group assisted thiolysis of dinitrophenyl ether strategy. J. Mater. Chem. B. **6**(47), 7916–7925 (2018). https://doi.org/10.1039/C8TB02218F
18. Wache, N., Scholten, A., Klüner, T., Koch, K.-W., Christoffers, J.: Turning on fluorescence with thiols—synthetic and computational studies on diaminoterephthalates and monitoring the switch of the Ca2+ sensor recoverin. Eur. J. Org. Chem. **2012**(29), 5712–5722 (2012). https://doi.org/10.1002/ejoc.201200879
19. Jean-Gérard, L., Vasseur, W., Scherninski, F., Andrioletti, B.: Recent advances in the synthesis of [a]-benzo-fused BODIPY fluorophores. Chem. Commun. **54**(92), 12914–12929 (2018). https://doi.org/10.1039/C8CC06403B
20. Ilhan, H., Cakmak, Y.: Functionalization of BODIPY dyes with additional C–N double bonds and their applications. Top. Curr. Chem. **381**(5), 28 (2023). https://doi.org/10.1007/s41061-023-00438-5

21. Jin, J.H., An, J.M., Kim, D., …, Kim, D.: Amino-SBBF (single benzene-based fluorophore) library: its synthesis, photophysical property, and cellular imaging application. Dyes Pigm. **221**, 111811 (2024). https://doi.org/10.1016/j.dyepig.2023.111811
22. Li, H., Wang, J., Kim, H., Peng, X., Yoon, J.: Activatable near-infrared versatile fluorescent and chemiluminescent dyes based on the dicyanomethylene-4H-pyran scaffold: from design to imaging and theranostics. Angew. Chem. Int. Ed. **63**(6), e202311764 (2024). https://doi.org/10.1002/anie.202311764
23. Ye, J., Wang, D., Yao, H., …, Yu, B.: Current status and future developments in NIR-II-emitting organic small molecule fluorophores for bioimaging and phototherapy. Small. **21**(27), e2409722 (2025). https://doi.org/10.1002/smll.202409722
24. Dumat, B., Chieffo, C.: Harnessing cyanine-like properties to develop bright fluorogenic probes based on viscosity-sensitive molecular rotors. Chem. Eur. J. **31**(9), e202404077 (2025). https://doi.org/10.1002/chem.202404077
25. Bai, L., Wang, C., Zhang, M., …, Bai, Y.: Structural and electronic properties of D-π-A coumarin dyes: a DFT/TD-DFT assisted design of low-molecular-weight, long-wavelength-emissive fluorescent dyes. J. Fluoresc. **35**, 9077-9092 (2025). https://doi.org/10.1007/s10895-025-04217-x
26. Shukla, A.K., Choudhary, S., Sisodiya, D.S., …, Bhattacharya, A.: Pyrrolo[1,2-a]quinoxaline: a combined experimental and computational study on the photophysical properties of a new photofunctional building block. J. Phys. Chem. B. **129**(23), 5842–5853 (2025). https://doi.org/10.1021/acs.jpcb.5c01810
27. Zhang, J., Wang, N., Ji, X., …, Zhao, W.: BODIPY-based fluorescent probes for biothiols. Chem. Eur. J., **26**(19), 4172–4192 (2020). https://doi.org/10.1002/chem.201904470; Zhang, J., Peng, F., Dong, X., Zhao, W.: Water-soluble BODIPY derivative as a highly selective "turn-on" fluorescent probe for hydrogen sulfide in living cells. Chem. Lett. **44**(11), 1524–1526 (2015). https://doi.org/10.1246/cl.150631
28. Ali, M., Zhou, C., Gao, Z., …, Wang, F.: Sensitive fluorescent probe for monitoring and bio-imaging biothiols in living systems. Dyes Pigm. **235**, 112650 (2025). https://doi.org/10.1016/j.dyepig.2025.112650
29. Fan, J., Lin, B., Tang, S., Han, Y.: A novel ratiometric BenzoBODIPY-based fluorescent probe for the detection and imaging of cysteine in living cells and zebrafish models. Talanta. **285**, 127332 (2025). https://doi.org/10.1016/j.talanta.2024.127332
30. Wang, Z., Gao, Z., Gao, Y., Pang, X., Xia, X.: A fluorescent probe with large Stokes shift for detecting biothiols in living cells and zebrafish. Tetrahedron. **179**, 134615 (2025). https://doi.org/10.1016/j.tet.2025.134615
31. Qiu, X., Jiao, X., Liu, C., …, Zeng, X.: A selective and sensitive fluorescent probe for homocysteine and its application in living cells. Dyes Pigm. **140**, 212–221 (2017). https://doi.org/10.1016/j.dyepig.2017.01.047
32. Lee, S.-C., Heo, J., Woo, H.C., …, Kwon, O.-P.: Fluorescent molecular rotors for viscosity sensors. Chem. Eur. J. **24**(52), 13706–13718 (2018). https://doi.org/10.1002/chem.201801389
33. Zhao, L., Chen, R., Jia, C., …, Cheng, T.: BODIPY based OFF-ON fluorescent probe for endogenous carbon monoxide imaging in living cells. J. Fluoresc. **34**(4), 1793–1799 (2024). https://doi.org/10.1007/s10895-023-03403-z
34. Zhang, M.-Q., Guan, W.-H., Wang, J., …, Wang, X.-B.: Engineering a swich-on fluorescent probe for the visual detection of HOCl in envirnomental and living system. Spectrochim. Acta A Mol. Biomol. Spectrosc. **330**, 125738 (2025). https://doi.org/10.1016/j.saa.2025.125738
35. Cosco, E.D., Bogyo, M.: Recent advances in ratiometric fluorescence imaging of enzyme activity in vivo. Curr. Opin. Chem. Biol. **80**, 102441 (2024). https://doi.org/10.1016/j.cbpa.2024.102441

36. Zhang, Y., Zheng, Y., Meana, Y., Raymo, F.M.: BODIPYs with photoactivatable fluorescence. Chem. Eur. J. **27**(44), 11257–11267 (2021). https://doi.org/10.1002/chem.202101628
37. Zhang, Y., Zheng, Y., Tomassini, A., Singh, A.K., Raymo, F.M.: Photoactivatable fluorophores for bioimaging applications. ACS Appl Opt Mater. **1**(3), 640–651 (2023). https://doi.org/10.1021/acsaom.3c00025
38. Li, H., Wang, J., Jiao, L., Hiao, E.: BODIPY-based photocages: rational design and their biomedical application. Chem. Commun. **60**(45), 5770–5789 (2024). https://doi.org/10.1039/D4CC01412J
39. Zhu, H., Fan, J., Du, J., Peng, X.: Fluorescent probes for sensing and imaging within specific cellular organelles. Acc. Chem. Res. **49**(10), 2115–2126 (2016). https://doi.org/10.1021/acs.accounts.6b00292
40. Cai, Z.P., Hagan, A.K., Wang, M.M., …, Voglmeir, J.: 2-Pyridylfuran: a new fluorescent tag for the analysis of carbohydrates. Anal. Chem. **86**(10), 5179–5186 (2014). https://doi.org/10.1021/ac501393a
41. Wilson, D.L., Kool, E.T.: Ultrafast oxime formation enables efficient fluorescence light-up measurement of DNA base excision. J. Am. Chem. Soc. **141**(49), 19379–19388 (2019). https://doi.org/10.1021/jacs.9b09812
42. Maujean, T., Wagner, P., Valencia, C., …, Bonnet, D.: Rapid and highly selective fluorescent labeling of peptides via a thia-Diels–Alder cycloaddition: application to apelin. Bioconjug. Chem. **34**(1), 162–168 (2023). https://doi.org/10.1021/acs.bioconjchem.2c00500
43. Yu, A., He, X., Shen, T., …, Wu, H.: Design strategies for tetrazine fluorogenic probes for bioorthogonal imaging. Chem. Soc. Rev. **54**(6), 2984–3016 (2025). https://doi.org/10.1039/D3CS00520H
44. Marshall, O., Sutherland, A.: One-pot synthesis of alkynyl-conjugated phenylalanine analogues for peptide-based fluorescent imaging. Org. Lett. **27**(29), 8023–8027 (2025). https://doi.org/10.1021/acs.orglett.5c02361
45. Shimomura, O., Johnson, F.H., Saiga, Y.: Extraction, purification and properties of aequorin, a bioluminescent protein from the luminous hydromedusan, aequorea. J. Cell. Comp. Physiol. **59**(3), 223–239 (1962). https://doi.org/10.1002/jcp.1030590302
46. Tsien, R.Y.: The green fluorescent protein. Annu. Rev. Biochem. **67**(1), 509–544 (1998). https://doi.org/10.1146/annurev.biochem.67.1.509
47. Pakhomov, A.A., Martynov, V.I.: GFP family: structural insights into spectral tuning. Chem. Biol. **15**(8), 755–764 (2008). https://doi.org/10.1016/j.chembiol.2008.07.009
48. Wachter, R.M., Watkins, J.L., Kim, H.: Mechanistic diversity of red fluorescence acquisition by GFP-like proteins. Biochemistry. **49**(35), 7417–7427 (2010). https://doi.org/10.1021/bi100901h
49. Follenius-Wund, A., Bourotte, M., Schmitt, M., …, Pigault, C.: Fluorescent derivatives of the GFP chromophore give a new insight into the GFP fluorescence process. Biophys. J. **85**(3), 1839–1850 (2003). https://doi.org/10.1016/S0006-3495(03)74612-8
50. Paul, D., Sahoo, P., Pradhan, A.K., …, Chatterjee, S.: Revealing the importance of the C(1) position to modulate the photophysics and Z-scan responses of o-locked GFP chromophores. J. Chem. Phys. **162**(23), 234308 (2025). https://doi.org/10.1063/5.0274692
51. Karasawa, S., Araki, T., Nagai, T., Mizuno, H., Miyawaki, A.: Cyan-emitting and orange-emitting fluorescent proteins as a donor/acceptor pair for fluorescence resonance energy transfer. Biochem. J. **381**(1), 307–312 (2004). https://doi.org/10.1042/bj20040321
52. Ando, R., Hama, H., Yamamoto-Hino, M., Mizuno, H., Miyawaki, A.: An optical marker based on the UV-induced green-to-red photoconversion of a fluorescent protein. Proc. Natl. Acad. Sci. U. S. A. **99**(20), 12651–12656 (2002). https://doi.org/10.1073/pnas.202320599

53. Shcherbo, D., Merzlyak, E.M., Chepurnykh, T.V., …, Chudakov, D.M.: Bright far-red fluorescent protein for whole-body imaging. Nat. Meth. **4**(9), 741–746 (2007). https://doi.org/10.1038/nmeth1083
54. Cranfill, P.J., Sell, B.R., Baird, M.A., …, Piston, D.W.: Quantitative assessment of fluorescent proteins. Nat. Meth. **13**(7), 557–562 (2016). https://doi.org/10.1038/nmeth.3891
55. Fatima, A., Bressan, G., Ashworth, E.K., …, Meech, S.R.: Substituent effects on the photophysics of the kaede chromophore. Phys. Chem. Chem. Phys. **26**(46), 29048–29059 (2024). https://doi.org/10.1039/D4CP03272A

Solutions 15

Abstract

Solutions to the questions of chapters 1-13, providing also more profound explanations.

15.1 Solutions Chap. 1

Answers

1.1 Yes, $K_a = K_w/K_b$

1.2 The reaction rate is, in principle, unrelated to thermodynamic parameters. Alkoxides are **more nucleophilic** than alcohols. In this case, the higher nucleophilicity (a kinetic parameter) also corresponds to a higher basicity (thermodynamically defined).

1.3 Cyclopentadiene is indeed cyclic, but not fully conjugated; the CH_2 group interrupts the conjugation. Moreover, it would not satisfy the Hückel rule, but this consideration only applies to fully conjugated cycles.

1.4 **F**alse: Resonance forms are only hypothetical limiting structures; all must always be considered. In some cases, a particular reaction can be especially well explained using a specific resonance form.

1.5 **T**rue

1.6 **T**rue

D. Krois, *Organic Chemical Methods*, https://doi.org/10.1007/978-3-662-73573-2_15

15.2 Solutions Chap. 2

Answers

2.1 Coded proteinogenic amino acids are those that occur in proteins and are specified by the DNA code. For the list, see Fig. 2.1.

2.2 Tryptophan (heterocycle: indole), histidine (heterocycle: imidazole). For formulas, see Fig. 2.1.

2.3 In Fig. 2.1, in the general formula for the L-configuration, you must replace the R group with CH_3. For the *R*,*S* nomenclature, the priorities of the four substituents on the asymmetric C atom must be assigned: NH_2 … a, COOH … b, CH_3 … c, and H … d. You should then view the asymmetric C atom so that substituent d is pointing away from you. The groups a > b > c are then arranged counterclockwise ➔ configuration *S*. If this is difficult, you can also look at it the "wrong" way (with group d pointing toward you), but then you must simply reverse the sense of rotation you obtain!

2.4 Because the N atom, according to the third resonance structure (which carries significant weight in amide bonds), has a positive charge (thus, overall, a partial positive charge).

2.5 The Fischer projection formula of D-lyxose is shown in Fig. 2.4. The conversion to the Haworth formula is shown in Fig. 2.5. This results in a five-membered ring with the OH groups at C2 and C3 pointing up, as well as the CH_2OH group at C4 pointing up. The new OH group at C1 points down in the α-anomer and up in the β-anomer.

2.6 Anomers are the two possible diastereomers that result from hemiacetal formation and which are in equilibrium in solution *via* the carbonyl form.

2.7 The glycosidic bond is a full acetal bond. Full acetals can only be cleaved by acids.

2.8 A nucleoside is a ribose (or 2-deoxyribose) linked via a β-*N*-glycosidic bond to a heterocyclic base (nucleobase).

2.9 The formula for ATP is obtained by attaching an additional phosphate group as an anhydride to the diphosphate group of 5′-ADP in Fig. 2.6.

2.10 Fats are the triesters of glycerol with three long-chain carboxylic acids.

2.11 A micelle is a spherical or ellipsoid-shaped compartment in aqueous solution that is polar and/or charged on the outside and lipophilic on the inside. It is formed by detergents in aqueous solution when a certain concentration (cmc … *critical micellar concentration*) is exceeded.

15.3 Solutions Chap. 3

Answers

3.1 In principle, there are at least two different possibilities:

a. Protect the N-terminus of alanine and the C-terminus of glycine, form the peptide bond using a coupling reagent, and then remove both protecting groups:

e.g., Boc–Ala–OH + H–Gly–OtBu --- DCC/HOBt ---> Boc–Ala–Gly–OtBu, followed by cleavage of the *tert*-butyl protecting group with TFA, or

b. Prepare the active ester Z–Ala–OSu from, e.g., Z-Ala–OH with DCC/HOSu, add free glycine (H–Gly–OH), and obtain Z–Ala–Gly–OH. After hydrogenolysis (PdO/H_2), the desired free dipeptide is obtained.

3.2 See Fig. 3.5

3.3 When using coupling reagents, both coupling partners are present in solution upon addition of the activating reagent, so all carboxyl and amino functions that are *not* to react must be protected. When using an active ester, however, only a specific carboxyl group is activated—other carboxyl groups cannot react and therefore do not need to be protected.

3.4 See Conclusion at the end of Sect. 3.1.3.2.

3.5 The coupling yield should approach 99% as closely as possible, otherwise too many "gap peptides" will accumulate, which are difficult to remove completely at the end.

3.6 Urethanes are half-esters, "half-amides" of carbonic acid (R′–N–CO–OR). They are suitable for peptide synthesis because, depending on the R group, they can be easily cleaved and the reaction proceeds essentially to completion. Gaseous CO_2 is formed. In addition, urethane-protected amino acids tend to racemize less upon activation (epimerization at the α-C atom).

3.7 A peptide library is an ensemble of many different peptides of the same length but different sequences. It is used for screening methods to determine affinities. Examples and methods, see Sect. 3.1.5 and Sect. 3.3.

3.8 HPLC, SFC, CE, gel electrophoresis, and MS, Sect. 3.2.1 and Sect. 3.2.2.

3.9 Amino acid analysis refers to the quantitative determination of amino acids in a mixture that has resulted from total hydrolysis of a peptide or protein. However, it can also refer to free amino acids in a solution, in food, or in a body fluid.

3.10 In Edman degradation, the N-terminal amino acid of a peptide is cleaved off in three reaction steps and identified as a phenylthiohydantoin derivative (Sect. 3.2.4).

3.11 X-ray crystallography, cryo-electron microscopy, and for proteins in solution, primarily NMR methods.

3.12 The method of choice for this is the polymerization of an NCA of an amino acid (Sect. 3.1.2.3). A catalyst is added for this—various catalysts have been reported in the literature. Such amino acid polymers are used in materials science and medical research. – Oligomers and polymers of a single amino acid can also be obtained as follows, e.g.: Z–Ala–OSu ---H2/Pd---> [H–Ala–OSu] → H(Ala)$_n$-OH (if stopped by addition of H_2O).

3.13 Of course, there is no solution for this—you can do this for several literature sources. In doing so, you will learn much more than is contained in the book.

15.4 Solutions Chap. 4

Answers

4.1 There are far more possible linkages (α- and β-glycosidic bonds and possibilities for branching).

4.2 Number as for glucose and mannose in Table 4.1: 20, namely for Gal*p*→Man [α(1→2), β(1→2), α(1→3), β(1→3), α(1→4), β(1→4), α(1→6), β(1→6)], for Man*p*→Gal [α(1→2), β(1→2), α(1→3), β(1→3), α(1→4), β(1→4), α(1→6), β(1→6)] and for Gal*p*→Man*p* [(α1→α1), (α1→β1), (β1→α1), (β1→β1)].

4.3 The glycosidic bond is a full acetal bond, and is therefore more or less labile under acidic conditions, but stable under basic conditions.

4.4 The glycosidic bond is directional: it always originates from the anomeric C atom (the carbonyl C atom), which is why it is indicated with an arrow (→).

4.5 Only the types are asked for, i.e., ester protecting group, ether protecting group, and acetal protecting group.

4.6 Sometimes two or even three orthogonal temporary protecting groups are needed on a monosaccharide when branches need to be synthesized.

4.7 Activation of the anomeric C atom (glycosyl donor) as a thioglycoside –SR′, as a trichloroacetimidate –O–C(=NH)–CCl_3, or as a phosphate –O–P(OR′)$_2$=O; in addition, an activator is usually added (in catalytic or stoichiometric amounts).

4.8 The acceptor HO group is attached to the solid phase, so the assembly proceeds from the reducing end (the end of the oligosaccharide with the free anomeric C atom, after cleavage from the solid phase). Therefore, the activated glycosyl donor in solution can also be used in excess—just as in SPPS.

4.9 *Capping* refers to the saturation of unreacted acceptor sites by an irreversible reaction at the end of each synthesis cycle, to prevent the formation of gap-oligosaccharides of similar length to the desired product. This would otherwise complicate the final purification.

4.10 See Sect. 4.3: In addition to chemical degradation methods, MS and NMR should definitely be mentioned. It should also be explained what each method can accomplish.

4.11 Not really, since both MS and NMR require at least several different experiments, which must then be interpreted together. Perhaps one day a method using nanopores will be developed to this extent—but that is still far in the future.

4.12 Alduronic acids are aldoses whose terminal primary hydroxyl group has been oxidized to a carboxylic acid. They occur in oligosaccharides; glucuronic acid is also "used" by the liver to make lipophilic substances water-soluble as conjugates.

4.13 **T**rue

4.14 **F**alse: There are only four diastereomers, since there are only three asymmetric C atoms. The general formula is $HOCH_2$–C(=O)–CHOH–CHOH–CHOH–CH_2OH.

4.15 **F**alse: The ratio of the two anomers is structure-dependent, for glucose, for example, about 2:1.

15.5 Solutions Chap. 5

Answers

5.1 Glycoconjugates are biomolecules that, in addition to the carbohydrate portion, also contain a peptide and/or lipid portion. The oligosaccharide chain (or the monosaccharide part) is often crucial for biological function, and this region is usually variable and heterogeneous. This increases the demands on both analytics and synthesis.

5.2 Alaninol is a derivative of alanine in which the carboxyl group has been reduced to a primary alcohol. Such amino alcohols are formed in nature by enzymatic reduction. Alaninol is found in the peptide part of the GPL from *Mycobacterium avium* (see Fig. 5.1).

5.3 Rha stands for rhamnose, which is a 6-deoxymannose (see Fig. 4.6).

5.4 Glc*p*A denotes a glucuronic acid in the pyranose form (see Fig. 4.6).

5.5 Trehalose is a non-reducing disaccharide composed of two glucopyranose units, linked ($\alpha 1 \leftrightarrow \alpha 1$) (see Fig. 5.2).

5.6 Oligosaccharide chains can be linked *O*-glycosidically to serine or threonine, and *N*-glycosidically to asparagine.

5.7 No—a prerequisite is that the amino acid two residues downstream is serine or threonine, i.e., the sequence –Asn–Xxx–Ser/Thr–. A rationale for this can be found in Sect. 5.2.4.

5.8 Enrichment of glycoproteins in a biological sample for MS analysis can be achieved by reaction with polymer-bound hydrazides or with boric acid derivatives (see Fig. 5.3).

5.9 The main problem in glycopeptide synthesis is the choice of protecting groups, since the glycosidic bond is stable under basic conditions but labile under acid catalysis, whereas peptides undergo rearrangements and epimerizations at high pH.

15.6 Solutions Chap. 6

Answers

6.1 The Maxam-Gilbert sequencing method is based on degradation of the single-stranded DNA (in contrast to all other sequencing methods, which use polymerases for synthesis) by four different reactions, where it is essential that the base-specific reaction occurs only with low yield.

6.2 The 2′-,3′-dideoxynucleoside triphosphates, when incorporated by a polymerase, terminate replication. It is essential that in each of the four reactions, a *limited* amount of *one* ddXTP is added alongside an excess of dATP, dGTP, dTTP, and dCTP, so that after each X, termination can occur—but only in a small fraction, so that many longer strands are also produced.

6.3 In these methods, only one specific dNTP (deoxynucleoside triphosphate) is added per cycle. If it is incorporated, a signal is generated, then another dNTP is offered, and so on.

6.4 In oligonucleotide synthesis, the 5′-OH group is temporarily protected, usually by a dimethoxytrityl protecting group (DMT).

6.5 A phosphoramidite is a compound of trivalent phosphorus with the formula R^1O–$P(OR^2)$–NR_2.

6.6 Oxidation occurs at the end of each cycle, usually with I_2 or tBuOOH.

6.7 Solid-phase oligonucleotide synthesis is simpler because there are only four different building blocks and the structural differences are far from the coupling site. In peptide synthesis, the coupling site is directly adjacent to the α-C atom with the most diverse side chains. The reaction at P(III) is also apparently much faster.

6.8 LNA (*locked nucleic acids*) make oligonucleotides more rigid and less susceptible to degradation. The ribofuranose ring is greatly restricted in its conformational flexibility by the bridging.

6.9 The backbone of Nielsen PNA consists of aminoethylglycine units, with the nucleobase attached via an amide bond to the glycine N atom: NH–CH_2–CH_2–N(CO-CH_2-B)–CH_2–CO.

6.10 The introduction of a third base pairing (in addition to A–T and G–C), which is faithfully replicated by natural polymerases, would enormously increase the information storage capacity of DNA. However, I also see the release of such replicating units into our world as problematic.

15.7 Solutions Chap. 7

Answers

7.1 The methods for preparing conjugates of biomolecules should be specific for a particular functional group. The reactions should be "bioorthogonal," meaning they should be applicable even in complex mixtures. In general, modification of the biomolecule should not affect its biological activity. Usually, a defined product is desired—but sometimes a mixture (e.g., where different lysines have been conjugated) is preferred.

7.2 For this, either active esters (such as *N*-hydroxysuccinimide esters) are used (➔ amide bond) or isothiocyanates (➔ thioureas) (see Fig. 7.2)

7.3 Cysteine is less common as an amino acid than lysine and is therefore better suited for defined conjugation products. In addition, thiols are more nucleophilic than amines and are not protonated at neutral pH. Reversible disulfide formation can also be exploited.

7.4 Thiols react as nucleophiles with electron-deficient double bonds; this type of reaction is called a Michael addition.

7.5 R–CH=O + NH_2-OR′ ➔ R–CH=N–OR′ (an oxime)

7.6 Aldehydes are formed by periodate cleavage from compounds with vicinal dihydroxyl groups (e.g., neuraminic acid) or from N-terminal serine (Sect. 7.9.1, Sect. 10.1). Enzymes can transform the amino acid cysteine (possibly also serine) into formylglycine (Sect. 7.9.2).

7.7 Sharpless referred to reactions as *click chemistry* that proceed rapidly and selectively. An example is the Cu(I)-catalyzed 1,3-dipolar cycloaddition of azides with alkynes.

7.8 Azides can form stable triazoles with alkynes (Sect. 7.4), form amides with phosphines and carboxylic acids (Staudinger ligation, Sect. 7.5), or, after photolysis, be used as nitrenes for photoaffinity labeling (Sect. 7.6.1).

7.9 See Fig. 7.12 and 7.13.

7.10 With such reagents (Sect. 7.7.2), the aim is to identify protein pairs that interact with each other (that form complexes or aggregates). With a suitable label, these covalently linked pairs can subsequently be enriched and, in MS/MS experiments, the total molecular mass and, after CID, the mass of the individual proteins (+ part of the linker) can be determined.

7.11 The reason is that 2-sulfhydrylpyridine immediately tautomerizes to 2-thiopyridone, thus shifting the equilibrium in this direction (Sect. 7.2.2)

7.12 Biotin is a small biomolecule that forms an extremely strong—non-covalent—bond with the protein streptavidin, and there are many streptavidin conjugates with a wide variety of labeling options and enzymes commercially available (Sect. 7.8).

15.8 Solutions Chap. 8

Answers

8.1 The peptide fragments must be linked without protecting groups on the side chains and in aqueous solution; therefore, methods for native chemical ligation (NCL) were developed. If all side chain protecting groups are retained, the solubility of such large peptide fragments is generally too low even in organic solvents, and there would usually be a very strong tendency to aggregate. For semisynthesis, when part of a protein is produced in microorganisms or cell culture, one must work without protecting groups anyway.

8.2 In NCL, a C-terminal thioester must react with an N-terminal free cysteine. Initially, a thioester is formed with the HS group of the cysteine, followed by an *S*→*N* acyl transfer and formation of the natural peptide bond.

8.3 Cysteine is not very common in proteins! In addition, the C-terminal amino acid with the thioester should be sterically unhindered—ideally glycine.

8.4 In this case, a peptide fragment with cysteine at the N-terminus is linked, and the cysteine is subsequently "desulfurized" to alanine. This is achieved by a radical reaction.

8.5 NCL also works very well with the corresponding selenium compounds. In general, these are even more reactive!

8.6 In STL, a salicylaldehyde ester is introduced at the C-terminus, which thus carries a free aldehyde. This aldehyde group can form a cyclic aminal with an N-terminal hydroxy amino acid such as Ser or Thr, which then, after elimination of salicylaldehyde by *O*→*N*-acyl transfer, leads to the peptide bond.

8.7 In Fig. 8.1c, d, and e, three methods are given. The simplest is the direct introduction of a hydrazide upon cleavage from the solid phase and its conversion via an azide into the thioester or, as in Fig. 8.3a, via a pyrazole.

8.8 In SPPS, a Boc-protected thiazolidine carboxylic acid can be directly introduced. The thiazolidine ring is actually a thioaminal of formaldehyde and can be cleaved to cysteine in acidic solution by trapping the formaldehyde as an oxime.

8.9 This is the coupling of an oligosaccharide, which has a terminal 1-amino group, with the β-carboxyl group of Asp using a coupling reagent.

8.10 Pseudoprolines (see Fig. 8.4) are "prolines" with an additional oxygen atom in the ring—that is, a 1,3-oxazolidine ring. At the same time, they are aminals that can be cleaved under acid catalysis into serine or threonine and acetone.

8.11 The ability to obtain uniform, defined glycoproteins is very important for understanding the significance of the various oligosaccharide variants that are normally present, and possibly also for their use as therapeutics. The difficulty lies in the lability of the glycan part under acidic and the peptide part under basic conditions (see Chaps. 3, 4 and 5). Therefore, the choice of protecting groups and the order of couplings is especially critical. This is very clear in the example in Sect. 8.2.2 when the original literature source [25] is considered.

15.9 Solutions Chap. 9

Answers

9.1 The Ras protein is activated by growth factors (GF) and is a key switch in the MAP kinase signal transduction cascade, which ultimately leads to increased cell division. It has an intrinsic deactivation activity (GTPase activity), which then switches off this signal. Point mutations can block this activity and thus lead to cancer formation (uncontrolled cell growth)—the Ras protein becomes an oncogene.

9.2 Four posttranslational modifications are required: farnesylation (farnesyl transferase), cleavage of the three C-terminal amino acids (specific protease), formation of the methyl ester at the C-terminal cysteine (methyl transferase), and palmitoylation at the penultimate cysteine (protein *S*-acyl transferase).

9.3 Two lipid anchors are necessary to ensure localization at the cell membrane (near the corresponding receptor tyrosine kinase RTK) and also at the membrane-localized Raf protein, which is to be phosphorylated by Ras.

9.4 Because these PGs can be cleaved under practically neutral conditions, the terpene residues are acid-sensitive, and the palmitoyl thioester is very base-sensitive.
9.5 With the knowledge from Chap. 3 (Sect. 3.1 and 3.2), you should be able to do this.
9.6 In Sect. 7.2, conjugate formation of cysteines.
9.7 Fluorophores can, due to their polarity properties or their size, lead to significant changes in biological activity, as was clearly shown in Ref. [8].
9.8 A bisubstrate inhibitor has similarities to the two substrates of an enzyme that are to be linked together, and will therefore exhibit special binding properties.
9.9 Upon binding of farnesyl pyrophosphate, the polarity in the region of the fluorophore is greatly changed, and this can be monitored in a concentration-dependent manner.
9.10 See Fig. 9.4 and Ref. [6].

15.10 Solutions Chap. 10

Answers

10.1 The causative agent of malaria is *Plasmodium falciparum* (among other Plasmodium species) and is a parasite, thus neither a bacterium nor a virus.
10.2 No—it is transmitted by the intermediate host, the Anopheles mosquito, through an insect bite (in principle, probably also possible through the blood of an infected person).
10.3 Antibodies are protein molecules (immunoglobulins) that recognize the antigen (a defined region, e.g., of a protein of the pathogen). They are produced by the so-called B cells of the immune system.
10.4 Usually, the variable regions are not functional, not important for attacking the pathogen, and antibodies against such regions therefore often do not confer protection; they do not neutralize the pathogen. In addition, they then do not recognize a related pathogen strain.
10.5 MAPs, *multiple antigenic peptides*, are dendrimers; they consist of a branching "core," which thus has several identical antigenic peptides covalently attached.
10.6 They are linked by oxime formation with the *core*, which carries a hydroxylamine function. The aldehyde is generated by periodate cleavage of a serine at the ε-amino group of the C-terminal lysine of the antigenic peptide (see Fig. 10.2).
10.7 Dimorphism refers to the exchange of individual amino acids in an otherwise strictly conserved region in different strains of the pathogen (see Table 10.2).
10.8 The antigen with which immunization was performed is adsorbed in the wells of a microtiter plate (*coating*). The antiserum is then added in various dilutions, washed, and incubated with an enzyme conjugate that binds the antibody (anti-antibody). After another wash, the enzyme substrate is added, the enzyme reaction is stopped after a certain time, and the concentration (of the antibodies

associated with the plate) is determined photometrically. The so-called antibody titer is determined from the sigmoidal concentration dependence.

10.9 Whereas in the "normal" ELISA the affinity between antibody and adsorbed antigen is determined, in the competitive ELISA this interaction is determined in solution between free molecules.

10.10 The CSP protein is presented to the immune system on a virosome (a virus-like particle) formed from a protein of the hepatitis B virus.

10.11 A tertiary amine.

15.11 Solutions Chap. 11

Answers

11.1 Difluorotoluene is isosteric to thymine; the F atoms perfectly replace the carbonyl O atoms. Although fluorine is more electronegative than oxygen, the H-bonds to F are much weaker. It is well known that fluorocarbons are nonpolar.

11.2 Since the UV spectra of nucleobases in a single-stranded DNA differ significantly from those in the double strand, the dissociation of the double helix can be very well monitored in the UV-Vis at a specific wavelength (e.g., 260 nm). A sigmoidal curve is recorded, from which the melting temperature can be read. The free binding energy ΔG can also be calculated from this.

11.3 The same spatial occupancy is not achieved, since =N– had to be replaced by =CH–, i.e., an additional atom (H) was introduced.

11.4 At an abasic site, the nucleobase is missing from the ribose. As a natural nucleotide triphosphate, dATP is most likely to be incorporated opposite. However, the designed pyrene derivative dPTP is incorporated with much better selectivity. Pyrene best fills the space of the pairing nucleobases (e.g., A–T). This was a nice proof of the steric model of polymerase selectivity.

11.5 *Molecular beacons* consist of two complementary sequences at the ends and a sequence in the *loop* that is complementary to the target. In the hairpin conformation, the fluorophore at one end comes close to the quencher at the other end, so no fluorescence occurs. Only upon binding to the template does the quencher lose its effectiveness. Therefore, unlike the FISH method, no washing step is necessary, and detection of DNA or RNA can be performed in solution.

11.6 They should be as stable as possible against nucleases, because cleavage would cause them to fluoresce. This would result in a high background signal, i.e., lower detection accuracy. An easy option is the use of PNA.

11.7 It is an S_N2 reaction, in which the S atom in the phosphothioester acts as a strong nucleophile.

11.8 As in the Wittig reaction, it is the formation of the phosphine oxide from the phosphine.

11.9 In principle, the fluorescence signal can be amplified in a cascade manner, because after successful autoligation and cleavage of the quencher, the fluorescent oligosaccharide can dissociate and the reaction on the template can repeat. However, the question always remains how fast the reaction and dissociation actually are.

11.10 At a specific site of the oligonucleotide, an unpaired electron (charge injection) is actually generated by the modified nucleobase. The modified thymine undergoes cleavage upon irradiation, forming a radical. The additional electron can then migrate from one thymine to the next in the oligonucleotide (*hopping*).

11.11 There is, of course, no solution for this task.

15.12 Solutions Chap. 12

Answers

12.1 It is one of the codons that signals the end of synthesis during translation, but it occurs very rarely and is therefore used for the incorporation of non-coded amino acids. In DNA it is TAG, and in mRNA accordingly UAG. The anticodon of the suppressor tRNA is (from 5′→ 3′) CUA.

12.2 In the method developed by P. G. Schulz in a 1995 review in Angewandte Chemie, the suppressor tRNA—shortened by two nucleotides at the 3′ end—is produced by molecular biology (*run-off transcription*). The synthetically produced dinucleotide pdCpA is then coupled to the N-terminally protected unnatural amino acid, enzymatically attached to the 3′ end of the truncated suppressor tRNA, and the N-terminal protecting group is then photolytically cleaved. Addition to the "UAG"-modified mRNA and the *in vitro* extract for ribosomal protein synthesis ultimately allows the synthesis of the modified protein. In the later developed *in vivo* method, on the other hand, the suppressor tRNA is loaded in the cell with the introduced unnatural amino acid by an aminoacyl-tRNA synthetase (AARS) developed through mutation and selection, and then the protein—read from the altered mRNA—is synthesized, all within the living cell.

12.3 Ras proteins with modifications at lysine 16 were produced. The question was whether the chain length had an effect on GTPase activity. Result: The distance of the amino group from the protein backbone does not appear to play a significant role, but replacing $-NH_2$ with –OH does; GTPase activity is completely lost.

12.4 If, instead of the quadruplet codon, a triplet codon is read and the corresponding incorrect natural amino acid is incorporated, the next triplet is a stop codon—the synthesis is terminated.

12.5 If Fl1 is excited at $\lambda = 490$ nm, and is sufficiently close to the N-terminus, FRET (*fluorescence resonance energy transfer*) to Fl2 occurs, which then emits at 575 nm. If the two fluorophores are too far apart, no transfer occurs via FRET, and the observed emission has a maximum at about 515 nm.

12.6 Grb2 transmits the signal from the receptor tyrosine kinase (EGFR) to the Ras protein.

12.7 The aim was to see which amino acid residues are in close contact with the activated EGFR.

12.8 pBpa carries a benzophenone group (Sect. 7.6.1), which is excited to a triplet state (i.e., a diradical) by UV light. A radical (electron) is transferred to spatially adjacent atoms, and through radical combination, a covalent bond is formed.

12.9 Photoaffinity labeling (*photoaffinity-labeling*). In Sect. 7.6.1 we encountered further examples: azides ($-N_3$) or diazirines ($-C{<}N_2$).

12.10 This refers to the photolytically cleavable 4,5-dimethoxy-2-nitrobenzyl protecting group (nitroveratryl protecting group) as an ether. A serine protected in this way cannot be phosphorylated. Upon irradiation with blue light, the hydroxyl group is released even in living cells. This allows observation of whether phosphorylation occurs and what effect it has.

12.11 It is visible under the fluorescence microscope because it is fused to the C-terminus with the *green fluorescent protein* (GFP).

12.12 From phenylselenocysteine by oxidation to the selenoxide with hydrogen peroxide and spontaneous elimination reaction.

12.13 It is a cell extract, but contains only the most essential elements of protein biosynthesis in purified form and offers particularly high efficiency for the synthesis of modified proteins—but is extremely expensive.

12.14 AARS are responsible for faithful protein synthesis according to the reading of the mRNA, as they are the ones that load the tRNAs with the corresponding anticodons only with the corresponding amino acid.

15.13 Solutions Chap. 13

Answers

13.1 SAR refers, for example, to the interactions between enzyme and substrate or receptor and ligand.

13.2 The inactive original gene product angiotensinogen is degraded to angiotensin I by the specific protease renin; the protease ACE (*angiotensin converting enzyme*) then further shortens it to angiotensin II. This is the active peptide, the ligand for the angiotensin II receptor.

13.3 All three inhibitors are transition state analogs of the cleavage site, resulting in strong binding to the enzyme, but since there is no peptide bond at that position (γ- or δ-amino acid), no cleavage occurs and the enzyme's active site remains blocked.

13.4 Primarily, this prevents the inhibitor from being degraded by peptidases, thereby increasing its residence time in the circulation.

13.5 Published synthesis: The correct configuration at the stereocenters had to be introduced via asymmetric syntheses, namely using the Evans auxiliary, the Schöllkopf method (which then worked without a chiral auxiliary), and finally two diastereomers had to be separated.

13.6 2002

13.7 These and others have an H-bond donor and acceptor site in close proximity (models by Shallenberger/Acre, Kier, Tinti/Nofre, and also Goodman).

13.8 Monosodium glutamate is recognized by the umami receptor, which exists as a T1R1–T1R3 heterodimer.

13.9 See "Definition" in Sect. 13.2.2.

13.10 See Fig. 13.13

13.11 Lactisol is an inhibitor of sweet taste, but also of umami taste, as it binds to the transmembrane region of T1R3, which is present in both receptors.

13.12 In principle, by *photoaffinity labeling* (such as benzophenone or diazirine conjugates) of the sweetener. However, for such small molecules, the modification by labeling is so significant that the SAR is almost certainly altered.

13.13 It has been discovered that the receptors are expressed not only in the mouth, but also in the intestine, pancreas, and even in the brain. This means that the organism is deceived by such sweeteners and may undergo unfavorable metabolic reactions.

GPSR Compliance

The European Union's (EU) General Product Safety Regulation (GPSR) is a set of rules that requires consumer products to be safe and our obligations to ensure this.

If you have any concerns about our products, you can contact us on ProductSafety@springernature.com

In case Publisher is established outside the EU, the EU authorized representative is:

Springer Nature Customer Service Center GmbH
Europaplatz 3
69115 Heidelberg, Germany

Batch number: 10513644

Printed by Printforce, the Netherlands